AGRIBUSINESS AND EXTENSION MANAGEMENT

By the Same Author

Extension Education for Human Resource Development/*Birendra Kumar* and *B.S. Hansra*

Modernizing Indian Agriculture in 21st Century: Challenges Opportunities and Strategies/*B.S. Hansra, G. Perumal* and *K. Chandrakandan* (eds)

About the Editors

B. S. Hansra, Assistant Director General (Agri. Extn.) at the Indian Council of Agricultural Research, New Delhi, was Professor and Head, Department of Extension Education, Punjab Agricultural University, Ludhiana (Punjab). He is a well-known scientist and has more than three decades of teaching, research, training and management experience. He has authored 14 books and more than 200 research and popular articles. He is a National Fellow of Indian Society of Extension Education. He is closely associated with many national and international professional societies. He is the Vice-Chairman of International Extension Forum.

K. Vijayaragavan is a Principal Scientist at the Division of Agricultural Extension, Indian Council of Agricultural Research, New Delhi. He is a well-known extension scientist and has more than 25 years of teaching, research, training and extension experiences in the areas of extension management and rural development. He was a visiting scholar at Ohio State University, U.S.A. during 1998. He is a prolific writer and has several publications to his credit in the form of research papers, books and manuals. He also served as an Editor of *Indian Journal of Extension Education* for about a decade.

AGRIBUSINESS AND EXTENSION MANAGEMENT

Editors

B.S. HANSRA
K. VIJAYARAGAVAN

CONCEPT PUBLISHING COMPANY, NEW DELHI-110059

ISBN 81-8069-021-0

First Published 2003

Published and Printed by
Ashok Kumar Mittal
Concept Publishing Company
A/15-16, Commercial Block, Mohan Garden
NEW DELHI-110059 (India)

Phones: 25351460, 25351794
Fax: 091-11-25357103
Email: publishing@conceptpub.com

FOREWORD

In the context of present agricultural scenario, agribusiness has become one of the most important components of rural development in India. The liberalisation of Indian economy coupled with recent advances in farm technology have opened up opportunities for agribusiness. The focus of agriculture is increasingly shifting from raising production to market orientation and value addition. The factors such as fast growing market for agricultural products at domestic as well as international arena are also responsible for increased commercialisation of agriculture. The recently drafted national agricultural policy has accorded priority for promoting agribusiness at different levels. In the above context agricultural extension system of our country has to reorient itself to meet the needs of farmers which will help them to improve their agro-enterprises.

In the whole process of agribusiness development use of knowledge will be a crucial factor. Farmers as well as the extension functionaries have to be aware of changes occurring in agro-food market at domestic and world level. The extension agency has to improve its linkage with knowledge institutions, both government and non-governmental agencies so that the recent technological advances are made available to farmers in time. A comprehensive programme at various levels for education and training in agribusiness has to be undertaken.

The basic task and responsibility of agricultural extension should go beyond mere increase in yield. Extension organisations should make efforts to promote diversification at farm level, value addition and production of products for international market. This calls for increasing farmers' technical competency as well as entrepreneurial skills.

In simple economic terms, globalisation of agriculture means agro-products crossing the borders and establishing global network for its promotion. The revolution in information and communication technology has made it possible to communicate globally at a cheaper cost and to establish a global network of business centres. Therefore, efforts should be made by extension system to improve access to internet by the farmers at village level.

Realising the need for reorientation of extension system to promote agribusiness, the International Extension Forum in collaboration with MANAGE, Hyderabad, organised a national conference on "Agribusiness and Extension Management: Status, Issues and Strategies" on 7th and 8th December, 2001. I am glad that the papers presented in the above conference have been edited and published in its present form. The contributors of different chapters and editors of this volume have done a unique service to the extension system by the way of analysing the scope, problems and extension strategy to promote agribusiness. I hope this book will be of great use to those who are involved in agribusiness development. My best compliments to all the contributors and editors of this volume.

New Delhi
September 2002

(A.G. SAWANT)
Chairman
Agricultural Scientists Recruitment Board

PREFACE

This book is the result of a National Conference on 'Agribusiness and Extension Management: Status, Issues and Strategies' held during December 2001 at MANAGE, Hyderabad (AP). The seminar was jointly organised by International Extension Forum and MANAGE, Hyderabad and attended by delegates from different parts of India and abroad.

Indian agriculture traditionally has been a subsistence sector aiming mainly at foodgrain production. However, in recent years it has witnessed a great transformation resulting in attaining not only self-sufficiency in food production but also increased commercialisation. It is a well-known fact that the focus of present day agriculture is increasingly shifting from raising the production to market orientation and value addition. The factors such as fast growing market for agricultural products at domestic and international arena coupled with globalisation of economy are responsible for the above situation. The recently drafted National Agricultural Policy has accorded priority for promoting agribusiness activities at different levels.

The agribusiness is not a new concept, it started gaining momentum since the beginning of 1990s due to liberalised policies in Indian agriculture. In fact during early 1990s taking advantage of new policies in seed development, some private entrepreneurs and multinationals made entry in agribusiness dealing with hybrid seed, horticulture, floriculture and meat processing to cater both domestic and foreign markets. Our experiences with agribusiness show that it has succeeded only in selected pockets while remaining as an untapped potential in many areas. There is a greater scope for agribusiness activities related to production of seeds, medicinal and aromatic plants, propagation of planting materials through tissue culture techniques, aquaculture, floriculture, production of fruits and vegetables for domestic and export purpose. What is needed is to identify the potentials of agribusiness under different agroclimatic regions and to find out its irrelevance to improve the farming situations of the country.

The changing scenario of agriculture as discussed above calls for reorientation of agricultural extension system so as to transfer knowledge based inputs, which are essential for agribusiness at various levels.

Further agribusiness extension activities with emphasis on value addition have to be promoted at farm level. One of the important measures that can be undertaken by extension organisations at village level is promotion of rural entrepreneurship. The existing entrepreneurship development programme of selected State Agricultural Universities and Krishi Vigyan Kendras are mainly concerned with establishment of cottage level or small business activities related to poultry, beekeeping, mushroom and other crop production activities. While we have to continue to support the above activities, there is a strong need to promote specialised agribusiness activities for domestic and export purposes. This will also involve encouragement of public/private cooperation such as ITC tobacco programme in Andhra Pradesh, which has been supported by the varietal development authority of Central Tobacco Research Institute (CTRI), Rajahmundry. Another example of public support for agribusiness is provided by the Punjab Agricultural University (PAU), Ludhiana which supplied hybrid seed to Pepsi on contract farming business. Since the government extension machinery is likely to face more financial crunches in coming years encouragement of agribusiness through joint venture will strengthen extension at field level through involvement of NGOs and young private entrepreneurs.

This book is an attempt to identify the scope of agribusiness and the role of agricultural extension system in promoting the same. We hope that this book will be of interest and use to those who are involved in fostering agribusiness. Our thanks are due to contributors who prepared and revised their contributions. We would like to thank Dr. Panjab Singh, Secretary, DARE and Director General ICAR; Dr. A.G. Sawant, Chairman, International Extension Forum, Director General MANAGE; Dr. G. Perumal and all others who were involved in the National Seminar. Their vision and encouragement have resulted in publication of this volume in the present form.

September 2002 — B.S. HANSRA
New Delhi — K. VIJAYARAGAVAN

INTRODUCTION

This volume consists of papers presented at the national conference on "Agribusiness and Extension Management: Status, Issues and Strategies" held during 7th and 8th December, 2001 at MANAGE, Hyderabad. In the context of present agricultural scenario, the issues related to agribusiness development and the role of extension system in promoting agribusiness activities are examined in this book.

We all know that agribusiness is not a new concept. The term agribusiness was introduced by Professor Ray Goldberg of Harvard University during mid-fifties. It encompasses a wide variety of activities related to production, processing and marketing of crops, livestock, and forest products. Accordingly to the United States Department of Agriculture as quoted by Surya Kumar (1999)* the following types of employment fall under agribusiness:

(i) "agricultural production and propagation of animals, animal products, plants, plant products, forests and forest products;
(ii) the provision of services associated with agricultural production and the manufacture and distribution supplies used in agricultural production;
(iii) the designing, installation, repair, operation, servicing of machinery, equipments and power sources and construction of structures in agricultural production;
(iv) any activity related to inspection, processing and marketing of agricultural products and primary by-products;
(v) any aspects of greenhouse, nursery landscaping and other ornamental horticulture operations;
(vi) the conservation, propagation, improvement and utilisation of natural resources; and
(vii) the multiple use of forest lands and resources."

Indian agriculture traditionally has been a subsistence sector aiming mainly at improving foodgrain production. However, in recent years Indian agriculture has witnessed a great transformation resulting in

* Surya Kumar, P.V.S. (1999). "Export Oriented Agriculture—The Power House of Agribusiness in India," *SBI Monthly Review*, June 1999.

achieving not only self-sufficiency in foodgrain production but also increased commercialisation. It is a well-known fact that the focus of present day agriculture is increasingly shifting from raising the production to market orientation and value addition. The structure of income from agriculture is also changing very fast. The share of livestock as well as fishery sector to agricultural income is increasing at a faster rate in recent years. This indicates a trend of diversification from crops in favour of economic activities related to animal husbandry and fisheries.

Agribusiness started gaining momentum since the beginning of 1990s in India due to liberalised economic policies. During early 90s, taking advantage of new policies, some private entrepreneurs and multinationals made entry in agribusiness dealing with seed-farm, horticulture, floriculture and meat processing to cater both domestic and foreign markets. Our experiences with agribusiness shows that it has succeeded only in selected pockets while remaining as an untapped potential in many areas. There is a greater scope for agribusiness activities related to production of seeds, crops, medicinal and aromatic plants, propagation of plant materials through tissue culture techniques, aquaculture, floriculture, production of fruits and vegetables for domestic and export purposes. What is needed is to identify the potentials of agribusiness under different agro-climatic regions and find out its relevance to improve the income and economy of farmers. For example, food processing offers a lot of scope for agribusiness as only about one to two per cent of our food production is processed at present as compared to more than 70 per cent in developed countries.

Agribusiness should aim both at development of micro and macro enterprises at village level. The micro enterprises will generate employment and income opportunities for small farmers and farm women, and offer a greater scope to fight rural poverty. Small farmers and farm women have to be empowered through training in the technical and entrepreneurial skills to start agro-enterprises. The rural youths have to be motivated to take up self-employment related to micro agribusiness activities. This calls for development of entrepreneurship capabilities among small farmers, farm women and rural youths.

The changing scenario of agriculture as discussed above calls for reorientation of agricultural extension system so as to transfer knowledge based inputs at various levels. Further extension activities with emphasis on value addition and commercialisation have to be promoted at farm level. The scope of extension services should go beyond transfer of technology and include activities related to development of entrepreneurial skills of farmers and formation of Self-Help Groups which will help them to eliminate middle men eating away their profits. The problems in development of agribusiness among small farmers and farm women are many. Some of the constraints which are to be tackled by agricultural extension organisations are as follows:

(a) Poor access to latest technology
(b) Inadequate market information
(c) Poor access to finance facilities
(d) Poor infrastructural facilities
(e) Lack of motivation among small farmers and farm women
(f) Poor entrepreneurial skills
(g) Non-availability of information related to quality of products.

At present agricultural universities and Krishi Vigyan Kendras are promoting establishment of cottage level small business enterprises related to poultry, bee-keeping, mushroom cultivation and other crop production activities. While we have to continue to support the above activities there is a strong need to promote specialised agribusiness for domestic and export purposes. This may involve encouragement of public-private cooperation such as ITC tobacco programme in Andhra Pradesh, which has been supported by the varietal development authority of Central Tobacco Research Institute (CTRI) Rajahmundry. Another example of public support for agribusiness is provided by the Punjab Agricultural University (PAU), Ludhiana which supplied hybrid seed variety for Pepsi contract farming business. Since the government extension machinery is likely to face more financial crunches in the coming years, encouragement of agribusiness through joint venture is needed to strengthen extension services at field level through involvement of NGOs and private entrepreneurs. In this context it is essential that every State Agricultural University takes up agribusiness services with the following mandates:

(a) to provide consultancy on agribusiness enterprises and prepare project proposal on demand. The specialised enterprises may be related to seed-farming, crop and animal production, medicinal and aromatic plants, farm machinery, fruits, vegetables, food processing, green house activities, ornamental horticulture etc.;
(b) to provide market intelligence to the farmers on agribusiness for domestic and export purposes;

(c) to establish effective linkage and promote joint ventures with Farmer Organisations, NGOs/private agencies for agribusiness training and consultancy; and

(d) to train farm graduates to undertake agribusiness enterprises who in turn may set up agro-clinics at block and village level.

This volume consists of five parts. The first part discusses the scope, strategies of agribusiness in the context of globalisation. It consists of chapters dealing with World Trade Organisation and its impacts on agriculture, emerging agribusiness opportunities related to crops, animal husbandry, fisheries etc.; and problems of agribusiness and strategies for its development.

The second part is on "Farmers' organisations and farmer-led marketing in promoting agribusiness." It consists of chapters which portray the roles of NGOs, farmer-led marketing organisations, and women Self-Help Groups.

The third part, "Export oriented agribusiness" consists of chapters which deals with opportunities and problems related to export of farm commodities with particular reference to cashewnut and flowers.

The fourth part is on reorganisation of extension management systems and role of information technology. It consists of chapters and discusses themes related to reorientation of extension to meet needs of agribusiness, value addition for farm produce, role of international technology in promoting agribusiness, dissemination of information through internet, and training needs for agribusiness.

The final section, "Socio-economic and environmental impact of agribusiness" consists of three chapters dealing with impact of agro-enterprises on economic development and environment.

Editors

September, 2002 B.S. HANSRA

New Delhi K. VIJAYARAGAVAN

CONTENTS

PART II — FARMERS' ORGANISATIONS AND FARMER-LED MARKETING IN PROMOTING AGRIBUSINESS

LIST OF CONTRIBUTORS

Anandraja, N., Ph.D Student, Department of Agricultural Extension & Rural Sciology, TNAU, Coimbatore (TN).

Arun Kumar, S., M.Sc. Student, Department of Agricultural Communication, G.B. Pant University of Agri. & Technology, Pantngar (Uttranchal).

Beeton, Bob, University of Queensland, NRSM, Gatton Campus, Qld 4343, Australia.

Bhardwaj, Neelam, Assistant Professor, Department of Agricultural Communication, G.B. Pant University of Agri. & Technology, Pantnagar (Uttranchal).

Bhosale, S.S., Junior Research Assistant, Department of Agricultural Economics, Dr. Balasaheb Sawant, Konkan Krishi Vidyapeeth, Dapoli, Distt. Ratnagiri (MS).

Bhosale, S.S., Research Assistant, Department of Agril. Economics, MPKV, Rahuri (MS).

Bonde, P.S., Department of Agricultural Economics, M.P.K.V. Rahuri (MS).

Boparai, B.S., Associate Director (Training), KVK, Gurdaspur (Pb).

Chahal, V.P., Senior Scientist (Agril. Exten.), Division of Agricultural Extension, KAB, ICAR, New Delhi.

Chamala, Shankariah, University of Queensland, Tropical Dairy Research Centre, Teowan.

Chandrakandan, K., Professor, Department of Extension Education, TNAU, Coimbatore (TN).

Chaturvedi, V.K., Technical Officer (T-6), Department of Agricultural Extension, IARI, New Delhi.

Chole, Vijaykumar, Maharashtra State Grape Growers Association, Pune (MS).

Cowan, Tom, University of Queensland, Tropical Dairy Research Centre, Teowan.

Gaikwad, Bhaskar, Chief Scientist, Krishi Vigyan Kgyhasn Kendra, Babhlaeshwar, Tal. Rahata, Distt. Ahmednagar (MS).

Gangadharappa, N.R., Associate Prof., Department of Extension Education, UAS, Bangalore (KT).

Ghuman, Parminder K., Associate Professor (Home Science), KVK, Gurdaspur (Pb).

Gill, S.S., Additional Director, Extension Education (Communication Centre), PAU, Ludhiana (Pb).

Gill, T.K., Assistant Professor, Department of Extension Education, PAU, Ludhiana (Pb).

Govind, Santha, Reader in Agricultural Extension, Faculty of Agriculture, Annamalai University (TN).

Gupta, V.P., Vice-Chancellor, RAU, Pusa (Bihar).

Hansra, B.S., Asstt. Director General (AE), ICAR, New Delhi.

Ingavale, S.S., Department of Agriculture Economics, MPKV, Rahuri (MS).

Ingle, P.O., Professor and Head, Department of Extension Education, Dr. PDKV, Akola (MS).

Jaiswal, A.K., Indian Lac Research Institute, Namkum, Ranchi (Jharkhand).

Jayachithra, N.K., Asstt. Manager, VFPCK. (TN).

Jha, P.N., Former Vice-Chancellor, Rajendra Agricultural University, Pusa (Bihar).

Jhamtani, Anita, Principal Scientist, Division of Agricultural Extension, IARI, New Delhi.

Joshi, R.M., Sr. Research Assistant, Department of Agricultural Economics, Dr. Balasaheb Sawant, Konkan Krishi Vidyapeeth, Dapoli, Distt. Ratnagiri (MS).

Kalavathi, S., Ph.D Scholar, Gandhigram Rural Institute, Gandhigram (TN).

Kale, P.V., Research Assistant, Department of Agril. Economics, MPKV, Rahuri (MS).

Kalra, Anjana, PAU, Ludhiana (Pb).

Kapase, P.M., Professor of Farm Management, MPKV, Rahuri (MS).

Kasar, D.V., Head, Department of Agricultural Economics, MPKV, Rahuri (MS).

Kaur, Rabinderjit, PAU, Ludhiana (Pb.).

Kokate, K.D., Director of Extension Education, Dr. Balasaheb Sawant Konkan Krishi Vidyapeeth, Dapoli, Distt. Ratnagiri (MS).

Krishnaveni, M., Head, Department of International Business, SNMV College, Coimbatore (TN).

Kumar, K.K., Indian Lac Research Institute, Namkum, Ranchi (Jharkhand).

Kumar, Kanwal, PAU, Ludhiana (Pb).

Latha, K. Madhavi, NAARM, Hyderabad (AP).

Makesar, A.D., Agricultural Assistant, Department of Extension Education, Dr. PDKV, Akola (MS).

Mali, B.K., Agricultural Research Officer, Department of Agril. Economics, MPKV, Rahuri (MS).

Malik, Arvind S., Senior Research Fellow, Department of Seed Science and Technology CCSHAU, Hisar (Haryana).

Manoharan, P. Muthiah, Prof. & Head, Department of Agril. Extension, AC&RI, Killikulam (TN).

Marawar, S.S., Associate Professor, Department of Agril. Economics & Stat., Dr. PDKV, Akola (MS).

Mishra, B.P., Training Associate (Agril. Ext.), KVK, Bhanjanagar, Distt. Ganjam (Orissa).

Nadre, K.R., Head, Department of Extension Education, Marathwada Agricultural University, PARBHANI (MS).

Nayak, N., Training Organiser, KVK, Bhanjanagar, Distt. Ganjam (Orissa).

Navadkar, D.S., Asstt. Statistician, Department of Agricultural Economics, MPKV, Rahuri (MS).

Nehru, S. Mothilal, Director-Project & Implementation, VFPCK. (TN).

Pagire, B.V., Assistant Professor, Department of Agricultural Economics, MPKV, Rahuri (MS).

Palande, R.S., Prof. of Extension Education, Department of Extension Education, MPKV, Rahuri (MS).

Patil, H.K., Head, Department of Agricultural Economics, Dr. Balasaheb Sawant, Konkan Krishi Vidyapeeth, Dapoli, Distt. Ratnagiri (MS).

Patole, S.D., Department of Agricultural Economics, M.P.K.V. Rahuri (MS).

Pawar, B.N., Sr. Research Asstt., MPKV, Rahuri (MS).

Philip, H., Department of Extension Education, TNAU, Coimbatore (TN).

Ponnusamy, K.A., Professor (Agricultural Extension), TNAU, Coimbatore (TN).

Prabu, J.V., Ph.D student, Department of Agricultural Extension & Rural Sciology, TNAU, Coimbatore (TN).

Prameela, K., Lecturer in Agricultural Extension, Faculty of Agriculture, Annamalai University (TN).

Prasad, T. Krishna, NAARM, Hyderabad (AP).

Ranganathan, T., Dean, Faculty of Agriculture and A.H., Gandhigram Rural Institute, Gandhigram (TN).

Rao, S.V.N., Professor & Head, Rajiv Gandhi College of Veterinary and Animal Sciences, Pondicherry.

Ravishankar, G.T., Department of Extension Education, UAS, Bangalore (KT).

Reddy, P. Venkataram Muni, NAARM, Hyderabad (AP).

Rout, S.K., Information Specialist, Directorate of Extension Education, OUAT, Bhubaneswar (Orissa).

Saini, S.K., Associate Professor, Department of Extension Education, PAU, Ludhiana (Pb).

Sale, Y.C., Jr. Res. Assitt. Department of Agricultural Economics, MPKV, Rahuri (MS).

Sarda, M.K., Agricultural Development Officer, State Department of Agriculture, Punjab, Chief Agriculture Office, Ludhiana (Pb).

Satapathy, B., Dean, Extension Education, OUAT, Bhubaneswar (Orissa).

Satapathy, C., Dean, Extension Education, OUAT, Bhubaneswar (Orissa).

Sawant, A.G., Chairman, Agricultural Scientists Recruitment Board, New Delhi.

Sawant, G.K., Director of Extension Education, MPKV, Rahuri (MS).

Sharma, K. Krishan, Indian Lac Research Institute, Namkum, Ranchi (Jharkhand).

Sharma, Poonam, Scientist (S.G.), Department of Agricultural Extension, IARI, New Delhi.

Sharma, Rachana, Research Associate, Division of Agricultural Extension, IARI, New Delhi.

Shete, V.R., Associate Professor, Department of Agricultural Economics, MPKV, Rahuri (MS).

Shinde, H.R., Jr. Res. Assitt., Department of Agricultural Economics, MPKV, Rahuri (MS).

Singh, A.K., M.Sc. Student, Department of Agricultural Communication, G.B. Pant University of Agri. & Technology, Pantngar (Uttranchal).

Singh, A.K., Scientist, IARI Regional Station, Indore (MP).

Solanke, V.P., Department of Agricultural Economics, M.P.K.V. Rahuri (MS).

Spies, Airton, Epagri/CNPq/UQ, C.P. 502, CEP 88034, Florianoplis, Brazil.

Srinivasan, N., Professor (Agri. Economics), Directorate of Research, TNAU, Coimbatore (TN).

Sunil, V.G., Ph.D Student, Division of Agricultural Extensions, IARI, New Delhi.

Suryawanshi, R.R., Department of Agriculture Economics, MPKV, Rahuri (MS).

Teixeria, Sergio Rustichelli, University of Queensland, Tropical Dairy Research Centre, Teowan.

Venkatasubramanian, V., Senior Scientist, KVK, CTRI, Rajahmundry.

Vikhe, V.S., M.Sc. (Agri.) Scholar, MPKV, Rahuri (MS).

Wadkar, S.S., Associate Professor, Department of Agricultural Economics, College of Agriculture, Dapoli, Distt. Ratnagiri (MS).

Wangikar, S.D., Associate Prof. Department of Extension Education, Marathwada Agricultural University, PARBHANI (MS).

Wegener, Mal, University of Queensland, NRSM, St. Lucia, Qld 4072, Australia.

LIST OF TABLES

LIST OF FIGURES

PART I

SCOPE, ISSUES AND STRATEGIES OF AGRIBUSINESS IN THE CONTEXT OF GLOBALISATION

1

AGRIBUSINESS AND EXTENSION MANAGEMENT

— *A.G. Sawant*

Agriculture in India has now blossomed into a mature industry, comprising the input supply system, the agricultural production system and the output marketing system. All these systems are integrated into one system termed as "Agribusiness". It includes all the business enterprises that buy from or sell to the farmers, the transactions related to a product, a commodity or a service. It encompasses productive resources like seed, fertilisers, equipments, and machinery etc., the agricultural commodities like all food and fibre, and facilitative services such as credit, insurance, marketing, storage, transportation, packaging, distribution etc. The growth of agribusiness, in its endeavour to meet the requirements of food and fibre to ever growing population, depends upon evolving comprehensive policies covering all the three systems. This national conference and its deliberations will hopefully help us in designing the policies for the promotion of agribusiness in general, and the management of extension system in particular.

Changing Policy Context

India's economic policies followed till 1991 were characterised by inward orientation, export pessimism and growing public sector investments. It was hoped that by giving protection to the infant industry, we can realise the economics of scale and that the industry will emerge strong and competitive. But this protection only resulted in a perpetually parasitic, outmoded and outpriced industrial sector. With the failure of this socialist model, the stage was set for a gradual reversal of the policies pursued over more than four decades. Resultantly, we had to shift to the new economic policies which were crystallised in the forms of devaluation, deregulation, disinvestments and decontrol of prices. Now, the net catchwords are liberalisation, privatisation and globalisation. The orientation has shifted from inward to outward. As an ultimate outcome of the new economic policy, the demand for agricultural products in the domestic and foreign markets has increased to a greater extent. So also, movement of capital and investments across countries has been facilitated. These new developments might benefit our country if we gear upto encase the emerging opportunities. For this, we will have to support the production efforts of the farmers. We will have to extend credit support to them on easy terms. We will have to ensure adequate and timely input supply. We will have to create facilities for storage, packaging and marketing of farm products. Under the new scenario, the most important thing is to provide the extension support to the farmers, especially in the market-oriented agribusiness. The farm scientists and the experts in agribusiness in general, and the extension professionals in particular, are expected to help and guide the communities engaged in agribusiness in their endeavour to improve their socio-economic conditions, by providing them opportunities to perform their agribusiness in more efficient manner. This calls for understanding the various facets of market oriented agribusiness and reorienting the extension system accordingly.

Commercialisation and Diversification

Indian agriculture has experienced spectacular changes in the recent period, manifesting large-scale commercialisation and diversification. They broadly include cultivation of new crops and varieties, increase in the share of area under cash crops, large scale spread of livestock activities and fisheries, pursuance of high-tech in the areas of aquaculture, bio-technology, horticulture, processing etc. In response to the new economic environment ushered in by the process of liberalisation and globalisation, many new agribusiness activities are cropping up. Now, we experience greater degree of diversification from crops and forestry in favour of allied activities like animal husbandry and fisheries. Under these circumstances, the extension system will have to be broad based, encompassing all the elements of commercialised and diversified agribusiness.

Consumers' Point of View

So far, the farmer has remained the focus of attention of the extension system. The time has come to take into consideration the consumers' point of view. The demand for agricultural products has been increasing at a faster rate as a result of influences of a large number of factors like growing population, rising income level of people, inelastic demand for food items, dynamic economic policy, World Trade Order, consumer subsidies, public distribution system, minimum support prices and above all, changing tests and preferences of consumers. While speaking of consumers' point of view, we have to bear in mind that the consumers' preference is an outcome of an complex interplay of the factors such as social value, social change, demonstration effects through external factors and multi-media influences and availability of product range, awareness about processed products, accessibility to outside products, product quality expectations and aesthetic value considerations. We, the extension professionals will have to include, therefore, consumer as an important element in our technology transfer programme. Our programmes should focus on both the farmers, as well as, the consumers. For this, we should develop our own market research methodologies suited to different locations and agro-based products and educate and help the farmers in applying these methodologies under their situation, plan and execute the production programme based on the results of such research and help farmers find out the outlets for their farm products. Now, our job will not end at increasing the production and productivity of the agricultural enterprises, but at the market places for their farm products. We shall have to design the modalities for this kind of work in collaboration with the agricultural economists, marketing experts and agencies within the country and abroad.

Developing the Human Resource

One of the major tasks before us is the building of human assets, which is as important a pre-requisite of economic prosperity, as is the growth of physical and financial assets. Several empirical studies have shown that the entrepreneurs as the human capital has made a larger contribution to economic development than non-human capital. So, there is a great need to make our farmers enterprising. This is an achievable proposition, since the characteristics of entrepreneurs are not inherited but can be developed through systematic motivational training. Keeping in view the changing scenario, we need to equip extension professionals with techniques of stimulating rural youth to initiate appropriate enterprises for meaningful self-employment and income generating activities. In addition to this, support from development agencies and management—

sustenance efforts are also equally important. Now, the question arises, how this can be done?

Entrepreneurship Development Programme aims to develop entrepreneurship through motivation development, providing support services and sustain the effort. So, firstly, awareness and motivation has to be developed not only in the farmers, but also in the field level extension workers. Once, this is done, they will be able to provide some thrust towards commercialising agriculture, as they are the one who are implementing government sponsored schemes. Secondly, EDP could be included in the mandate of KVKs. The staff of KVKs should be trained in this concept, so that it reaches the ultimate users, i.e., the farmers. Similarly, entrepreneurship development should be included in the UG and PG courses at the SAUs. The Agricultural Finance Officers in banks, especially those who are working in rural areas should be exposed to this concept. The entrepreneurship development centres in the States should widen their area to include activities in agriculture.

Besides identifications, selection, training and support, the extension organisations must equip themselves to provide guidance and counselling for modernisation, expansion, diversification, and rehabilitation, if needed, to the prompting enterprises.

Emerging Issues and Future Course of Action

It is maintained that the efforts aiming at exploitation of potential advantages of agribusiness on one side and removal of limitations in efficient management of agribusiness on the other, will result into accelerating the growth of agribusiness in our country. Such efforts, however, encompass many activities closely related to management of the enterprises and extension system.

First of all, the extension organisations will have to make concerted efforts for bringing diversification at the farm and regional levels, depending upon the suitability of agri-climatic conditions and other factors related to production and marketing. For this purpose, educating the farmers to induce specialisation at the farm level by economising the source use and expertise is necessary. It is also important to motivate the farmers to adopt high-tech agricultural production systems/technologies for raising crop productivity and reducing unit cost of production. The measures like effective technology transfer programmes, supply of farm inputs including credit, strengthening of market infrastructure, etc. will strengthen the farmers' abilities to effect crop shifts and adopt new production technologies.

The question is, are we capable of strengthening the farmers' abilities to withstand the challenges ahead? In my opinion, yes, we are capable of doing this. After fifty years of experience in extension, experimenting with various approaches, on way successfully meeting different challenges, we have been able to emerge into a sound extension system. We have proved time and again, that we can rise to any challenging occasion with the help of the appropriate technologies generated by the farm scientists, and positive response of the farmers to our plea. The new economic order has once again posed a challenge to us; let us face it with full confidence. This confidence will come to us only if we are able to manage our own work and our organisation effectively and efficiently. We have a strong network of extension professionals from national to village level; we have the sophisticated means of communication for quick and effective transfer of technologies; we have excellent institutional support for training the human resource; and above all, we have millions of innovative farmers ready to cooperate with us in our endeavour to promote sustainable agribusiness. The only thing is to manage these resources in a cost effective way.

As we know, efficient management calls for economic use of men, material, money and time. In the present scenario of globalisation let us discuss these issues in the context of management of extension system for agribusiness development. Let us think as to how we can make a small marginal resource poor farmer an entrepreneur, let us discuss as to how we can convert an average farm into a remunerative enterprise, let us talk on the modalities that would help in expansion of the agribusiness in our country, and let us focus our attention on the technological and other interventions being made by the countries all over the world.

2

IMPACT OF WTO ON AGRIBUSINESS

— *K. Prameela* and *Santha Govind*

The establishment of World Trade Organisation is an important landmark in the history of international trade which reiterates the objectives of GATT. The WTO formally came into being on January 1, 1995.

The WTO Agreement on Agriculture, is stipulated to bring agribusiness effectively for the first time under the discipline of a multilateral trading system. The WTO Agreement on Agriculture did not deliver what is promised, but the shortfall in performance can largely be attributed to the fact that the developed countries managed to take advantage of some loopholes of ambiguities of the agreement to maintain their protection level. There is considerable misconception about what WTO is all about and how it will affect the agribusiness as a whole. A more than cursory look at the Agreement on Agriculture is required to understand how international competition would impact production, domestic market, prices and farm income. It is now time to draw a balance sheet of the gains and losses that accrued from the implementation of the WTO's Agreement on Agriculture. It is also imperative on the part of economists to analyse whether or not the WTO is in the interest of the Indian farmers rather than politicians who are known to be speaking different languages at different times.

All elements of India's food security policy are being dismantled under pressure from the World Bank and the World Trade Organisation (WTO).

The new threats to the food security faced by the food producers came from the following:

Increasing Cost of Inputs

The first is the shift from internal input sustainable farming systems to external purchased inputs like seeds, fertilisers, pesticides which drain farmer's income and lock peasants into debt. As the subsidies that made Green Revolution have been withdrawn and the input sector has been deregulated, farmers expenditure on purchasing seeds and chemicals has increased, pushing peasants into debt and penury. The epidemic of farmers' suicide is a reflection of this crisis of the rising costs of inputs.

Decline in Food Production

The second threat to food security is the shift from staples to cash crops. From 1960-61 to 1998-99, the area under coarse grains has gone from 45 million hectares to 29.5 million hectares, area under cotton has increased from 7.6 to 9.3 million hectares and area under sugarcane increased from 2.4 to 4.1 million hectares. During 1999-2000, 2000-2001, food production has gone down from 208.9 million tonnes to 196.1 million tonnes, a 12.8 per cent decline.

Dumping of Imported Subsidised Products

The dumping of imported subsidised commodities and removal of import restrictions (Quantitative Restriction) is another dimensions of the erosion of entitlements of agricultural procedures in India. Price of coconut has fallen from Rs. 10 per piece to Rs. 2, coffee price has collapsed from Rs. 60 per kg to Rs. 26 per kg, pepper price has fallen from Rs. 19,055 per quintal to Rs. 10,550 per quintal.

The destruction of food entitlements by dumping is an unfair trade practice which is totally legal in the asymmetric and unbalanced WTO rules. For example, U.S. soya beans are cheep not because of cheap production but because of subsidies. The price of soya bean is $155 a ton. And this low price is possible because the U.S. Government pays $193 a ton to those farmers who would not otherwise be able to stay in production given the low commodity prices. This Government support is not really a farmer subsidy, it is an indirect corporate subsidy. As heavily subsidised soya bean flooded Indians domestic market, prices crashed by more than two-thirds.

While the conditionalities from Global trade and financial institutions are preventing the Government from supporting the poor to have access to adequate and nutritious food, they are promoting the diversions of subsidies from people to corporations. Exports increases while people starve. Corporations are subsidised while people's food subsidies are withdrawn.

Crop-wise Analysis of Trade Issues

Rice

Rice imports and exports are largely canalised through Food Corporation of India (FCI). Rice also does not come under Open General Licence provision. The import traffic rates for rice is fixed at 80 per cent. In South Asia, North Africa, non-EU West Europe and Caribbean Islands, mean tariff grains is more than 80 per cent. India basmati rice attracts less tariff in the major export markets indicating further scope for any large dumping rice in India. Since India is not a major rice importing country and an export competitor, the market access problem from Vietnam and Thailand may not arise.

Pulses

As far as pulses is concerned, there will be any adverse impact by imports for Indian farmers. Rather it is criticised that the introduction of 5 per cent import duty may have negative effects on consumption. It is, therefore, argued that import of pulses under OGL must be continued.

Canada is the largest supplier of peas and pulses to India with a market share of over 20 per cent. Main varieties of pulses imported from various sources include green peas (Canada, USA, Hungary, New Zealand, France and Australia), kidney beans (China, Myanmar), chick peas (Kabli gram—Turkey, Syria, Canada, Australia, Mexico), lentils (Turkey, Syria), green beans (Myanmar, China, Tanzania), chick bean (Pesigram, Australia, Canada), and black mapte and Pilea peas (Myanmar). The import unit value for pulses was US $334.03 per tonne or 0.33 dollar per kg during 1990s. At Rs. 47 per dollar exchange rate, per kg of pulses works out to Rs. 16, which was almost at par with domestic prices.

Sugar

The country have surplus sugar production to meet domestic demand and exports. The nation's sugar production is estimated at 182 lakh tonnes (1999-2000). We exported 2.05 lakh tonnes in the same year. At present sugar attracts 60 per cent import duty. Low global prices of sugar are reportedly due to subsidies extended by the exporting countries. Our domestic price is higher than world market prices and also some of the export competitive measures like Net Protection Coefficient and

Domestic Resources Cost indicate that sugar in India is a protected crop and have comparative disadvantage in the international market. Announcing export subsidies may bring in export competitiveness and sizeable sugar can enter into world market at competitive price.

Oilseed

India is a net importer of edible oils. Due to the efforts taken by Technology Mission on Oilseeds, oilseed production increased gradually from 11.27 million tonnes in 1986-87 to 21.62 million tonnes in 1999-2000 with a peak of 25.21 million tonnes in 1988-89. Inspite of increase in production India's import is increasing manifold. Import of sunflower oil at a cheaper price is affecting the groundnut and sunflower farmers. The refined sunflower oil is priced very low compared to even the unrefined groundnut oil. Due to lower prices on health grounds, consumers in the upper middle class and above are using more sunflower oil substituting groundnut oil. In this situation import tariff to sunflower oil could be raised to protect the farmers and will not trouble the elite consumers.

Cotton

India has turned to be a net importer of cotton from a net exporter. In 1995-96, India imported only 50,000 bales and exported eight lakh bales. During 1996-97 the imports were only 30,000 bales as against an export of 16.82 lakh bales. The situation reversed during 1999-2000 as the import of cotton was 2.36 lakh bales against the export of nearly 16,000 bales. Taking into account the maximum quantity of import, it is suggested that domestic production of cotton has to be increased to nearly 7 or 8 times. Quality production has to be ensured.

Impact of WTO on Dairy Sector

Dairy is one of the sectors affected by WTO. During the negotiations in 1985, we failed to bargain and agreed to allow import of milk products under zero per cent bound duty. Subsequently, our new EXIM policy which brought some essential milk products under Open General Licence, further facilitated import. Initially there was no threat from import, as the international price for milk and milk products was high compared to the prevailing price in India. Subsequently many developed countries were able to exert pressure on their governments to provide subsidy to dairy farmers which helped them to lower their price of dairy products. Furthermore, new technological inventions and improved management practices helped the farmers in the Western countries as well as in New Zealand to bring down the cost of milk production. Thus the price for dairy products in the international market fell far below the price paid by consumers in India. This provided good opportunities for the traders to import cheaper milk products and thereby earn high profit at the cost of Indian dairy farmers.

This is the background of WTO and its impact on our farmers. As long as we continue to remain a member of WTO, this threat will continue to bother us. In 1999, Indian traders imported 10,000 metric tonnes of milk powder and in 2000, we were threatened by the arrival of fresh milk in Mumbai from New Zealand at the landed cost of Rs. 9. Fortunately, the Government of India in its budget of 2001 has imposed heavy duty on milk and the problem has been halted temporarily. This duty will have to be abolished before the year 2006, as per the WTO agreement. Hence we have 5 years to gear ourselves for international competitions. This is probably the last chance for our farmers to organise themselves and convert this challenge into an opportunity.

The main problems are high cost of milk production, high cost of milk processing, marketing and poor quality milk due to unhygienic milk handling. Hence we need to address these problems on priority.

Conclusion

It is time for a re-think on the impact of world trade on food and evolve a new agenda for food security. This liberalisation and globalisation of food through WTO threatens food security and health everywhere while dividing producers and consumers. Such a system will affect the agribusiness as a whole. For agricultural products developing countries would open avenues and India will have an opportunity to boost up her exports. For this there is strong need to put in place appropriate policies to safeguard the immediate and long-term interest of farmers. The latest EXIM policy has taken a number of measures to make import of agricultural goods difficult. To avoid the infiltration of pest and diseases, Sanitary and Phyto-sanitary permits to be issued by Government of India. Farmers must improve the qualities of their agricultural products to achieve the international standards, because in the competitive market quality cannot be compromised.

REFERENCES

Ajjan, N., Raveendaran, N., and Ramasamy, C. (2001). "Impact of World Trade Organisation on Tamil Nadu Agriculture and Agro Based Products," paper presented at National Seminar on Implications of World Trade Organisation on Agriculture in India, conducted by CARDS, TNAU, Coimbatore on November 26-27, 2001.

Hegde, N.G. (2001). "WTO Challenges for Indian Dairy Farmers," *Yojana*, Vol. 45, December 2001.

Shiva, Vandana (2001). "At the Mercy of Globalisation," in *The Hindu*, 23rd September 2001.

Yadav, Subah Singh (2001). "WTO and Its Impact on Indian Agriculture," *Yojana*, Vol. 45, December 2001.

3

WTO AGREEMENT ON INDIAN AGRICULTURE
Scope, Issues and Strategies of Agribusiness

— *P. Muthiah Manoharan* and *M. Krishnaveni*

General Agreement on Trade and Tariffs (GATT) is an international trade organisation that came into being in 1945, in response to predatory and protectionist trade in the 1930s. After over seven years of negotiations the Uruguay round multilateral trade negotiations were concluded on December 15, 1993 and were formally ratified in April 1994 at Marrakesh, Morocco, and WTO was formed on January 1, 1995.

WTO is the operational secretariat of GATT, an international body to administer the new provisions which are designed to reduce tariff and non-tariff barriers in international trade. Trade barriers are instruments used by policy-makers all over the world to restrict or prohibit imports from particular countries in order to protect the domestic industry or to manage balance of payment position or to give most favoured nation treatment (MFN) to other countries with whom they have common interests. There are various tariff and non-tariff barriers like duty variations, calculations of duty, quota restrictions, foreign exchange regulations, technical and administrative regulations, consular formalities and the like.

The WTO Agreement on Agriculture (AoA) was one of the many agreements negotiated during the Uruguay round. The implication of the AoA started with effect from 1.1.1995. As per the provisions of the Agreement the developed countries would complete their reduction commitments within 6 years, *i.e.*, by the year 2000, whereas the commitments of the developing countries would be completed within 10 years, *i.e.*, by the year 2004.

The World Trade Organisation (WTO) has set in motion a substantially expanded multilateral trading system. It places special emphasis on removal of non-tariff trade barriers and low level of tariffs. This has set in motion a tremendous amount of opportunities as well as extensive threats to Indian agriculture. The reduction in tariffs and removal of trade restrictions has exposed India's agriculturists as well as the agricultural industry to unexpected challenges.

Domestic markets are being increasingly threatened by lowering of tariffs leading to free entry of foreign goods. World markets are opening up due to lowering of tariffs and dismantling of other restrictions in developed and developing countries. The Indian farmers and agro-industry have to face a competitive environment in their international business. Export market will become tougher because of competition among developing countries.

The AoA is perceived as likely to create opportunities for our agricultural exports. For that to happen the industrialised countries have to substantially reduce their subsidies and provide increased market access. Reduction of their subsidies will raise the prices of agricultural products in the world market and this will make India's export more competitive. Liberalisation measures in agriculture world-wide will create market openings, which will be available to use provided we rise up to make use of the opportunities. Hence, this present study envisages:

(i) to study the *salient features of AoA*,
(ii) to gain insight into, the *implementation status of AoA* and the demands of India in this area, and
(iii) to comprehend, India's position in the *upcoming "Doha" negotiations.*

Salient Features of AoA

India being a signatory to the Uruguay round of GATT is also one of the original members of WTO. The AoA (Agreement on Agriculture) under the WTO provides a framework for the *long-term reforms* of agricultural trade and for domestic policies to be tuned towards the *market orientation* of agricultural trade.

The obligations and disciplines incorporated in the AoA relate to four aspects, namely,

(i) Agreement on domestic support
(ii) Agreement on export competition/subsidy

(iii) Agreement on market access
(iv) Agreement on sanitary and phytosanitary measures.

Domestic support and export subsidy are to be reduced in a phased manner over an agreed period. The special clauses of sanitary and phytosanitary measures aim to take care of the food safety, animal and plant health regulations among the WTO members.

Domestic Support Policies

Domestic support to farmers is extended in various ways. These include market support price, subsidies on inputs like fertiliser and electricity, lower interest rates on loans and the like. Government expenditure on agricultural research, extension services, assistance given for pest control and environmental protection measures are also a part of domestic support given by the government to the farmers.

The provisions on domestic support measures of the AoA, are meant to (i) identify acceptable measures of support to farmers and (ii) curtailing unacceptable trade distorting support to farmers.

Categories of Domestic Support

Domestic support is divided into two categories, namely, (i) support with 'no' or 'minimal', distortive effect on trade, often referred to as "Green Box" and "Blue Box" measures, and (ii) trade distorting support, often referred to as "Amber Box" measures.

Amber Box Supports: Subsidies, placed in the Amber Box, are considered as trade distortionary. These subsidies directly influence the market prices; they influence the farmers' decision about the choice of crop to be produced and quantity to be produced. Farmers who are enjoying these kinds of subsidies can sell their products at a lower price than other competitive farmers of the same products. Subsidies like input subsidies for fertilisers electricity, subsidies in the form of lower interest rates, market price support fall under Amber Box category.

Green Box Supports: The 'Green Box' policies include a variety of direct payment schemes that supplement farmer's income. These subsidies are not deemed to have major effect on production and trade. The subsidies given through the Green as well as Blue Box are *decoupled supports*. Decoupled means they are not linked to the quantity of production.

Green Box support includes assistance given through:

- Certain type of rural infrastructure
- Services such as research, training and extension
- Regional assistance programmes
- Resource (land) retirement programmes
- Certain types of investment aid
- Environment protection programmes
- Producer retirement programmes
- Marketing information

The subsidies under Green Box are not only excluded from any reduction commitments, but they are not subjected to any upper limit.

Blue Box Supports: Like Green Box support, supports under Blue Box are also exempted from any reduction commitment as they are also decoupled and hence considered to be least or minimally trade distorting. However, it differs from the Green Box support. The Blue Box has an upper limit.

These subsidies include direct payment given to farmers in the form of deficiency payment. In certain countries like United States, the difference in the government's minimum support price and the market price is paid directly to the farmers. This is called deficiency payment.

Direct payments under production limiting programmes are also part of the Blue Box policy. Here the farmers are paid so that they do not produce (by setting aside land) beyond a certain level.

Aggregate Measurement of Support (AMS): The *trade distorting domestic support* is measured in terms of what is called the "Total Aggregate Measurement of Support" (Total AMS).

(i) It is expressed as a *percentage of* the *total value* of *agricultural output.*
(ii) It includes *both* product specific and non-product specific support.
(iii) The AoA stipulates a *reduction commitment* of total AMS by 20 per cent for developed countries in 6 years (1995-2000) and by $13^1/_3$ per cent by developing countries in 10 years (1995-2004), taking 1986-88 as the base period.
(iv) *'de-minimis' level:* It is also essential to note that no reduction is necessary if the AMS value in the base period is below the threshold limit, which is 5 per cent and 10 per cent of value of production respectively for the developed and developing

countries. This level is called as 'de-minimis' level. Also, if support for any commodity is less than 5 per cent (for developed countries) and 10 per cent (for developing countries) of the value of that commodity, this support is not considered while estimating the AMS of that year.

Indian State of Affairs on Domestic Support

In India, the product-specific support is negative, while the non-product specific support, that is, subsidies on agricultural inputs, such as, power, irrigation, fertilisers and the like, is well below the permissible level of 10 per cent of the value of agricultural output. Therefore, India is under no obligation to reduce domestic support currently extended to the agricultural sector.

Export Subsidies

In many developed countries farmers are given subsidies to export their products. The need for export subsidies arises due to huge surpluses. The domestic support measures encourage farmers to produce more. The demand for agricultural products in developed countries is inelastic to the changes in prices unlike in the developing countries where demand increases when prices fall. So the huge surpluses are sold in the international market, with the help of export subsidies. The heavy export subsidies depress the world market prices. So this becomes a trade barrier for the competitive farmers from other countries where level of export subsidies is low or zero. These depressed international prices resulting from export subsidies compel some countries to raise the tariff levels in order to protect their farmers. Thus the effect of trade distorting in one area has cascading effect on the other area making the trade all the more difficult.

According to the AoA, developed countries have to reduce, over a period of 6 years (the base period being 1986-90), the volume of subsidised exports by 21 per cent and the corresponding budgetary outlay for export subsidies by 36 per cent. For developing countries, these reductions are 14 per cent in volume terms and 24 per cent in budgetary outlays over a period of 10 years.

Indian State of Affairs on Export Subsidies

(i) Export subsidies of the kind listed in the AoA, which attract reduction commitments, are not extended in India.

(ii) Developing countries are free to provide certain subsidies, such as subsidising of export marketing costs, internal and international transport and freight charges. India is making use of these subsidising in certain schemes of Agricultural and Processed Food Products Export Development Authority (APEDA), especially for facilitating export of horticulture products.

Market Access Restrictions

These aims at protecting local producers from international competition. Access to domestic market is restricted through imposition of tariffs, variable levies, import quotas and other non-trade barriers like complicated licensing procedure and such other measures.

Tariffs or import duties are one of the most important measures to protect the domestic producer. Like tariffs, variable levies or variable import duties is another measure where in, the import duty is linked to the difference in the world market price and the domestic price.

Tariff Rate Quota (TRQ) is another effective way of restricting the market access. This method employs two levels of tariffs, if a country applies a general tariff of 80 per cent on a particular commodity and then allows a limited quantity say 30,000 tonnes to be imported at a lower rate say for 15 per cent, then 15 per cent is in quota tariff and 80 per cent is tariff for imports beyond 30,000 tonnes.

Other than tariffs and import quotas other non-tariff barriers like complicated, time consuming bureaucratic and restrictive licensing procedures also serve as impediment to trade. These measures of restricting the market access deny the consumers access to commodities at lower prices and discourage the efficient, competitive producer who can sell these commodities at lower price.

Now, the AoA says that there can be no restrictions on farm trade except through tariffs. Market access for agricultural products is to be governed by a "tariffs only" regime. All other quantitative restrictions (QRs) like quotas, import restriction through permits, import licensing etc., have to be done away with.

QRs are to be replaced by tariffs and it is to be followed by progressive reduction of tariff levels. According to AoA, tariffs resulting from the "tariffication process" are to be reduced by a simple average of 36 per cent over 6 years in the case of developed countries and 24 per cent over 10 years in the case of developing countries. However, developing countries like India who had not converted their QRs into tariffs, were allowed to have *ceiling bindings* that were not subjected to these reduction commitments.

The Method to Reduce the Barriers to Trade

The three main clauses, (a) market access, (b) domestic support commitment, (c) commitment for export subsidies, to reduce these trade barriers forms the main components of the AoA. The method adopted to reduce these barriers is as follows:

(i) 1986-1988 is selected as base period for reducing market access barriers and domestic support commitment.
(ii) 1986-1990 is the period for reduction of export subsidies.
(iii) The average values of the tariffs, domestic support and export subsidies in this base period are considered as the bound level of the support.
(iv) A reduction is to be achieved in this bound level of support over a fixed period and by a fixed percentage.
(v) This method is in accordance with the difference in the levels of development among the member countries. The least developed countries are not required to reduce their support to agriculture. Developing countries also are given differential treatment. The period over which reduction is to be completed and the percentage by which they are to reduce is smaller than that of the developed countries.

Reducing the Restriction on Market Access

(i) *Bound Tariff Level:* The first phase is fixing the tariff level in the base period, by tariffication. Tariffication means conversion of the non-trade barriers (NTBs) into equivalent tariffs. The objective is to achieve transparency. NTBs as barriers are not obvious. So these NTBs are made transparent by converting them to equivalent tariff levels. The resulting tariff level is considered to be the bound tariff level.

(ii) *Reduction in Tariff Levels:* After establishing the value of the base period tariff through the tariffication of NTBs, countries are committed to reduce these as follows:

(a) For developed countries, by an average of 36 per cent. The minimum reduction of tariff for each product has to be of 15 per cent. The reduction has to be accomplished in a period of 6 years.

(b) For developing countries, the commitments are 24 per cent and 10 per cent respectively, and the implementation period is 10 years.

(c) For least developed countries there are no reduction commitments.

(iii) *Tariffs Bound at the Final Level:* At the end of the implementation period all tariffs are bound at the final level and in future may not, except under specific circumstances be raised above these levels.

(iv) *Minimum Access:* It was also mandatory to maintain minimum access where there was no binding on tariffs. This minimum access was to be maintained through tariffs. This minimum access was to be maintained through tariff rate quota. These quotas are to be maintained for commodities where minimum access was less than 3 per cent of domestic consumption in the base period of 1986-88. Minimum access was to be gradually increased to 5 per cent of base period consumption.

Indian State of Affairs on Market Access Restrictions

India has *bound its tariffs* at 100 per cent for primary products, 150 per cent for processed products and 300 per cent for edible oils except for certain items comprising of about 119 tariff lines, which were historically bound at a lower level in the earlier negotiations. Out of these low bound tariff lines, bindings on 15 tariff lines which included skimmed milk powder, spelt wheat, corn, paddy, rice, maize, millet, sorghum, rape seed, mustard oil, fresh grapes etc., were successfully negotiated under GATT Article XXVIII in December 1999 and the binding levels were suitably revised upward to provide adequate protection to the domestic producers.

India has also not undertaken any commitment to provide *minimum market access* opportunities which other countries who had scheduled their QRs had to undertake to the extent of 3 per cent of its domestic consumption going upto 5 per cent at the end of the implementation period.

Though India is not entitled to use the *special safeguard mechanism* of the Agreement, which can be used only by countries which had scheduled, yet it can take safeguard action under the WTO 'Agreement on Safeguards', if there is a surge in imports causing serious injury or if there is a threat of serious injury to the domestic producers.

Sanitary and Phyto-Sanitary (SPS) Measures

SPS measures is a general term for a variety of regulations on trade in food stuffs and feed and other natural products to protect human, animal and plant health in the importing country. The agreement recognises that governments have the right to take sanitary and phyto-sanitary measures to the extent necessary to protect human, animal or plant life or health. It should not arbitrarily or unjustifiably discriminate between members where identical or similar conditions prevail.

The European market is becoming increasingly difficult for developing country exporters to access, because of SPS requirements and the way in which they are being enforced.

Classification of SPS Measures: There are different kinds of SPS measures. They can be classified in terms of policy instruments, scope and regulatory goals.

(a) *Policy Instruments*

- (i) Import bans: Total bans, Partial bans.
- (ii) Technical specifications; process standards (*e.g.*, quarantine), product standards (*e.g.*, maximum residue level), and packaging standards.
- (iii) Information remedies; labelling requirements, controls on voluntary claims.

(b) *Scope*

- (i) Uniform: apply to all products, domestic or imports.
- (ii) Border universal: applies to all imports—most common for maximum residue levels on imports where pesticide is banned in importing countries.
- (iii) Border specific: for certain imports—to mitigate different levels of risk from different sources.

(c) *Regulatory Goal*

There are biological and toxicological risk reducing measures and measures which do not reduce risk but promote other goals such as conservation, quality, and compatibility with domestic production systems. There are three broad social objectives:

(i) protecting the economic interests of producers.
(ii) protecting the health and economic interests of consumers.
(iii) protecting the environment.

Problems Faced by India and Other Developing Countries

In practice, SPS measures create many problems for developing country exporters. There is a gulf between their international rights and capacity to enforce and utilise these rights. Developing countries are 'not well-positioned to address' issues such as whether SPS measures are inconsistent with the SPS Agreement and unfairly impede the flow of agricultural trade.

In general developing countries face difficulties as a result of:

(a) the nature of the regulations imposed by importing countries and how they are enforced, and
(b) their own capacity to implement the regulations on their exports.

Projects aiming to increase capacity of developing countries in this area include:

- Strengthening Research and Policy-Making Capacity on trade and environment in developing countries, being undertaken by UNCTAD (http://www.unctad.org/trade-env/docs/Issues1.doc) and
- The World Bank's Integrated framework for trade-related assistance to least developed countries (http://www.ldcs.org/index.htm).

Implementation Issues of AoA

The AoA stipulates in detail the dismantling of barriers by the developing and the developed nations. Whether the signatories have stood by the provisions, is the critical issue to be addressed.

Current Assessment

Talking in general terms, it has been realised that the following situation exists, as per various studies conducted by WTO, OECD and FAO.

- It is now an established fact that the Uruguay round did not bring about *trade liberalisation* in agriculture to the desired extent.

- There were *no significant reductions* in domestic support as well as export subsidies by the developed countries.
- Although AoA achieved a great deal by *defining rules* for international trade, its achievement in terms of immediate market opening has been limited.
- The *anticipated gains* from agricultural trade liberalisation, therefore, have eluded the developing countries till now.
- It was expected that the contemplated fair trading regime would help the efficient producers in *realising higher prices* for their products. On the contrary, prices of most agricultural commodities are declining in the world markets.
- It was anticipated that due to the reduction in domestic support in developed countries, cereal *production* would *shift* from developed to developing countries. It has been now realised that there has not been much change in the pattern of world cereals production and export.
- At best the policies in many developed countries have only been cosmetically altered by *shifting the support* from one "box" to another. The continuation of the high domestic support to agriculture in many developed countries is a cause of concern as they encourage over-production in these countries leading to low levels of international prices.
- *Market access* in the developed countries is also hampered by their maintaining high tariffs on products of interest to developing countries, in addition to other non-tariff barriers, in a recent study of 14 countries, Food and Agriculture Organisation (FAO) concluded that there are little change in the volume exported or in diversification of products and destination.
- *Tariff peaks* continue to block exports from developing countries to the developed world. Tariffs still remain very high in certain sectors, specially, in cereals, sugar and dairy products.
- *Tariff escalation*, that is, increase in tariff with successive stages of processing, block exports of value-added products from developing countries to the developed countries.

India's Demands

- Benefits to developing countries in terms of increasing their exports will only occur after complete elimination of export subsidies and substantial reduction in domestic support in the developed countries has been effected. In this context, India has been demanding a substantial reduction in the trade distorting domestic support and elimination of export subsidies by developed countries.

- Stringent Sanitary and Phyto-Sanitary (SPS) measures continue to be a major barrier in diversifying exports in horticulture and meat items. Fresh commitments have, therefore, to be negotiated to substantially improve market access for products of particular interest to developing countries.
- Since entry of new comers is difficult in the existing Tariff Rate Quota (TRQ) regime, India is demanding substantial expansion of TRQs pending their eventual abolition. It is also essential that administration of tariff rate quotas should become more transparent and equitable.

Opportunities for Better Management of Indian Agriculture

As far as India is concerned, it has been possible to maintain without any hindrance, the domestic policy instruments for promotion of agriculture or for subsidised targeted supply of foodgrains.

The domestic policy measures like the operation of the Minimum Support Price (MSP), the Public Distribution System (PDS) as well as provision of input subsidies to agriculture have not in any way been constrained by the Agreement.

In fact, certain provisions contained in Annex 2 of the Agreement, popularly known as the "Green Box", give India the flexibility to provide support for, research and extension services, pest and disease control, marketing and promotion services, infrastructure development, payments made for relief from natural disasters, payments under regional assistance programme for disadvantaged regions and payments under environmental programmes.

Indian Government has put this to advantage, through its recent budget announcement, a tax holiday for five years and 30 per cent deduction of profits from income for the next five years to the enterprises engaged in the integrated business of handling, transportation and storage of foodgrains.

Future Outlook—Forthcoming WTO Ministerial Meet, November 2001

As agriculture constitutes a vital segment of the Indian economy, finding greater market access for India's agricultural products, especially in the developed country markets, has to be one of the important issues during forthcoming negotiations.

(i) Food security for the people of the country, (ii) protection of the interest of domestic farmers and their livelihood, as well as (iii) the need for export maximisation, will be the guiding principles during the upcoming negotiations.

State of Affairs based on Seattle Meet, Nov.-Dec. 1999

A basic querry in all minds is that, will India with other developing countries be successful in reducing the subsidies as well as the tariffs of the developed countries during the coming trade negotiations?

The situation in the negotiation arena (based on December 1999 positions) was that, the developed countries were divided on the issue of trade liberalisation. CAIRNS group, comprising of developing as well as developed countries like Canada, Argentina, New Zealand, was in favour of unrestricted free trade. This group comprises of very competitive countries. The domestic support given to farmers is these countries is very less as compared to US and especially EU. Members of the CAIRNS group have naturally endowed agricultural producers and are totally opposed to the trade distorting subsidies and the protectionist regime practiced by EU, US and Japan.

It is the members of EU who have been distorting the trade with heavy subsidies and for this their policies have come under severe criticism by the CAIRNS group, the developing countries as well as the US.

US is in favour of furthering trade liberalisation. Elimination of provision of export subsidies is their major demand. However, US is silent about the issue of Blue Box. In general, it was felt that countries like EU and Japan are on one side defending their trade distorting policies. CAIRNS, the developing countries and US are on one side demanding further liberalisation of the international trade. Many economists in India were of the opinion that India's interest lies in joining the CAIRNS group.

Informal Ministerial Meet at Singapore, October 2001

The launch of a new round of negotiations is now almost a certainty. Working together, the US and the EU have been able to organise a compromise on a very substantial part of the agenda for a new round that could well be launched next month.

All that now seems possible for India and the other like minded countries, mainly the Africa group and the Least Developed Countries (LDC) which are still against new talks, is to contain the agenda to the extent possible.

The developing country's demand of first settling the "implementation issues"—the shortcomings in the existing WTO agreements—before launching a new and comprehensive round has weakened on account of two reasons. First, India is the only large developing country that holds on to this position; others like Egypt, Indonesia and the latest, Pakistan, have gone over to the other side. Second, the developing countries' freedom to manoeuvre in the WTO preparations have so far mainly come from the divisions within the developed world. The significance of the Singapore meeting is that it saw a cementing of most of these cracks.

On Agriculture: A compromise agenda on trade talks in this sector has been more or less reached between the liberalisers, the CAIRNS group of farm exporters (Australia, Thailand, Brazil and others) and the US on the one side and the "subsidisers", the EU and Japan, on the other. The key was the CAIRNS group, anxious about market access for its products dropping its insistence on "elimination" of export subsidies.

Environment: The EU's demand to put food safety, the "precautionary principle" and related issues on the agenda had the rest opposed to the proposal. But an issue that the EU had called "a deal breaker" will see another compromise: another time-bound study programme that will be followed by negotiations for a new agreement.

Other Issues: The agenda will include talks to reduce customs duties as well as draw up new agreements on trade facilitation and transparency in government procurement. On all these issues, India has been one of the few large countries still opposed to negotiations.

The challenge now confronting India, which is in a corner, is how to make the best of a bad bargain.

Upcoming Doha Meet

On the eve of the Doha Ministerial Conference, the World Trade Organisation has suggested a major negotiating agenda for a new round of trade liberalisation talks that India and most other developing countries will find difficult to digest, even as only marginal progress has been made in addressing their long-standing demands to correct the problems that have surfaced in implementation of past WTO/GATT agreements.

A *draft ministerial declaration* that was released on 22.10.2001 in Geneva put forward "a board and balanced work programme" that could lead to more than a dozen new WTO agreements in areas ranging from import tariffs on industrial products to entirely new subjects such as foreign investment and national competition policies. Other issues include trade in agriculture, transparency in government procurement and a possible review of the agreement on anti-dumping duties.

The agenda now includes a specific programme on agriculture trade liberalisation. In another controversial area—food safety— which the E.U. has previously called "a deal breaker", the proposed agenda is very modest and makes on reference to the European demand to frame rules that will allow countries to keep out agricultural imports if there is a slightest hint of danger to health, a demand that developing countries and the U.S. have called protectionism in disguise. Not listed explicitly in the agenda is one that India has long asked for—talks to provide protection to "geographic appellations" like Basmati rice or Alphonso mangoes; only its possibility is mentioned.

On the implementation issues, Doha will be asked to approve proposals to modify some 40 clauses in the existing agreements in order to benefit developing countries. But the more substantive implementation demands—numbering more than 50 and including, for example, a proposal to permit developing countries to impose export obligations on foreign industry—has been pushed back to talks in the next round.

Pointers for Policy-Makers

From the above discussions it can be deduced that,

- Being a signatory, India has to adhere to the stipulations of AoA, as applicable. So, the policy-makers have to carry out the necessary *internal reconstruction* of the agricultural sector as per the norms of AoA.
- *Case by case dealing* must be undertaken regarding each agricultural product that comes into threat due to the dismantling of trade barriers.
- *Prove in the WTO forum*, the true picture of so-called dismantling of trade barrier by the developed nations and fight for the right of developing nations to capitalise on the global market that opens up.
- In the event of gaining access to more of global market, is our Indian agribusiness sector competent to make use of this

widened market? Towards this, efforts should be *geared up*. if not for all agricultural commodities, at least a group of agri-products must be identified and *a basket of export commodities* must be concentrated upon.

Scope of Further Research

- *Product-wise study* on the impact of WTO provisions of particular product markets will help to clearly plan for remedial and developmental actions.
- International organisations working for the betterment of the agrarian society should take up updated *studies on the exact position of dismantling of barriers* undertaken by the signatories to AoA. The very purpose of AoA would be lost if this is not carried out.
- Rather than allowing misleading misconceptions to float among the public, it would be better if true status and picture of the *actual efforts taken by the government* are brought to the common knowledge of the public.
- Research and extension efforts should be taken for this proper dissemination of knowledge in this area of WTO an Indian Agriculture.
- A basic study may be conducted about the *tariffication process* and ceiling bindings framed by the Government of India.
- A *comparative study on tariffs* in developed and developing countries can be undertaken based on the bound tariff levels to be maintained on the basis of the "base period" tariffs. This would give a clear picture of how 'better-off' the developed and the developing nations have been placed as per the provisions of the AoA.
- A study on *sanitary and phyto-sanitary measures* insisted by developed countries and their impact on the trade of the developing countries would be a good insight to be gained.
- A study into the "*Agreement on Safeguards*" is highly necessary for helping our policy-makers.
- A world-wide data collection on *market access commitment* of various developed countries would be helpful for the developing nations to gear up to make use of the expanded market opportunities.

Conclusion

Agriculture is a bone of contention for developed and developing countries. WTO watchers feel that the negotiations are likely to continue for a few years, as there are serious differences amongst the major players. Food security is not only of great economic relevance but also a very important socio-political concern in large agrarian economies like India, and as such it has to be addressed upfront in the ongoing negotiations on agriculture.

The WTO regime will benefit those countries that show wits and skills in the ongoing dialogue. The concepts of liberalisation of international trade, de-regulation and privatisation of internal economy have now been strengthened and legalised under WTO. The new multilateral trade regime under the WTO offers excellent scope for growth and development, provided we are in a position to meet global competition.

The countries that have understood this have moved swiftly in fine tuning their domestic and international trade policies creating a winning environment for their businesses. Those who are still debating the issue or are still in the stage of bewilderment will help neither themselves nor their businesses. Hence it is the right time that emphasis be given to various aspects, namely domestic reforms, infrastructure development, the development of technology and human skills, cost reduction and quality improvement through modernisation and upgradation.

REFERENCES

Chakravarthi, R. (2001). "Fewer options at the WTO", *The Hindu*, dated 31.10.2001.

Rammanohar Reddy, C. (2001). "A New WTO Round Now Almost Certain", *The Hindu*, dated 17.10.2001.

Rammanohar Reddy, C. (2001). "WTO for Major Agenda in New Round", *The Hindu*, dated 29.10.2001.

Ramesh Chand (2001). "Import Liberalisation and Policy Implication", *International Extension Forum Newsletter*, Vol. 7, No. 3 & 4, pp. 2-4.

Vittal Rao, S. Yadav and Ansari, M.R. (2001). "GATT 1994: The Conceptualised Hopes and Fears of Agriculture Provisions", *Journal of Extension Education*, Vol. 12, No. 1, pp. 3027-3032.

Web sites http://www.wto.org
http://www.fao.org
http://www.oecd.org
http://www.codissia.com
http://www.zdh-connect.com

4

AGRIBUSINESS
Value Addition and Marketing

— *T.K. Gill* and *S.K. Saini*

Agribusiness includes all operations involved in the manufacture and distribution of farm supplies. These are production activities on farm and storage, processing, and distribution of farm commodities. In the modern context, on-farm activities are called farming sector of agriculture and off-farm activities are called as agribusiness sector of agriculture (Chandrakandan, 2001).

The ultimate purpose of agribusiness is of catering/satisfying the food/nutritional, fibre and floriculture needs of the consumer, his style of living, his society's changing value structure in the given political and economic priorities of the government and it should be done in an efficient, acceptable socially as well as globally desirable manner. So it has to be both (i) domestic, (ii) international in scope and must keep in view some critical variables like social, economic, political and technological changes. As India is one of the original members of WTO, so it becomes obligatory to adjust its trade and other policies as envisaged in the GATT accord which also covers agriculture sector.

Promoting agribusiness in the context of globalisation is not an easy task because majority of the cultivators belong to small and marginal categories. They are unorganised, illiterate and live in perpetual debt. They have negligible contribution to marketable surplus and insignificant market power on account of their poor financial position. They remain conservative and need much education and protection against risks to accept any change because farmers are reluctant to deviate from their traditional role. They view agriculture and business as two separate sectors. So there is need to create awareness about agribusiness and to promote it in the country and to be competitive internationally, appropriate steps should be taken as follows.

Increasing Productivity

India is the world's second largest food producer and has the potential to be the world's number one (Bhattacharya, 1997) producing over 600 million tonnes of food products. It is first in cereals, livestock population and milk and second in fruits and vegetables (Khan, 2001). The population of the country is increasing and to feed it India has to dramatically increase the yield. Though India has 60 per cent more arable land than China, it produces 40 per cent less output. Yields in India for most key products are 25 to 40 per cent of world levels. With noticeable exception of Punjab, wheat yields are just 32 per cent of world levels, rice yield are 33 per cent, maize yields are mere 7 per cent of world levels. Milk production of Indian cattle is just 500 kg a year compared to yields in USA, that is 7,000 kg/year (Bhattacharya, 1997). Reasons for low productivity are:

- low inputs—Indian farmer uses far less soil nutrients
- low quality seeds—over 95 per cent of the seeds used in planting are grown by the farmers themselves.
- high cost of inputs
- low level of mechanisation
- inefficient plan protection measures.

To overcome these problems agribusiness firms can help to increase agricultural production by supplying inputs like HYV seeds, fertilisers, insecticides/pesticides and enhancing mechanisation to decrease labour inputs.

Produce can also be saved by reducing post-harvest losses. Post-harvest losses in India are too high, *i.e.*, about 65 million tonnes which is more than the consumption in UK. The extent of these losses depends on the nature of food raw materials, *i.e.*, whether it is perishable, semi-perishable or durable (*e.g.*, about 10% of foodgrains, 25-30% of horticultural crops and 10-15% of animal and fishery products) are lost in various operations. Of these losses about 50 per cent can be reduced by adopting the post-harvest processes/technologies developed and advocated by different research organisations (Khan, 2001).

Stressing Diversification

Reducing area under cash crops like wheat and rice and shifting towards dynamic commodities occupying progressively large space in the international market can be another step to promote agribusiness. Due to piling up of the foodgrain surpluses—cereals, there is need to identify and exploit the international market potential. Agriculture should shift towards the production floricultural products, live plants, bulbs, tubers, cutflowers, dried flowers, herbs, vegetables, fruits and other products in demand. Even in the domestic market people are moving from subsistence foods to basic foods, because per capita income is rising and it will enable a large number of people to move out of pure subsistence levels and increase expenditure on protein rich foods (Bhattacharya, 1997). Due to vast diversity in country's agro-climatic conditions—a variety of products can be produced. Plastic culture can also be of great help for the farmers can turn towards earning more profits through commercial production. They can grow off-season vegetables fetching handsome price in the market. The vegetable production in early, late and out of its normal season is only possible if some of the growth phases of the plants are completed by providing shelter, creating a congenial environment for the plants. The role of plastic culture has gained significant importance for the production of vegetables out of its normal season. Nursery of summer vegetables can be raised under plastic cover to provide protection against frost and winter injury. Plastic culture plays an important role in modifying effectively both the predictive and non-predictive elements of weather to take advantage of off-season production by creating optimal environmental conditions. The crops can be raised much more early than their normal growing season which will fetch remunerative prices in the market (Khan, 2001).

Value Addition

Value addition to raw produce in the country is only seven per cent, compared to as much as 23 per cent in China, 45 per cent in the Philippines and 188 per cent in UK (India Grains, 2001). Difference between the farmer's realisation and final consumer's price is very high in our country even in case of fresh produce. So if the farmers sell their produce after adding value at a reasonably higher price, he can earn more profits. Liberalisation has opened new vistas for growth in the value addition.

Value can be added at two levels:

(a) Value addition in raw material.
(b) Value addition in processed foods.

Value Addition in Raw Material

Value addition to agricultural produce starts at the post-harvest level. It includes even the primary processes like sorting, grading, cutting, seeding, shelling, packaging etc. Farmers, if become processors of raw produce, can sell it at a higher price and earn profits. But with the coming of WTO regime in the country, requisite quality standards of grading of packaging etc. should be met in order to compete with imported products in the domestic market.

Value Addition in Processed Foods

It is expected that consumption of processed foods will grow at a fast rate in future due to following reasons:

- Today's consumer is spending less time in home food preparations and its recent trend is the increase in consumption of food away from home.
- Rise in consumer's income is enabling him to shift from food for sustainability to food they wish to eat and the basic foods, *i.e.,* for nutrition. Food consumption patterns of rural India are also following the cities, with a time lag of fifteen to twenty years (Bhattacharya, 1997).

The growth of food processing industry can bring immense benefits to the economy of the producer as well as that of the country. According to Bhattacharya (1997) it will:

— raise agricultural yield as food processing industries will provide HYV, pesticides, fertilisers etc., to the farmers. Examining the impact of a number of food processing projects it was revealed that yield increased by 50 to 250 per cent.

— enhance employment opportunities as creation of 50 lakh jobs has been estimated by 2005.

— raise standards of living of a number of people, farmers and rural people. It has been found that income increased by an average of 100 to 200 per cent through the impact of the projects.

— directly or indirectly triggers growth in a number of other industries such as transport, refrigeration, pesticides, fertilisers etc. In India the multiplier for food industries is 2.4 which is much higher than that for industries as power and telecom.

In addition to above food processing is also beneficial because:

— post-harvest losses of perishable and semi-perishable products can be reduced as their shelf-life is increased.
— food items processed from produce during the peak seasons can be used in lean seasons.
— farmers can get higher cost instead of disposing their produce at throw away prices in peak seasons.

Marketing

Marketing is the key link between the two forces in agribusiness, *i.e.*, the agricultural producer and the consumer. An efficient marketing system minimises the cost of marketing, increases the share of the producer, minimises the effects of seasonal and geographical differences and handles the farm surpluses efficiently. But the marketing system in the country is inefficient. It comprises a hierarchy of intermediaries, grading packing, storing and transportation etc. Here are some suggestions to make the system effective.

Decreasing Intermediaries

A large number of intermediaries are involved in the marketing system of the country as compared to the developed countries. In case of fresh produce, 6 to 7 intermediaries are involved as compared to two or three in UK (Bhattacharya, 1997). To reach at consumer's gate from the farmer's the following chain is formed:

Farmer → Consolidator → Commission agent → Trader → Commission agent → Whole saler → Retailer → Consumer

As a result of this chain the farmers receive a much lower share of the consumer's price. Moreover so many intermediaries required to deliver the produce from farmers to consumers are the significant cause for the inefficiency of the system. There is no utilisation of profits earned on the infrastructure except the cooperatives, which are very few in number.

Grading

Uniform food product quality grades and standards contribute to both operational and pricing efficiency in the agribusiness. The farmers should be encouraged to sell their produce after grading. Grading is a must because of the following reasons:

— lowers the cost of transportation because higher valued products can be transported to bigger markets whereas lower valued products to the local/nearer markets.
— reduces the market spoilage by separating the poorer quality products from the higher quality products.
— contributes to market competition among the sellers and reduces extraordinary profits.
— provides incentives for producers to adjust to changing consumer's preferences. If the producer receives average price of all grades of products, there would be little incentive to produce high quality products.
— gives specific information for consumer's preferences.
— contributes to consumer's satisfaction and increased producer's returns.

Packaging and Storing

Another important aspect of agribusiness is packaging and storing. The domestic market are flooding with the imported products having an attractive, hygienic and good quality packing. So in order to compete them packaging of the domestic products should be done mechanically so that its cost is reduced, is hygienic and of good quality. Moreover, the packaging of products should be attractive and presentable.

Raw produce is handled roughly at the farm. Damage is compounded when perishable products are piled into large cane baskets or to the trucks or trolley beds without cushioning and packaging. There is lack of infrastructure for storing specially for perishable items because, there need cooling/refrigeration system. Even the durable produce is not being properly stored. Millions of tonnes of foodgrains are lying in open exposed to sun and rain. Storing is also necessary for co-ordination of demand and supply in the peak and lean seasons, otherwise the farmer has to dispose of their produce at throw away prices.

Transportation

Shortage of transportation facilities during the harvest seasons result in falling of cash prices. India has inefficient transportation system. High wastage in farm produce can be attributed to transportation facilities such as poor quality and rough surface roads, more time taken to reach the produce at the retailer, lack of refrigeration during transportation, exposure to high temperature etc. Transportation is a key link in agribusiness connecting geographically specialised farmers and the urban consumer population. Improved transportation can help in:

— linking the food production centre and consumer centres.
— increasing the area/size of the market.
— affects the location of market/food processing units.
— compensates for the inflexibility of geographically specialised agriculture.
— affects the adequacy and cost of farm product.
— affects the storage cost of the producer.

Government Policies

The government should devise elaborative policies to implement above mentioned measures. Here are some suggestions:

— Attitude of farmers should be changed and they must be made aware about various aspects of agribusiness.
— Incentives should be provided to increase production.
— Diversification should be encouraged to meet the domestic and foreign demands.
— Development of untapped potential especially in the north-eastern region.
— Quality consciousness should be generated among farmers/processors and international standards should be adopted.
— Cost of production should be reduced by promoting mechanisation.
— Marketing and transportation system should be improved.
— As the farmers have small land holdings, co-operatives, contract farming and lease farming should be encouraged.
— Subsidies should be rationalised.
— Marketing fee, sales tax, surcharge etc., should be reduced. High taxes encourage unethical practices. Food items should not be subjected to more than one tax and processed items should be considered as luxury items.
— Revenue collected should be used appropriately for infrastructure, *i.e.*, roads, covered sheds, platforms, storage space etc.

— Financial assistance should be provided to producers/processors. Loan procedures should be simplified. Banks should be directed to provide adequate and timely funds at economical interest rates and priority basis.
— Co-operatives should be organised as they are well suited to provide financial services to rural people as they operate at the grass-root level among people who know each other well, a basic condition for trust.
— Agricultural research and extension system should be revamped for transferring innovative technologies from laboratories to the fields. Adequate funds should be provided for the purpose.

So if India has to promote agribusiness and become internationally competitive, the government must not lose time in taking appropriate measures to support it through improving the productivity, stressing diversification and value addition keeping in view, the quality of produce, packaging, storing and evolving an efficient marketing system.

REFERENCES

Bhattacharya, U. (1997). "India is the World's Second Largest Producer", in *Modernising the Indian Food Chain*, Food and Agriculture Integrated Development Action CII, McKinsey and Company Inc., New Delhi, pp. 13-20.

Bhattacharya, U. (1997). "Food Consumption Patterns are Changing", in *Modernising the Indian Food Chain*, Food and Agriculture Integrated Development Action CII, McKinsey and Company Inc., New Delhi, pp. 21-30.

Bhattacharya, U. (1997). "Unlocking the Opportunities through Integration", in *Modernising the Indian Food Chain*, Food and Agriculture Integrated Development Action CII, McKinsey and Company Inc., New Delhi, pp. 40-50.

Chandrakandan, K. (2001). "Agribusiness Management Basic Issues," *International Extension Forum, Newsletter*, 7 (3-4): 2.

Janaiah, A. (2001). "Overflowing of the Indian Grain Bowls: Looking at Beyond the 'Poverty' Factor", *India Grains*, 3(7): 20-23.

Khan, M.J. (2001). "Indian Agricultural Exports: Strengths and Shortcomings", *Agriculture Today*, 5(2) : 34-35.

Khan, M. J. (2001). "Post-harvest Processing: In a Climate of New Trends", *Agriculture Today*, 5(2) : 43-46.

Khan, M. J. (2001). "Plastic Culture: Boon for Raising Vegetables", *Agriculture Today*, 4(5) : 9-10.

Shaji, T.K. (2001). "Indian Crucially Awaits Agripreneurship", *Agriculture Today*, 5(2) : 10-11.

Singh, P. (2001). "The National Food Processing Policy—Agro Processing Committee", *India Grains*, 3(4) : 3-13.

Vardarajan, S. (1989). *Marketing Farm Products*, Country paper presented in Seminar on Marketing. Farm Products in Asia and the Pacific held on March 3-14, 1986, Asian Productivity Organisation, Tokyo, pp. 247-55.

5

IMPACT OF WTO ON INDIAN AGRICULTURE
Issues and Options

— *N. Anandraja, K. Chandrakandan* and *H. Philip*

The WTO came into existence on 1st January, 1995 to replace GATT (General Agreement on Trade and Tariffs) with the following aims:

- To promote open, fair and rule-based flows of trade without distortion of competition,
- To encourage development and economic reforms in developing countries,
- To encourage ever increasing participation of transition economics in the international trading system,
- To provide garter time and flexibility to the least developed countries for implementing certain WTO provisions,
- To encourage industrial countries to assist developing countries by liberalising trading conditions and by not expecting reciprocity in trade concession negotiations,
- To protect the environment and to promote sustainable development,
- Progressive liberalisation and elimination of tariff and non-tariff barriers to trade in goods,
- Rejection of all forms of protectionism,
- Elimination of discriminatory treatment in international trade relations,
- Integration of developing and least developed countries and economics in transition into the multilateral system, and
- Maximum possible extent of transparency.

Functions of WTO

- To facilitate the implementation, administration and operation of the Uruguay round agreements,
- To provide a forum for negotiations among member-countries concerning their multilateral trade relations,
- To administer the trade dispute settlement procedures,
- To review national trade policies, and
- To co-operate with other international institutions, especially the IMF and the World Bank in order to ensure more meaningful compatibility in global economic policies.

Impact on Indian Agriculture

Subsidies

The aggregate measurement of support is to be brought at the level of less than 10 per cent by the developing countries. While the estimates made in this regard have been already shown that Indian agriculture has a support of much less than this level. The subsidies available on fertilisers, irrigation, electricity, credit, pesticides, etc., together accounted for 3 to 6 per cent of total cost of production in different commodities during 1986-88 period. However, the developed countries are supposed to lower such subsidies to a level of 5 per cent or less. According to an estimate, Japan gives a subsidy of 72.5 per cent, EEC at 37 per cent and USA at 26 per cent. Therefore, they will have to bring down subsidies drastically making our agricultural products much more competitive in the world trade.

Calculations according to GATT methodology show that the non-product specific agricultural subsidy worked out to about Rs. 4,582 crore, while the level of product specific subsidies calculated during the base period worked out minus Rs. 2,442 crore, the world prices being higher than the national prices. This means that at current subsidy level, the government can increase subsidy by a further Rs. 19,860 crore.

Trade Related Invest Measures (TRIMs)

Investment in *sin-quo-non* for the development of an economy. Most of the developing countries are capital starved and thus have high marginal productivity of this scarce resource. The scope of investment under trade related investment measures (TRIMs) provided in the draft can help to boost up the development process. The priority areas are to be identified and specified by the government for making such investment. Thus, it prohibits the investment measures that are inconsistent in the national interest. The developing countries have been given time of 5 years period to phase out the inconsistent TRIMs. The agreement does not impose any obligation to provide access to all or any particular sector for foreign investments.

Trade Related Intellectual Property Rights (TRIPs)

It provides norms for copyrights, trade-marks, geographical indications, signs, patents etc. On patents, the basic obligations is that of product and process patent in all areas including agriculture. On plant varieties there is a provision of protection by patents or by *sui generis* system. India has opted for *sui generis* systems, which is being developed taking into account the national requirements.

Social Clause

This Social Clause aimed at nullifying the price advantage enjoyed by the developing countries like India due to low labour cost. Such goods thus would be subject to import duties by the developed economies. As a result of this—

- The multinational firms interested to make investment in developing countries may find it unattractive as a result of this clause.
- The migration of labour to advanced countries may be checked resulting in still higher rate of unemployment and misery of developing economies.
- The child labour, which forms an important part of working population in some specialised enterprises like carpet industry, is supposed to be taken out causing severe effect on such industries.

This cause has yet been kept out of the agreement and India has opposed it along with most of other developing nations.

The Negative Effects

A large chunk (nearly two-third) of our population is dependant on agricultural sector. For the developing economy, a steady shift in population from agriculture to secondary and tertiary sectors is of vital importance. As a result of this change, the small-holdings may not find suitable places in the agricultural sector.

The product quality improvement is a must to come at par with the international standards. Therefore, the clause relating to sanitary and phyto-sanitary measures is likely to be exploited against our interest as it happened during the outbreak of plague and malaria in 1994 causing serious impact of India's exports. Further, the dishonesty, corruption and favouritism need to be rooted out in this connection which seems to be difficult task in the short run. The multinational companies are bound to exploit the research and the natural resources of the developing countries to get maximum in the short run without bothering for environment of the country. The effect of the agreement is to bring in specification rather than diversification of occupations in the economy. The specialisation in crop enterprises may be disastrous in situation of natural calamities. The social factors, which have been an attraction for the people towards developing countries, may also receive a setback. The social norms, taboos, values customs etc., are likely to be translated into economic parameters causing a decline in the overall social welfare.

Strategies to be of Concern

- Efforts should be made to educate the farmers about recent developments in the field of agriculture including WTO, such as policy subsidies on inputs, non-tariff measures, parent, law and IPR etc.
- There is a great need to improve the quality of agricultural products being exported by maintaining standards on per international norms to compete with MNCs of the world.
- Government should take into account the interests of small and marginal farmers. Separate strategies are also required to make the small farmers to participate effectively in the globalisation process.
- Need to explore and develop human resources for hi-tech farming, computer aided extension services for providing information on diving forces to farmers and export.
- All the state governments should amend the land ceiling.
- Agricultural sector should be given the status of an industry in terms of product price fixation and provision of budgetary support, credit, power, transport and export facilities.

- Export procedure should be nationalised and simplified in the light of the world trade agreement.
- Biotechnology should be promoted on a massive scale.

REFERENCES

Gulati, Ashok and Anil Sharma (1997). *Subsidies and Investments in Indian Agriculture*, paper presented at Rajiv Gandhi Institute for Contemporary Studies, New Delhi, July 1997.

Hansra, B.S., *et al.* (1999). *Globalising Indian Agriculture: Policies and Strategies*, Classical Publishing Company, New Delhi.

Thakur, S.P. (1999). "World Trade Organisation: Operation and Impact", *Civil Service Chronicle*, January 1999, pp. 12-14.

6

AGRIBUSINESS OPPORTUNITIES IN ORISSA

— *C. Satapathy* and *S.K. Rout*

With change of time and situation, the farmers of Orissa are in search of profitable enterprises in farming sector. The government as well as non-government organisations have been trying to change the ways of farming to a profitable enterprise. The concept of agribusiness have come to play a vital role among scientists, farmers and extension agencies. But unfortunately the structure, function and organisational behaviour do not support these efforts to accomplish goals. In the context of globalisation of agriculture the farmers of Orissa have been responding to challenges in different degree and ways.

The farming structure of Orissa consists of different situations. These are: rainfed and irrigated agriculture, tribal and non-tribal agriculture, old and modern agriculture and after all non-progressive and progressive agriculture. Keeping the consideration in view to motivate farmers for agribusiness is not an easy job. The response of the farmers to emerging agribusiness is directly proportional to the motivational inputs that we put in system to operate. The socio-psychological factors, economic advantage, market network and sound Government policy are key elements for the concept of agribusiness into operation in the State of Orissa. With significant proportion of small and marginal farmers, Scheduled Castes and Scheduled Tribes groups, rainfed areas and ill marketing structure, sometimes it becomes very difficult to receive positive response of the farmers. Research and extension experiences have clearly proved in tribal areas that rainfed agriculture will take a long time for transmitting into agribusiness. Even in agricultural tracts our experiences are not very positive in sensitising farmers for agribusiness.

The Enterprise and Decision-Making

Extension studies in Orissa have enlisted certain enterprises, which are emerging and have potential growth in the State of Orissa. These are seed production, off-season vegetable, tissue culture banana, mushroom cultivation, broiler farming, floriculture, raising of seedling and grafts. Before advising farmers to take up these enterprises there is need to make a SWOT analysis of these enterprises. 'SWOT' is an acronym stands for Strength, Weakness, Opportunity and Threat relating to various problems, programmes, plan of work and so also of an organisation. On the basis of field experience and interaction between farmers, scientists and traders through SWOT analysis some of these aspects have been studied which are described herewith.

Paddy-cum-Pisciculture

Orissa has a considerable large low lying and waterlogged areas. Due to waterlogging and poor management practices, there is limited scope to increase productivity. Similarly the fish production from available sources not sufficient to meet the requirement of the State. Considering these aspects, of the low lying areas having rice monocropping system is put to paddy-cum-pisciculture, there will be harvest of two commodities simultaneously. The SWOT analysis of paddy-cum-fish culture is analysed below.

Strength

1. Huge waterlogged areas available in the State particularly in coastal areas and Hirakuda command areas.
2. Rice-fish farming system increases the productivity of the land and provides better income than sole crop of rice.
3. Lowland and deep-water areas could be successfully utilised for rice-fish farming against crop failure.
4. Interested farmers have easy access to take training and advisory service from KVKs, OUST and CRRI.
5. Inputs required for rice-fish culture is available.

Weakness

1. Plant protection measures could not be taken for rice crop for the safety of fishes.

2. Since it is mainly rainfed farming, availability of water depends upon the rainwater for the growth of the fish.
3. For paddy-cum-fish culture, bonding of the area is necessary as the fish may move to other areas.

Opportunity

1. Low lying and waterlogged area could be successfully utilised and more income be generated through paddy-cum-fish culture.
2. Two crops can be raised with same cost and same space with pooled return.
3. There is heavy market demand of fish. The farmers can sell the fish in villages and nearby market. Local traders will also purchase fish from the spot itself.
4. Fish could be successfully grown under submerged condition of rice field for 4-6 months.

Threat

1. During harvesting period, depth of water becomes less; predator, animals and birds will easily eat the fishes.
2. The fish tracks in the rice field cause irrigation for the subsequent crops.

Raising of Vegetable Seedling and Fruit Plants

Only ten per cent of total gross cropped area of Orissa are covered under vegetables and fruits. Annual vegetable and fruit production is not sufficient to meet the requirement. The SWOT analysis for raising vegetable seedlings and fruit plants are detailed herewith.

Strength

1. Raising of planting materials is a good source of income and can be raised throughout the year in rural areas.
2. Nursery could be raised in a small piece of land and all categories of landholders could take up the enterprise.
3. Much demand exists for vegetable seedlings and fruit plants of better genotype.
4. Unemployed youths could be fully employed and take it as a vocation.
5. More profitable than other farm enterprise within a very short span of time.

Weakness

1. Regular surveillance against pest and diseases is essential.
2. Failure of germination of seeds/stocks incur loss in business.

Opportunity

1. It is good way of employment for the educated youths and school dropouts.
2. More investment is not required since the materials disposed of within a short period.
3. Necessary knowledge and skills could be acquired from the training at KVK or other training institution.

Threats

1. Prolonged bad weather prevents nursery raising and good growth.
2. Transit nursery may be required if not disposed of in seasons.

Tissue Culture Banana

Banana cultivation is a well-known remunerative enterprise. The plantlets from successors are late fruiting, irregularly bunched and are affected by various diseases. This restricts extension of banana in Orissa. Tissue culture banana has the advantage of true to the type, disease free, early fruiting, uniform bunching with average of 140 to 150 fingers per plant and high yielder. SWOT analysis on tissue culture banana enlight different aspects of it so that interested farmers can do for it.

Strength

1. Agro climatic condition is suitable for tissue culture banana cultivation.
2. Bunches are uniform, big finger size and average 140-150 fingers per bunch.
3. Easily available with affordable price.
4. Financing institutions are also interested to finance for commercial cultivation.
5. Uniform harvesting make easy to dispose to the traders in a reasonable price.

Weakness

1. It requires more investment, more irrigation, protected fencing for which resource poor farmers could not afford to that extent.
2. It is relatively sophisticated technology than suckers and farmers need to be well equipped with knowledge and skills.
3. Sufficient and frequent irrigation is necessary during dry seasons.

Opportunity

1. Heavy market demand with good price.
2. Plantless are sufficiently available with reasonable price.
3. The entrepreneur could get subsidy and credit facilities through Krishi Sahayak Kendra (APPICOL is financing such projects).
4. Crop insurance is there, if there is any loss due to unnatural situations.

Threats

1. There will be heavy damage to the plant if there is cyclone and hailstorm.
2. Complete watch and ward is necessary for discriminates for storing bunches.
3. Heavy winter during bunch initiation period reduces the bunch size and occurrence of throat choke disease.

Seed Production

Out of 64,20,000 ha total cultivated area, cereals cultivated in 75.20 per cent area, pulses 7.66 per cent, oilseeds 7.66 per cent, fibres 0.75 per cent and other crops 2.17 per cent area. While analysing the yield, the statistics indicate that, per hectare yield of cereals is 18.53 qtls., pulses 4.09 and oilseeds 7.11 qtls./ ha only which is much below the national average. One of the important factors for the low yield is the non-use of quality seeds. The farmers usually use their own seeds year after year. So far the State Government has been able to replace up to 10 per cent seeds. The government. is now encouraging private seed growers to meet the requirement. The farmers are purchasing vegetable seeds from different private firms without guarantee. The procurement price of different types of seeds are also quite high in comparison to non-seeds.

Therefore, seed production is more profitable and could be considered as a vocation by the young farmers. Production of seeds/seedling at panchayat level can be taken up. This will open new horizon for the farmers of Orissa.

Strength

1. There is much demand for quality seed.
2. Price of seed is quite remunerative.
3. Crop insurance facilities now available in the State.
4. Credit facilities available.

Weakness

1. Seed cost is high for which resource poor farmers cannot afford.
2. It requires assured irrigation facilities.
3. It is labour intensive; require timely recommended inputs for which cost of cultivation is high.
4. It requires good storage condition to maintain viability.

Opportunities

1. Farmers can easily acquire the knowledge and skills.
2. Processing of seeds done by the O.S.S.C. in its nearby processing plant with nominal price. Most of all DDA range has this provision.
3. After processing, rejected grains could be used for consumption purpose.
4. There is also demand by the private traders, input dealers, as well as farmers for purchasing seeds from the farmers.

Threat

1. Seeds cannot be certified, if there is unfavourable conditions like cyclone, pest and disease calamity, drought situation etc.
2. If not disposed of within time limit, the viability will be deteriorated and may not be used for seeds.

Broiler Poultry Farming

Broiler poultry farming is a very good profitable enterprise compared to different agro-based enterprises, India is a exporting broiler birds to south Arabian countries. Broiler farming can be taken up in large scale as well as in small numbers for which different types of farmers can go for it. The SWOT analysis of Broiler farming reveal the following.

Strength

1. Broiler birds are quick growing and disposed within 6-8 weeks.
2. On an average, a net profit of Rs. 10 per bird is obtained.
3. It could be reared from small scale to large scale commercial basis.
4. Returns obtained within a short span period which will be invested in a cyclic order.
5. One of the good enterprise for self employment.
6. Heavy market demand.
7. Family labour will be utilised.

Weakness

1. Specific shed required for protection against weather and predators.
2. Requires little more investment.
3. Regular visit by Veterinary Doctor is required.

Opportunity

1. Quality chicks of different strains are readily available.
2. Technologies are available and simple.
3. Easy access to demonstration units in KVKs. Farmer's condition for exposure and conviction.
4. Insurance facilities are available.

Threat

1. A couple of fatal diseases pose all time threat to broiler farming.
2. No alternative against market glut due to over production and incur loss.
3. Too hot or too cool climate restricts growth of the bird.

7

PRODUCTION, PROCESSING AND MARKETING OF LAC
Concerns and Remedial Measures

— *K. Krishan Sharma, A.K. Jaiswal* and *K.K. Kumar*

Lac is a natural resin secreted by a tiny insect which requires plant hosts for its propagation. Besides resin two other products namely, dye and wax are also produced by the insect. India produces about 15,000-20,000 metric tonnes of lac annually, which is about 50 per cent of total world lac production. Jharkhand, Chhattisgarh, West Bengal, Orissa and Maharashtra are major lac growing States. Jharkhand alone accounts for about 60 per cent of the lac produced in the country followed by Chhattisgarh (23%) and West Bengal (10%). 80-85 per cent of the total lac produced in our country is exported to over 100 different countries mainly in the form of seedlac, shellac and buttonlac. At present India earns about 70-80 crore of foreign exchange per annum through export of lac.

Lac cultivation has traditionally been carried out mainly by tribals inhabiting Chhotanagpur region of Jharkhand and adjoining areas of Chhattisgarh, West Bengal and Orissa. Lac culture has got considerable importance in the tribal economy. The importance of the lac for the region can be judged by the results obtained from recent studies which reveal that income from lac cultivation contributes about 27 per cent of their total agricultural income. It can be safely assumed that majority of the households dependant on agriculture in Ranchi, Palamau and Singhbhum districts of Jharkhand happen to be growing lac as a subsidiary source of income. More than three million families are engaged in cultivation of lac and about one million man-days per year are generated by 150 odd lac processing units which provide employment opportunity to the weaker sections of the society.

What ails Lac Cultivation and its Marketing?

The significance of the increased lac production lies in the fact that it can be obtained without divesting land from other crops. Production of lac, however, has not evoked a steady response from its growers owing to fluctuating prices. Similarly it is the lack of remunerative returns from lac which has stood in the way of popularisation of improved techniques of lac cultivation. Low levels of literacy has a significant bearing on lac production.

Production of lac in India over the years has shown a consistently declining trend. During sixties the mean value of lac production was 32.57 thousand tonnes per annum. It came down to 20.37 thousand tonnes in seventies and to an all time low of 16.54 thousand tonnes in eighties. Large scale deforestation, lack of appropriate measures by the agencies concerned and price instability are some of the factors responsible for this decline leading to the shrinkage in lac producing areas of the country. Some of the important lac growing States like Orissa, Uttar Pradesh and Assam now, have a negligible share in lac production. Because once the lac cultivation is abandoned, it becomes difficult to reintroduce it due to non-availability and high prices of broodlac. As a result un-exploited lac-host trees are cut for timber and fuel purposes, thus further eroding the lac cultivation area.

Some of the remedial measures to address the concerns afflicting the lac production, processing and marketing for creating a favourable impact are listed in Table 7.1.

Role of Indian Lac Research Institute

The Indian Lac Research Institute, Namkum, Ranchi; now a constituent unit of Indian Council of Agricultural Research, New Delhi was founded on 20th September, 1924 to serve as National apex body on lac, with a charter to not only promote R & D activities in biological, chemical and engineering sciences but also to promote and assist lac industry in all spheres. In more than 75 years of its existence, the institute has fulfilled many of its objectives, attained an international status and has gone on further to develop improved lac cultivation technologies for boosting the production and for diversified high-tech uses of lac and lac based products.

Lac production technologies developed at the institute for sustainable yield on conventional hosts, *palas* (*Butea monosperma*), *ber* (*Ziziphus mauritiana*) and *kusmi* (*Schleichera oleosa*) offer 50-100 per cent increase

TABLE 7.1: The Remedial Measures to Concerns Afflicting the Lac Production, Processing and Marketing for Creating a Favourable Impact

Sl. No.	*Concern*	*Suggested Remedial Measure*	*Expected Impact*
(1)	(2)	(3)	(4)
A. *Lac Production*			
1.	Exploitation of idle lac host-plants	* Scheme for extensive survey of lac-hosts should be started to create the base information. * Free distribution of broodlac in new areas should be taken up. * Integration of lac-culture with agriculture.	* It will bring un-exploited trees under lac cultivation. * It will also check deforestation as farmers will prefer sustained income from lac cultivation to one time income from sale of timber. * It will enhance per unit area returns.
2.	Revival of existing broodlac farms	* The existing broodlac farms owned by State Forest Departments or Institute of Forest Productivity (IFP) need to be revived and managed by trained personnel. * Establishment of new *kusmi* and *rangeeni* broodlac farms in different areas. * The task should be taken up by the State Forest Departments and/or by IFP.	* It will ensure steady supply of much needed broodlac to lac growers during period of scarcity and also to those interested in starting lac cultivation afresh.
3.	Mechanisation of lac cultivation	* Lac cultivation requires climbing on to trees for carrying out various operations. Available implements need to be popularised and more need to be developed to reduce drudgery.	* The farmers will be able to undertake lac cultivation on a larger scale. * Dependence on labour force will be reduced. * Overall lac productivity will be increased.
4.	Minimum support price for lac to lac growers	* Need to take immediate action for fixation of minimum price of raw lac as it relates to the poor of the poorest section of the society. * TRIFED and such other agencies should be given more government support for effective intervention in the market when lac price is very low.	* The lac growers will be prompted to take more interest and thus bring more area under cultivation of lac. * Price fluctuation would be checked.

(*contd.*)

TABLE 7.1 (*contd.*)

(1)	(2)	(3)	(4)
5.	Lac crop insurance scheme	* Since lac crop is subject to the vagaries of weather, introduction of lac crop insurance scheme will help the lac growers for taking risk in lac cultivation.	* This will help, particularly small farmers, who are dependant mostly on lac cultivation for their livelihood. * The farmers will be motivated to avail credit facility for raising lac crop.
B. *Lac Marketing*			
6.	Free trading (marketing) of lac produce	* Remove restriction on movement of broodlac and sticklac from one State to another.	* Will help lac growers of those States where lac processing units are not located and in restarting the lac cultivation where crop failed due to adverse climatic conditions. * Help the farmers to get remunerative price of their produce.
7.	Formation of Lac Growers' co-operatives	* Lac growers' co-operatives at block/village level in various lac growing. * Encourage inter-State co-operative marketing system.	* Development and usage of common facilities for cultivation and post-harvest operations. * They will be able to market their produce during favourable periods thus preventing distress sale. * Value addition before marketing for better returns. * Will help the farmers in neutralising the middlemen to a large extent.
8.	Minimise the role of middlemen	* The farmers may be motivated to form "self-help" groups for collective marketing of their produce.	* A favourable impact on the pricing of the commodity. * Better returns to the growers.

(*contd.*)

TABLE 7.1 (*contd.*)

(1)	(2)	(3)	(4)
9.	Liaison activity and awareness	* A nodal information agency with its network in different States on lac should be established, in order to help and provide information on lac to the needy.	* It will help lac growers and entrepreneurs in keeping abreast with recent developments and also procurement and selling of broodlac, sticklac etc.
C. *Lac Processing*			
10.	Modernisation of lac processing units	* Lac processing/product units should use latest technologies/machines available.	* It will prevent processing losses and reduce drudgery and production cost. * It will help attain consistent quality which is very critical for sustaining the demand.
11.	Incentive to the lac growers/ processing units	* Need for a policy decision to provide incentives and attractive loans to lac growers particularly tribals who cannot afford to buy the broodlac from the market. * For increasing the domestic consumption, the lac-based unit in the country should be given tax concessions. * Low interest loans should be made available to entrepreneurs interested in establishing lac-based industries.	* It will help to reinitiate the cultivation of lac by the farmers who have abandoned its farming. * It will help in stabilising lac price because at present it is governed by overseas buyers. * It will enhance the domestic consumption of lac.

in productivity as well as 100-400 per cent in profitability. The cultivation technology package developed for quality kusmi broodlac and sticklac production offers a return of Rs. 2,500 per tree per annum as against Rs. 666 under conventional method. Mixed plantation technique accommodating fast growing 1,200 *galwang* (*Albizia lucida*) and 3,600 *bhalia* (*Flemingia macrophylla*) bushes per hectare along with intercrops increases the scope of lac cultivation on waste land offering the profitability upto Rs. 33,000/ha/annum. Similarly, a new fast growing plant *Akashmani* (*Acacia auriculaeformis*) when used in alternation with *kusmi* trees provides a net profit of Rs. 43,000/ha/annum.

Integrated pest management has resulted in lac yield increase to 160 per cent over the traditional method. R&D efforts of ILRI has led to establishment of 25 industries in recent past for manufacturing of aleuritic acid, isoambrettolide, gasket shellac compound, shellac bond powder and bleached lac. Recently a lac scraper-cum-grinder has been developed for lac growers to earn more profit, if they do the processing and refining of lac at their own level for selling the semi-refined product.

The institute has laid emphasis upon the transfer of technology activities so that the research findings are passed on to the farmers, grass-root level workers, entrepreneurs so as to enable them to have more profit. The technology transfer is effected through arrangements which include training-cum-demonstration, consultancy and advisory services and quality control maintenance. The institute plays a major role in fixing national and international standards for lac and lac based products by virtue of being a permanent member of CHD 023 committee of Bureau of Indian Standards. The testing laboratory of the institute is internationally recognised by number of importing countries and provides services for quality testing of lac and lac based products. The institute is also collaborating with a number of government organisations (GOs) and non-government organisations (NGOs) in its Transfer of Technology (ToT) activities and has registered a tremendous growth in the participation of farmers/lac growers in the training programmes in recent years.

8

RAISIN PRODUCTION MANAGEMENT
An Economic Venture

— *R.R. Suryawanshi, S.S. Ingavale* and *P.M. Kapase*

In Maharashtra, grape growing in recent years has become a commercial proposition. However, the grape producers facing difficulties in getting reasonable prices for their fresh grapes. Processing of grapes helps in better utilisation of agri-produce. It also helps in value enhancements employment generation, substantial export earning, increase storage life, time utilisation, etc. As such raisin production enterprise in western Maharashtra has got vital importance in modern days especially with changed economic scenario under WTO regime. This will facilitate to explore technological options required for the promotion of agro-enterprise for rural prosperity, to analyse appropriate management system desired for sustainable development of agribusiness and to discuss opportunities and constraints in agriculture production, procurement, processing and marketing systems in the context of emerging raisin production enterprise.

The present study on production and marketing management of raisin enterprise was conducted during 1999-2000 in Tasgaon region of Sangli district of Western Maharashtra covering 96 raisin producers spread over eight villages. Further these producers were classified into three major groups.

Group I — Grape growers who prepared raisins from the grapes of their own grape gardens.

Group II — Grape growers who prepared raisins from the grapes of their own grape gardens as well as purchased grapes from the other grape gardens.

Group III — Grape growers who prepared raisins only from the grapes purchased from the other grape gardens.

A sample of 32 raisin producers from each group who adopted dipping oil method for raisin production and marketed the raisins were selected purposively.

Results

Production and Disposal Pattern of Grapes

The information regarding the average production of grapes and its disposal is presented in Table 8.1.

TABLE 8.1: Per Farm Average Production of Grapes and their Disposal Pattern *(figures in quintals)*

Sl. No.	*Particulars*	*Size Groups* I	II	III	*Overall*
(1)	(2)	(3)	(4)	(5)	(6)
1.	Total production grapes	285.86 (100.00)	303.31 (100.00)	—	295.58 (100.00)
2.	Disposal pattern				
	(i) Sale of fresh fruit	57.59 (20.11)	51.84 (16.86)	—	54.32 (18.38)
	(ii) Used for raisin making	226.63 (79.28)	250.16 (82.47)	—	239.4 (80.99)
	(iii) Used for home consumption including gifts to relatives and friends	1.72 (0.61)	2.00 (0.67)	—	1.86 (0.63)
3.	Variety-wise groups used for raisin making	226.63 (100.00)	335.12 (100.00)	570.18 (100.00)	443.97 (100.00)
	(i) Thompson Seedless	155.18 (68.47)	225.26 (42.10)	120.00 (21.05)	166.80 (37.57)
	(ii) Tas-A-Ganesh	55.00 (24.26)	290.28 (54.24)	450.18 (78.95)	265.15 (59.72)
	(iii) Sonaka	6.87 (3.04)	5.83 (1.09)	—	4.23 (0.95)
	(iv) Manik Chaman	9.58 (4.23)	13.75 (2.57)	—	7.77 (1.76)
4.	Raisin production pattern				
	(i) Total output of raisins	56.42 (100.00)	138.65 (100.00)	143.09 (100.00)	112.72 (100.00)
	(a) Grade—I	47.43 (84.00)	118.03 (85.12)	121.59 (84.97)	95.69 (84.79)
	(b) Grade—II	8.77 (15.08)	19.34 (13.94)	20.09 (14.04)	16.07 (14.26)

(Contd.)

TABLE 8.1 (*Contd.*)

(1)	(2)	(3)	(4)	(5)	(6)
	(c) Grade—III	0.52 (0.92)	1.28 (0.94)	1.41 (0.99)	1.07 (0.95)
	(ii) Recovery percentage	24.89	25.91	25.10	25.39

Note: Figures in the parentheses are the percentages to the respective totals.

It is revealed from Table 8.1 that, at the overall level, the average per farm total production of grapes was about 295.58 quintals of which quantity sold as fresh fruit and used for raisin making was about 18.38 and 80.99 per cent respectively. Among the size groups, grape growers from size Group II diverted maximum quantity of grapes for raisin production.

As regards to variety-wise quantity of grapes used for raisin production, it is seen that at the overall level, the proportion of Thompson seedless and Tas-A-Ganesh varieties was 37.57 and 59.72 per cent respectively in total quantity of grapes used for raising making.

Among the size groups, it was noticed that size Group I, used larger quantity of Thompson seedless variety of grapes (68.47%) was used for raisin making, while in Group III Tas-A-Ganesh variety grapes (78.95%) was used. Recovery percentage in raisin production was about 25.39 per cent at the overall level.

Cost of Processing/Production

It could be seen from the data of Table 8.2 that at the overall level, on an average per kg cost of raisin processing was worked out to be Rs. 56.52 of which proportionate sharer of variable and fixed cost were 86.16 and 10.84 per cent respectively. Among the variable cost, cost of grapes used (raw material cost) was major item (79.25%).

TABLE 8.2: Per kg Cost of Raisin Making

(*figures in Rs.*)

Sl. No.	*Particulars*	*Size Groups*			
		Group I	*Group II*	*Group III*	*Overall*
(1)	(2)	(3)	(4)	(5)	(6)
I.	*Variable Cost*				
	1. Labour cost	1.93 (3.39)	2.29 (4.09)	2.38 (4.28)	2.27 (4.02)
	2. Chemicals	1.42 (2.50)	1.40 (2.50)	1.40 (2.49)	1.40 (2.47)

(*Contd.*)

TABLE 8.2 (*Contd.*)

(1)	(2)	(3)	(4)	(5)	(6)
3.	Packing material	1.31 (2.31)	1.34 (2.39)	1.29 (2.30)	1.31 (2.32)
4.	Electricity charges and other charges	0.20 (0.35)	0.35 (0.63)	0.36 (0.64)	0.34 (0.59)
5.	Interest on working capital	0.27 (0.48)	0.29 (0.52)	0.29 (0.53)	0.29 (0.51)
6.	Raw material (Cost of grapes used)	43.50 (76.58)	44.60 (79.75)	45.50 (81.00)	44.79 (79.25)
7.	Total variable cost	48.63 (85.62)	50.27 (89.88)	51.24 (91.23)	50.40 (89.16)
II.	*Fixed Cost*				
A.	Depreciation on shed and equipments	3.45 (6.08)	3.00 (5.37)	2.55 (4.54)	2.88 (5.11)
B.	Interest on fixed capital	4.72 (8.30)	2.66 (4.75)	2.38 (4.23)	3.24 (5.73)
C.	Total fixed cost (A+B)	8.17 (14.38)	5.66 (10.12)	4.93 (8.77)	6.12 (10.84)
	Total cost	56.80 (100.00)	55.93 (100.00)	56.17 (100.00)	56.52 (100.00)

Note: Figures in parentheses are percentages to total cost.

Methods of Sale

Regulated market for raisins was established at Tasgaon in Sangli district in the year 1994, wherein auction method of sale has been generally followed. The wholesalers and traders from the different places in the country take part in open auction sale in the market day. The raisin producers before their sale of produce, first contacts with the commission agents in the markets and after his consent, handover the sample of their produce for auction. The prices are given according to the quality and grade of the raisins. The raisin producers also market their raisins at Delhi market.

Marketing Channels

The information of marketing channels observed for sale of raisins in different markets is as follows:

1. Tasgaon market
   Producer—Commission Agent—Wholesaler—Retailer—Consumer

TABLE 8.3: Group-wise per kg Average Cost of Marketing of Raisin in different Markets

Sl. No.	Items of Marketing Cost	Tasgaon Market				Delhi Market				Overall			
		Group I	Group II	Group III	Overall	Group I	Group II	Group III	Overall	Group I	Group II	Group III	Overall
1.	Grading	1.15 (25.79)	1.05 (21.64)	0.88 (19.76)	1.01 (21.52)	1.43 (13.19)	0.98 (8.98)	0.95 (8.60)	1.12 (10.13)	1.17 (24.63)	1.04 (18.34)	0.92 (18.61)	1.01 (19.40)
2.	Packing	1.29 (28.74)	1.26 (26.07)	1.21 (27.40)	1.27 (27.10)	1.93 (17.80)	1.81 (16.58)	1.75 (15.86)	1.81 (16.33)	1.31 (27.73)	1.34 (23.59)	1.29 (26.21)	1.31 (25.27)
3.	Transport charges	0.16 (3.65)	0.15 (3.22)	0.12 (2.78)	0.14 (3.10)	2.41 (22.25)	2.36 (21.56)	2.28 (20.66)	2.35 (21.17)	0.25 (5.36)	0.45 (8.02)	0.23 (4.62)	0.32 (6.25)
4.	Hamali	0.02 (0.59)	0.02 (0.55)	0.02 (0.58)	0.02 (0.57)	0.10 (0.92)	0.10 (0.90)	0.09 (0.86)	0.09 (0.88)	0.03 (0.62)	0.03 (0.64)	0.03 (0.61)	0.03 (0.62)
5.	Commission charges	1.32 (29.32)	1.34 (27.65)	1.27 (28.63)	1.32 (28.36)	4.36 (40.20)	4.55 (41.64)	4.50 (40.67)	4.53 (40.83)	1.44 (30.33)	1.77 (31.31)	1.47 (29.87)	1.59 (30.59)
6.	Cold storage charges	0.43 (9.64)	0.93 (19.33)	0.87 (19.51)	0.84 (17.80)	0.51 (4.75)	1.06 (9.65)	1.36 (12.27)	1.08 (9.77)	0.43 (9.19)	0.95 (16.80)	0.92 (18.77)	0.85 (16.43)
7.	Other charges	0.10 (2.27)	0.08 (1.54)	0.06 (1.34)	0.08 (1.55)	0.10 (0.89)	0.08 (0.69)	0.12 (1.09)	0.09 (0.89)	0.10 (2.14)	0.08 (1.32)	0.06 (1.31)	0.08 (1.44)
	Total marketing cost	4.47 (100.00)	4.83 (100.00)	4.43 (100.00)	4.68 (100.00)	10.85 (100.00)	10.94 (100.00)	11.06 (100.00)	11.08 (100.00)	4.73 (100.00)	5.66 (100.00)	4.92 (100.00)	5.19 (100.00)

Note: Figures in parentheses indicate percentages to total marketing costs.

2. Delhi market
   Producer—Commission Agent—Wholesaler—Retailer—Consumer

Marketing Cost

It is apparent from the data in Table 8.3 that at the overall level, the average per kg cost of marketing worked out to Rs. 5.19. The major items of the cost of marketing were commission charges, packaging cost, grading cost and cold storage charges which constituted 30.59, 25.27, 19.40 and 16.43 per cent respectively in the total cost of marketing. Further, it was observed that at the overall level, the marketing cost in Group II was relatively high as compared to Groups I and III. It was due to the high commission charges and more transportation cost.

Price Spread

The price spread in marketing of one kilogram of raisins in Mumbai market had been worked out based upon the information given by traders in Tasgaon market. The results so obtained are present in Table 8.4.

TABLE 8.4: Per kg Price Spread in Marketing of Raisins

Sl.No.	*Particulars*	*Rupees per Kilogram*
1.	Gross price received by the producers	68.14 (64.89)
2.	Market expenses incurred by the producers	4.73 (4.50)
3.	Net price received by the producers	63.41 (60.39)
4.	Commission received by the commission agents @ 2%	1.36 (1.30)
5.	Expenses incurred by the wholesalers	8.12 (7.73)
6.	Margin of the wholesalers	4.00 (3.81)
7.	Expenses incurred by the retailers	5.38 (5.13)
8.	Margin of the retailers	18.00 (17.14)
9.	Price paid by the consumers in market	105.00 (100.00)

Note: Figures in the parentheses are the percentages to final price paid by the consumers.

It is seen from Table 8.4 that the average per kg consumers price was Rs. 105.00 in Mumbai (during March 2000) market while, the gross price received by the raisin producer in Tasgaon market was Rs. 68.14 per kg. As such the producers share in consumers rupee was about 65 per cent.

Cost and Returns per kg

The group-wise per kg cost and returns obtained in raisin processing are shown in Table 8.5.

TABLE 8.5: Group-wise per kg Cost and Returns from Raisin Making

(figures in Rs.)

Sl.No. Particulars	*Size Groups*			
	Group I	*Group II*	*Group III*	*Total*
I. *Costs*				
1. Variable cost	48.63	50.27	51.24	50.40
2. Fixed cost	8.17	5.66	4.93	6.12
3. Marketing cost (exclusive of cost of packing material)	3.41	4.32	3.63	3.79
4. Total cost (1+2+3)	60.21	60.25	59.80	60.31
II. *Production of raisins (kg)*	1	1	1	1
III. *Returns from raisins*				
1. Gross returns	66.08	68.49	68.63	68.14
2. Returns at variable cost	17.45	18.22	17.39	17.74
3. Returns at fixed cost	57.91	62.83	63.70	62.02
4. Net returns	5.87	8.24	8.83	7.83
IV. *Benefit-Cost Ratio*	1.097	1.136	1.147	1.129

The data in Table 8.5 indicate that at the overall level, on an average per kg total cost of raisin processing and marketing together computed at Rs. 60.31. Whereas, gross returns fetched through sale of raisin were worked out to Rs. 68.14. After deducting the total cost of processing and cost of marketing, thereof from the receipts of sale proceed, net returns realised by the raisin producers were estimated at Rs. 7.83. This revealed that raisin making was economically profitable venture.

The Benefit-Cost Ratio (BCR) for raisin making was computed at 1.13 which is greater than unity. Thereby showing that raisin processing is economically viable venture.

Constraints in Production of Raisins

The problems in production of raisins faced by the producers were enormous in number. However, from numerous problems, few which are of serious nature which need special and immediate attention are described below.

The most important problems in production of raisins perceived by the raisin producers were: (1) non-availability and high charges of skilled labourers; (2) adverse climatic conditions affecting drying, process; (3) non-availability of standard chemicals at cheaper rate; and (4) high cost of establishment of raisin processing unit. The number of raisin producers who reported the above problems constituted 89, 75, 65 and 54 per cent respectively of the total sample producers.

The other problems faced by the sample raisin producers were (1) non-availability of required quality grapes for (45.75%) raisin production; (2) high cost of raw material, *i.e.*, grapes (36.46%); (3) non-availability of grapes of required variety (46.89%); and (4) lack of technical knowledge (53.12%) etc.

Constraints in Marketing of Raisins

The problems in marketing of raisins faced by the producers were also enormous in number. The information about the major problems faced by the majority of raisin producers in marketing is presented as below.

The most important problems in marketing of raisins were: (1) delay in payments by commission agents; (2) loss of produce in sampling; (3) high cost of packing material; (4) high commission charges; (5) non-availability of labour in time; (6) fluctuation in prices etc.

Suggestions by the Raisin Producers

The raisin producers in the selected area suggested many suggestions for improvement in production and marketing of raisins. The important measures suggested by the raisin producers were: (1) payment must be made immediately after sale of produce; (2) need of proper control of market committee over the commission agents; (3) mechanical packing and grading facilities required to be made available; (4) reduction in commission charges; (5) minimum support price need to be declared; (6) need of quality packing material at subsidised rate etc. The number of raisin producers who reported the above suggestions constituted 83, 78, 73, 71 and 68 per cent respectively of the total samples producers.

The other suggestions of the raisin producers pertained to (1) standard chemicals need to be provided at subsidised rates; (2) provision of technical knowledge; (3) quality grapes of required variety need to be produced; (4) reduction in cold storage charges; (5) finance at cheap interest rate; (6) imports of raisins from other countries need to be restricted etc.

Conclusions

The present investigation intends to depict the picture of raisin production enterprise in Sangli district of Western Maharashtra.

From the above study the following conclusions are drawn:

It was observed that on an average per kg cost of raisin production and marketing together computed to Rs. 60.31. Whereas, gross return worked out to Rs. 68.14, with 7.83 as net return resulting in benefit-cost ratio of 1.13. The study concludes that raisin production is economically viable venture.

The major constraints faced by the raisin producers in production were, non-availability of good quality grapes of good variety at cheaper rate, non-availability of standard quality chemicals at subsidised rates, non-availability and high charges of labour, lack of technical know-how, high cost involved of establishment etc. The major marketing constraints were high cost of packing material, high charges of cold storage, high commission charges, delay in payments and fluctuations in prices etc.

The raisin producers suggested for improvement in production and marketing efficiency for which supply of standard chemicals, technical knowledge, quality packing material at cheaper rates, mechanical packing and grading facilities need to be developed. They expressed need for immediate payment, proper control over the commission agents and minimum support prices for raisins etc.

9

DAIRY BUSINESS
Perspectives for Privatisation

— *Y.C. Sale, B.V. Pagire, H.R. Shinde* and *V.R. Shete*

Livestock rearing and dairying are the important components of mixed farming supporting the sustainable Indian agriculture since long back. The dairying has been providing a strong support to the economy of the rural farm families. The dairying is considered as the prime subsidiary enterprise which not only gives additional income from milk but also provides required bullock power for cultivation of crops. The by-products of crop husbandry as well as the surplus labour and other resources can be gainfully utilised for milk production. Thus, dairy enterprise is a regular source of income and employment.

The Central and State Governments have been taking very constructive steps in increasing the milk production. The Government of Maharashtra has been implementing various dairy development schemes, programmes since last two decades. All the efforts have resulted into an increased milk production of 5,192.8 thousand MT in 1997-98 from 1,062 thousand MT in 1970-71. There is a greater demand for good quality milk and milk products in the country. And for fulfilling the consumer demand, number of dairy units comprising co-operative as well as private agencies are entering into the activity involving collection, processing, distribution of milk and also manufacturing and marketing of milk products. In the co-operative and private agencies, there is a cut-throat competition in procurement of sufficient quantity of raw milk from the producers. The success of this business managed under different management systems, therefore, depends largely on the abilities in the milk marketing activities right from collection to distribution.

The present study was undertaken with a view to examine whether private or the co-operative agencies have the perspectives in the marketing of milk.

Methodology

The present study was based on the information of marketing of milk through private and co-operative agencies in Phaltan tehsil of Satara district. One private and one co-operative agency were selected for the study. The data were collected for the year 1998-99. From the two selected dairy units the data on the aspects covering milk procurement, processing and distribution and costs such as fixed and variable in each of these were compiled and analysed with suitable statistical tools.

Results and Discussion

Management Cost

For efficient management of the business, the dairy units had to incur costs on account of several items. The cost included all costs of general administration, taxes, rent, salary, meeting and travelling allowances, postage, benefits, etc. The detail break-up of management cost of milk for these two units (agencies) is presented in Table 9.1.

The proportion of the fixed cost was 27.70 per cent and 20.21 per cent in the total management cost in case of co-operative and private agency, respectively. The share of variable cost however, was higher, *i.e.*, 79.78 per cent in case of private dairy unit whereas it was 72.30 per cent for co-operative unit. The high fixed cost in case of co-operative was due to high cost on interest whereas, the high variable cost in the private unit was due to higher cost on account of rent, taxes, fees, electricity, office expenses etc.

The management cost was estimated at Rs. 23.50 per 100 litres of milk in the case of the co-operative unit and Rs. 11.62 per 100 litres of milk for the private dairy unit. The high management cost observed in case of the co-operative dairy unit was mainly due to the salaries and benefits of the increased number of workers.

Gross Returns and Net Returns

The annual gross returns from the business and the net returns accrued therefrom were estimated for both of these units operating. The estimates

TABLE 9.1: Item-wise Management Cost

Sl. No.	*Items of Costs*	*Co-operative Dairy Unit*		*Private Dairy Unit*	
		Rs. (in lakhs)	*Per cent*	*Rs. (in lakhs)*	*Per cent*
(1)		(2)	(3)	(4)	(5)
A.	*Fixed costs*				
	1. Land rent	0.31	1.18	0.15	1.78
	2. Depreciation	3.09	11.79	0.69	8.20
	3. Interest	3.86	14.73	0.86	10.23
	Sub-total	7.26	27.70	1.70	20.21
B.	*Variable costs*				
	1. Salary and benefits	14.94	57.00	3.47	41.26
	2. Office expenses	0.16	0.61	0.27	3.21
	3. Postage, telegram, telex, etc.	1.41	5.38	0.80	9.51
	4. Rent, taxes, fees, electricity, etc.	0.30	1.15	0.87	10.34
	5. Meeting and travelling allowances	0.25	0.95	0.67	7.97
	6. Miscellaneous charges	1.89	7.21	0.63	7.49
	Sub-total	18.95	72.30	6.71	79.78
	Total cost	26.21	100.00	8.41	100.00
	Qty. of milk procured (lakh litres)	111.50	—	72.38	—
Management cost					
	I. Rs./100 litre of milk	23.50	—	11.62	—
	II. Rs./litre of milk	0.24	—	0.12	—

of profitability of the business were worked out by deducting the total costs from the gross returns. The details of these are presented in Table 9.2. The annual gross returns of the co-operative unit were Rs. 988.70 lakh and Rs. 778.47 lakh for the private unit. The total costs were Rs. 1,035.68 lakh and Rs. 701.90 lakh for the co-operative and private unit, respectively, which indicated that the size of the business of the co-operative unit was greater than that of the private unit. The net returns over total cost, for the private unit was Rs. 76.57 lakh, whereas the co-operative unit was having a loss of Rs. 46.98 lakh. The co-operative unit was also having a net loss of Rs. 37.70 lakh over the variable costs while the private unit was in profit with Rs. 96.22 lakh. The per unit analysis brought out more clear picture for the loss and profit. The co-operative unit incurred a loss of Rs. 42.13 per 100 litres of milk handling while the

TABLE 9.2: Profitability of the Dairy Units

Sl. No.	*Particulars*	*Co-operative Dairy Unit*			*Private Dairy Unit*		
		Quantity (lakh lts.)	*Rates (Rs./lit.)*	*Amount (Rs. in lakhs)*	*Quantity (lakh lts.)*	*Rates (Rs./lit.)*	*Amount (Rs. in lakhs)*
I.	Gross returns						
	1. From sale of milk	111.50	7.91	882.16	72.38	10.75	778.13
	2. Commission from Govt.	—	0.95	105.93	—	—	—
	3. Other (sale of tenders etc.)	—	—	0.61	—	—	0.34
	Total returns	—	—	988.70	—	—	778.47
II.	Costs						
	1. Variable cost	—	—	1,026.40	—	—	682.25
	2. Fixed cost	—	—	9.28	—	—	19.65
	Total cost	—	—	1,035.68	—	—	701.90
III.	Net returns over variable cost	—	—	–37.70	—	—	96.22
IV.	Net returns over total cost	—	—	–46.98	—	—	76.57
V.	Profit (Rs. per 100 litre)	—	—	42.13	—	—	105.78

private dairy unit earned the profit of Rs. 105.79 per 100 litres of milk handled. The co-operative unit incurred a loss of Re. 0.42 per litre of milk and the private unit earned the profit of Rs. 1.05 per litre of milk handled in the annual business.

Break-even Analysis

The break-even analysis was carried out in order to estimate the optimum quantity of milk to be handled for efficient organisation and carrying out the business. The results of the break-even analysis are presented in Table 9.3.

TABLE 9.3: Break-even Analysis of Dairy Units

Sl. No.	*Particulars*	*Co-operative Dairy Unit*	*Private Dairy Unit*
1.	Annual output (quantities of milk handled) in lakh litres	111.50	72.38
2.	Annual fixed cost (Rs. lakh)	9.28	19.65
3.	Annual variables costs (Rs. lakh)	1,026.40	682.25
4.	Total costs (Rs. lakh)	1,035.68	701.90
5.	Price per litre of milk (Rs./litre)	7.91	10.75
6.	Annual gross returns (Rs. lakh)	988.70	778.47
7.	Estimated break-even quantities (lakh litres)	118.64	14.89
	(a) Quantities short fall (lakh litres)	7.14	—
	(b) Quantities excess (lakh litres)	—	57.49
8.	Break-even quantities per day (lakh litres)	0.33	0.04
9.	Percentage of actual quantities to the break-even quantities	–6.02	386.10

From the break-even analysis, it was observed that the estimated quantity of milk to be handled annually were 118.64 lakh litres and 14.89 lakh litres for the co-operative and the private unit, respectively. It clearly indicated that the co-operative dairy unit should handle 118.64 lakh litres of milk per annum and the private unit should handle 14.89 lakh litres of milk per annum in order to run the business on no profit no loss basis. The co-operative unit was handling less quantity of milk (7.14 lakh litres) whereas private unit was processing higher quantity of milk (57.49 lakh litres) and, therefore, was in profit.

Conclusions

From the following discussion based on the study thus carried out the following conclusions could be drawn:

(1) Among the two dairy units studied, the private dairy unit ranked first in running the business efficiently and economically; (2) The per unit profitability of the business was relatively low (actually in loss) in case of the co-operative dairy unit on account of excessive procurement and management cost; (3) Inefficient management and under utilisation capacity were observed the reasons for the loss in case of the co-operative dairy unit, however, the reverse was the situation in case of the private dairy unit; (4) The profitability over total cost was higher in case of the private dairy unit (Rs. 105.78 per 100 litres of milk handled) as compared to that of the co-operative dairy unit (*i.e.*, Rs. 42.13 per 100 litres of milk handled).

It could be stated that with the changed scenario in the liberalised economy of the country, the competition in all types of product marketing is inevitable and it is very much rightly applicable to the dairy business. And the private agencies, organisations are having more perspectives in the marketing activities and especially in the agriculture dairy products.

10

AGRIBUSINESS MANAGEMENT
Issues and Strategies

— *Neelam Bhardwaj, A.K. Singh* and *S. Arun Kumar*

Indian agriculture has gone through seachange in terms of foodgrains production, *i.e.*, 51 million tonnes in early fifties to 206 million tonnes at the end of the century. Although the country has achieved self-sufficiency in foodgrains production, agriculture is still considered as an unrewarding profession due to low remunerative returns. The grim situation in agriculture sector is due to shrinking farm holdings, stagnation in productivity and lack of research support to diversified farming. Presently, Indian Farmers' economy is under lot of pressure with global forces affecting agricultural marketing.

The present chapter has focussed on issues that are of paramount importance for agribusiness management. Those are quality consciousness, grading, standardisation of agro-products, and market information. As agricultural sector is central to the socio-economic life of the country and is a source of livelihood to two-third country's population, to boost country's economy emphasis has to be laid on empowering local farmers and farmwomen for micro entrepreneurship.

Scenario of Agricultural Research in India

In post-independence years research was not a priority and centered around land reforms, extension, irrigation and credit. In mid-sixties, ICAR and State Agricultural Universities assumed major responsibility for supporting agriculture research both at the national and regional level. The All India Co-ordinated Research Projects (AICRP) was another significant effort by ICAR which brought the State system in the main stream. Today, the national agricultural research network consists of 97

ICAR units, 31 SAUs, 98 AICRPs and 313 Research Stations, 120 Zonal Research Stations and 200 Sub-stations (Jha, 2001). This well developed research system initiated several research programmes such as High-Yielding Varieties Programmes (HYVP), the dryland research project and initiated to address rainfed agriculture problems. Though, the research institutions made a noteworthy contribution but inbuilt lacunae prevailed in the system. Mostly scientists adapted the material and methods from the West. Scientists made research recommendations based on their own research experiences and neglected the local farmer's needs, resources and socio-economic constraints.

At the outset of the new millennium, the research system in agriculture has to be visualised in a broad spectrum. Under the new economic compulsions of GATT and WTO the agribusiness has come under the strong and direct influence of the international markets. Indian farm sector, mainly made up of the small and marginal farmers is now required to produce farm goods of quality at competitive unit costs for the international market. Therefore, the agribusiness should be reoriented both to the changes that have occurred in the past and to the possible global challenges envisaged. Agribusiness has to be revitalised by streamlining the research issues which needs immediate attention.

Need for Participatory Research

There is necessity to make collaborative research efforts by research scientists, extension workers and farmers. Selection of the enterprises and farming system components like foodgrain crops, horticulture, animal husbandry, sericulture, apiculture and mushroom growing etc., should be based on agro-ecological and socio-economic aspects of farming systems. Agribusiness management research should follow a micro-approach by analysing the problems, utilising the available resources and potentialities at the local level by making the farmers participate in research process of analysing the market needs and development of agro-based products. The organised groups of farmers at community level can participate in post-harvest technologies research with the scientists.

Quality Management of Agro-Products

Standardisation of products should be ensured to compete with international markets. There is a need to adopt internationally accepted quality management systems and establish a credible system of registration of accredited laboratories for quality certification of agro-products. The major responsibility lies with the SAUs which can have separate research cells for quality improvement and standardisation of agro-products. They have to develop an on-going counselling system for solving production problems and have standardisation and certification facility for the agro-based products.

Strong Institutional Linkages

In agriculture sector small farmers constitute the majority. The hope of generating prosperity in rural areas largely depends upon making their operations more viable through strong institutional linkages with research organisations, banks and corporate sectors. Research organisations can help the farmers to carry out important post-harvest activities successfully. This can be ensured through additional support for credit, expert guidance for marketing information and media organisations for commercial advantage. The research system should forge linkages with some of the organisations which are taking care of agro-industries at the National level like National Horticultural Board, National Bank for Agriculture and Rural Development (NABARD), Forward Market Commission (FMC), National Dairy Development Board (NDDB) and Indian Institute of Packaging. Linkages are equally important with Small Farmers' Agribusiness Consortium formed at the grass-root level. This will enable them to join the main stream of free market economy.

Promotion of Market Research in Agriculture

Farm entrepreneurs have failed to do any strategic market planning due to lack of information of market research. Most of the media organisations are profit making organisations hired by the private agencies to popularise their products. These agencies are conducting research on consumerism so that they can derive maximum benefits by promoting their sale. No organised efforts are visible for the promotion of agro-based products. The area-specific consumer research should be emphasised for the promotion of the agribusiness.

Research in Diversified Agriculture—Cases of G.B. Pant University

Project: Processing of Soybean for Diversified Uses and Its Socio-economic Aspects, Department of Food Science & Technology, College of Agriculture.

Soybean contains about 40 per cent good quality protein and 20 per cent oil rich in poly unsaturated fatty acids which are good for the prevention of cardiac diseases. It can be processed into different food products. The department is working on the conversion of soybean into food products suitable for human consumption. Soybean products such as soymilk, soycandy, soynuts, soypaneer, have been developed from whole soybean. Solvent extraction for oil is one of the main industry for processing soybean and a by-product of this industry called deoiled soy cake is used as cattle feed. However, this department has worked extensively on the utilisation of food grade soymeal for human consumption in different food products. Defatted soymilk has been successfully incorporated into traditional food products such as chapatti, jalebi, samosa, puri, parantha, sem, mathari, papad, bhujia at levels ranging from 10-80 per cent. Besides the work on the utilisation for human consumption, the work in extraction of isoflavones for value addition, is also being carried out because isoflavones are known to possess anti-cancer, anti-diabetic and hypocholesterolomic activity. These are also useful to provide relief to women during menopause.

Project: AICRP on Agroforestry, Department of Gentics & Plant Breeding, College of Agriculture

Project initiated from 1990 on biodiversity research on agroforestry. Major objective was to enhance genetical diversity in nature to avoid monoculturing in tree components of agroforestry system. Genetical diversity studies in *poplar* and *shisham* are being recommended to the farmers, foresters including plantation sector. Incorporation of genetical diversity in plantation has provided a high tolerance against biotic and abiotic stresses. This increased plantation has put a positive impact on industrial sectors as a large number of plywood, board, and compressed wood factories has come up in recent times. Adoption of genetic diverse clones has ensured the supply of raw material to the industrial sector as per their needs.

Project: Processing of Millets for Value Addition and Development of Health Food, Department of Foods and Nutrition, College of Home Science.

Millets are also termed as nutri-cereals because they provide good nutrition. These are hardy rainfed dryland crops. The common small millets are finger millet, barnyard millet and foxtail millet.

Objective: Preparation of ready-to-eat health foods such as diabetic foods, high fibre foods and utilisation of millets in bakery products and upgradation of traditional foods to viable commercial scale.

Work done: Millet based food products have been developed for diabetics. These include *dhokla, uppuma, laddu, barfi* and *khichdi*. All these food products have been tested for hypoglycemic effect in non-insulin dependent diabetics and have low glycemic index. The range of glycemic index is from 26 to 36. Clinical trials of feeding *laddus* to diabetic patients for 6 weeks showed decrease in blood glucose and blood cholesterol level.

Millet based recipes using malted millet flour upto 60 per cent have been developed, *e.g.*, Weaning mixes for children, ready to eat mixes for *laddus*. Other products like *sevian*, convenient mixes to prepare 4 food products from a single mix have been developed, tested by 100 people in Kumaon hills for acceptability and are ready to be used by budding entrepreneurs. Efforts are being made to introduce weaning mixes in ICDS in Uttranchal.

Project: Value Addition of Agro-Animal based Fibres, Department of Clothing & Textiles, College of Home Science.

Indian flora and fauna have abundance of natural dye source to offer for commercial exploitation. Due to German ban on synthetic dyes, European countries are looking for natural dyes for textile industry and Indians can take advantage of this opportunity. Bhotia tribe communities in Garhwal hills of Uttranchal still uses natural dye sources, namely dolu, bexleris walnut bark etc. for colouring wool yarn, for carpet making. AICRP, HSc team of Clothing & Textile Unit identified many dye yielding plants in Kumaon and Garhwal hills for commercial exploitation. The optimisation of dyeing conditions for the application of these identified dyes with eco-friendly mordants for better shades and variety of shades are also one of the area of research. Production of dye in the crude form could be one of the income generating activity envisaged for the people of hill region. The agro-based fibres such as Bhimal, Patsun, Jungli Bhindi and animal based fibres such as sheep wool, mountain goat hair and angora rabbit hair are commonly used by the people in hill region. These fibres could be made into suitable handicraft items and woollen garment of internationally accepted standard under proper quality control guidance. The research work is focussed in these lines to extract and process these fibres for preparing valuable handicraft items. Intensive intervention programmes as application of natural dyes and value addition of agro and animal based fibres using proper techniques have been conducted as part of the research programme.

Looking into the type of efforts being done by the research institutions, agriculture sector can become major enterprise in national economy. Some of the points for developing an appropriate strategy for agribusiness management are described below:

- It requires increase in investment for R&D sector, developing of infrastructural facilities and strengthening of linkages between research, extension and farming system.
- Intervention programme can be designed for providing training to farmers and farm women by using multidisciplinary approach.
- Information technology can play a promising role in domestic agricultural marketing. A design framework for IT based Agricultural Marketing Network (AGRIM-NET) was suggested by Mandavia in 'Design, Development and Implementation of AGRIMNET (1992). Microcomputers should be installed at grass-root level for creating databases on market and trade information with facility of electronic data interchanges. Private organisations have already taken a lead. Indian Tobacco Company (ITC) started SoyaChapaul.Com where farmers are trained to use internet facility and developing their own website. Information on soya crops are put in electronic form.
- To replace bureaucracy with professionalism, selective privatisation in the areas of agricultural machinery, bio-agents, bio-pesticides processing, and value addition for research and extension will be of healthy, competitive presence for the public research system.
- Research on simple and efficient ways to increase the shelf life and avoid the spoilage should be taken up and the technologies made available to the growers at the earliest.
- A symbiotic model in which the research organisations, corporate sector, small and marginal farmers are partners will ensure the success of any diffusion of innovations. Also, the networking within the farming community should be ensured for efficient production handling, storage, processing and transport of products for liberalised markets.

REFERENCES

Jha, Dayanatha (2001). "Agricultural Research and Small Farms—Presidential Address", *Indian Journal of Agricultural Economic*, 56(1): 1-23.

Mandavia, U.K. (1992). *Design, Development and Implementation of Agricultural Marketing Network (AGRIM-NET)*, workshop on Information Systems for Agricultural Marketing: Present Role and Future Challenges, October, 1992, Jaipur.

Nayyar, Harish and Ramaswamy, P. (1995). *Globalisation and Agricultural Marketing*, Rawat Publications, New Delhi.

Singh, Ajit (2001). *Problems of Small and Marginal Farmers in Marketing Fruits and Vegetables*, Books India International, New Delhi.

11

AGRO-ENTERPRISES
Sensitising Tribal Farmers

— *C. Satapathy* and *B. Satapathy*

Tribal farmers form a significant proportion of our population, national as well as State plans and programmes have given due importance to bring tribal farmers into the main stream of Agricultural Development. During last 55 years of our independence, a number of projects have been initiated to bring a change in their life style, income pattern and farming system. But all these efforts have not helped to yield desirable results.

The agricultural scenario is changing very fast. It has added more dimensions with inception of WTO in 1995. The trade of agro-product in WTO has received a special attention, comparing tribal farmers with that of non-tribals, it can be said that former lags behind. With emphasis on equalisation of status and living, we shall have to put extra efforts for tribal farmers to bring them into the folds of development in farm sector. The S.T. population of India is about 6.78 crores which constitutes 8 per cent of our country's population (1991). The distribution of S.T. population is found to be 55 per cent in central region, 28.11 per cent in western region, 12.02 per cent in north-eastern region, 4.22 per cent in southern region leaving rest in remaining regions of the country. In fact, the tribal farmers of our country has to be geared up to make them equal with that of non-tribal farmers who have switched over to modern farming system.

The position of Orissa with regard to S.T. farmers needs special attention. The S.T. population of the State is about 22.21 per cent (1991). Among 30 districts of Orissa; Bolangir, Deogarh, Gajapati, Jharsuguda, Kalahandi, Kandhamal, Keonjhar, Koraput, Malkangiri,

Mayurbhanj, Nawarangapur, Nuapara, Rayagara, Sambalpur and Sundergarh constitute majority of tribals of different kinds.

Agro-Enterprises

The tribal dominated districts except Sambalpur depend on rainfall for farming. The farming situations of these areas are identified by the amount of rainfall, distribution of precipitation and soil type. Rainfall being erratic, these districts experience droughts very frequently. The production, productivity and profit per hectare is comparatively low. Depending on cereals, pulses and oilseeds grown in these districts, the farmers cannot derive much benefits. The diversification in farming system now has been advocated based on past experiences. These districts have enough scope to have variety of land-based enterprises.

The enterprises are:

1. Seed production
2. Pisciculture
3. Goat rearing
4. Duck rearing
5. Floriculture
6. Off-season vegetable production
7. Nursery raising of fruit crops
8. Broiler farming
9. Grafts and seedling production
10. Agro-service centre

The tribal youths can be motivated to take up these enterprises provided they are made aware of the advantage. The strength of these enterprises in relation to market demand, local consumption pattern, technical back-up and government's encouragement need to be examined.

Strength of Enterprises

Enterprise	*Strength*
I. Seed Production	1. There is much demand for quality seed. 2. Price of seed is quite remunerative. 3. Encouragement by the State government by supplying Breeder and Foundation seeds for production of foundation and certified seeds at the farmer's level.

Enterprise	*Strength*
	4. Crop insurance facilities now available in the State.
	5. Credit facilities available.
	6. Required technology for seed production is available.
II. Pisciculture	1. Huge water-logged areas available in the State for fish-culture.
	2. People have strong affinity for fish consumption.
	3. Technologies developed by Central Institutes are available.
	4. Provides better income to the people.
	5. Inputs required for fish-culture is available.
	6. Technical expertise are available in Research Stations.
III. Floriculture	1. Soil type is suitable for flower cultivation.
	2. Resource poor farmers can easily afford cultivation with good return from a small piece of land.
	3. Market demand is in increasing trend.
	4. Technology is very simple and can be easily applied.
	5. It is a very good source for self-employment.
IV. Off-season Vegetable Cultivation	1. Off-season vegetables fetche very good market price.
	2. Hybrid seeds of vegetables are available and already accepted by the farmers.
	3. Hybrid varieties can be grown in any season of the year.
	4. There is not much difference in cost of cultivation between seasonal and off-season vegetables.
	5. Credit facilities available for large scale production.
	6. Disposal of off-season vegetables is comparatively easy.

Enterprise	*Strength*
V. Nursery raising for Fruit Crops	1. Nursery can be raised in a small piece of land and all categories of land holders can take up the enterprise. 2. Family members can do the work and utilise their leisure time. 3. Much demand exists for fruit crops. 4. Unemployed youths could be fully employed. 5. Latest technology is available.
VI. Broiler farming	1. Broiler birds are quick growing and disposed within 6-8 weeks. 2. On an average, a net profit of Rs. 10 per bird is obtained. 3. It could be raised in backyard without compromising with living space. 4. Weaker-sections are habituated with poultry keeping. 5. Returns obtained within a short span period which will be re-invested in a cyclic order. 6. It is one of the good enterprise for self-employment 7. Heavy market demand.

Sensitising Techniques

The farmers in general like to stick to the practices adopted by their forefathers. They need to be motivated through suitable extension techniques. The farmers of tribal areas need special attention because of their tradition, culture, belief, faith and norms. The social system, social norms and social structure decide the rate of movement of new technology in tribal communities. In Orissa, the tribal farmers maintain their identity in terms of group-action, leadership style and community planning. The sensitising techniques if based on these values will definitely yield better results.

In the process of adoption, when an individual moves from the stage of awareness to final adoption, he is being influenced by number of sources providing information. The steps to sensitise farmers of a particular locality depends upon the local custom, belief, faith and norms. For sensitising tribal farmers of Orissa, a variety of methods are available. But their use is location and tribe-specific. The major tribes of Orissa are Kishan, Bathudi, Omanatya. Bhumiya, Binjhal, Bhumij, Mirdha, Mahali, Kotya, etc.

The sensitising methods are individual, group and mass methods. It is established through research that group-approach has more credibility over other methods to motivate farmers of tribal areas. Group methods are to be used depending upon previous experience of the farmers, size of the group, composition, meeting place, leadership pattern and method of feed-back. Choosing of methods to sensitise the tribal groups of farmers can be thought of and described as such.

Methods of Sensitising Tribal Farmers

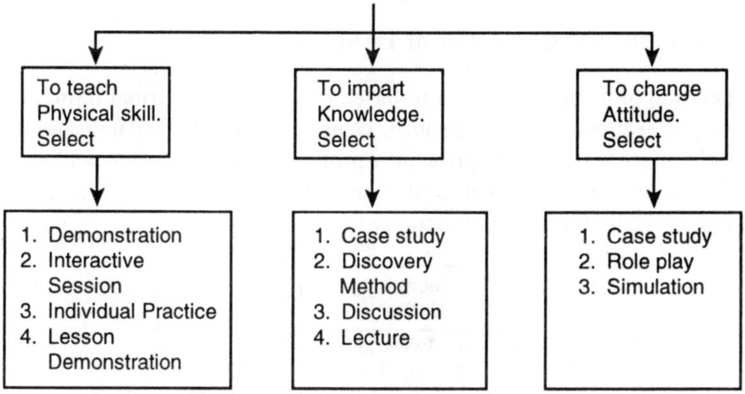

The above illustration provides a broad frame-work to apply appropriate method for motivation.

Matrix for Group Techniques

Use of group-method requires special technique and commitment to serve tribal farmers. The sensitiser should select the appropriate method keeping the group in view. Research has shown that group-method yields better learning experience and understanding among the tribal farmers of Orissa.

Use of group methods and their success are not independent of interest, level of understanding, need and SWOT of enterprises. Careful analysis has shown that the characteristics of agro-enterprises and their demands in market help to select right type of methods.

TABLE 11.1: Matrix for Use of Sensitising Methods

Sl.No.	*Methods*	*Large Group*	*Small Group*	*One View Point*	*Several View Points*	*Simulation*	*Problem Solving*
1.	Brain storming		✓		✓		✓
2.	Buzz session	✓			✓		✓
3.	Case study		✓	✓		✓	
4.	Demonstration		✓	✓			✓
5.	Interactive lecturette	✓			✓		✓
6.	Field trip	✓			✓	✓	✓
7.	Individual contact		✓	✓		✓	
8.	Film show	✓			✓		✓
9.	Field day	✓	✓		✓		✓
10.	Visit to institute	✓			✓	✓	✓

Tribal Norms and Generation of Technology

Generation of technology for agro-based business for the tribal farmers need special attention of the scientists. Generation, transfer and utilisation of technology to make enterprise more profitable requires understanding of these processes. The illustrated model provides insight into the situation of tribal farming community.

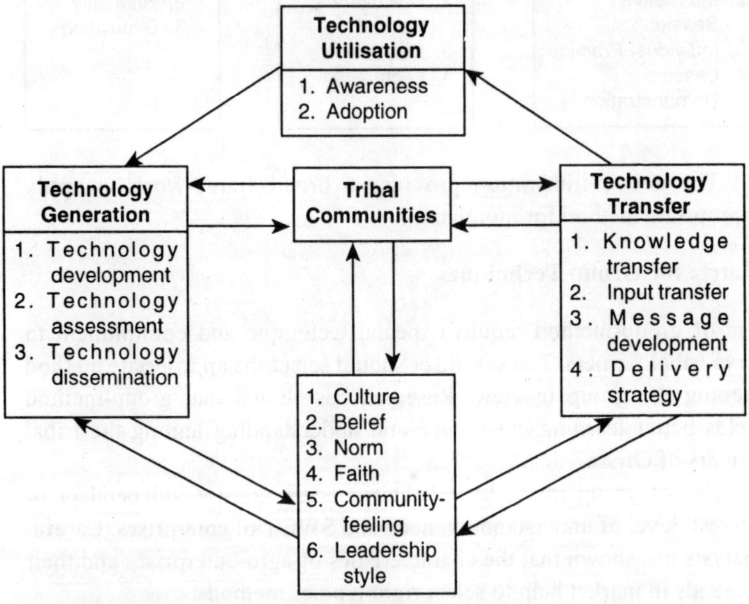

FIG. 11.1: Generation and Transfer of Technology in Tribal Community

12

AGRO-BASED ENTERPRISES OF PUNJAB FARMERS
Some Management Aspects

— *Kanwal Kumar, Anjana Kalra* and *Rabinderjit Kaur*

The output of any enterprise mainly depends upon the management practices followed by the entrepreneurs. The economy of our country is dependent on agriculture as more than 80 per cent of the population is employed in this occupation. Emphasis is, therefore, being given to develop this occupation in such a way, that maximum profits are gained by the farmers. For this purpose it is very important to educate the farmers regarding the modern practices to be followed for making this occupation more remunerative. The agricultural scientists are doing their best efforts in this direction and the farmers are fully associated from time to time in planning the future research, based on their felt needs in the adoption of latest agricultural technologies.

The Punjab Agricultural University has taken a lead in this direction and has constituted a number of farmers' committees to get first hand information regarding the agricultural problems being faced by them. The selected progressive farmers scattered all over the State are the members of these committees. The present study has been planned to know the types of agro-based enterprises taken up by them; sources of their technical information and financial assistance and also to know their mode of marketing of their produce.

Scope of Study: The findings of the study will help to know as to which agricultural enterprises the progressive farmers have taken up at commercial level in addition to crop growing. The results will also reveal the sources of their technical information and financial assistance. The other important aspect which will come to light would be the mode of their marketing the agricultural produce. The study will also bring out the problems being faced by them. These findings will help to educate the fellow farmers of the State to strengthen the enterprises already taken up by them and the enterprises which can be taken up by them at commercial level.

Research Methodology

Locale of the Study: The study was conducted at Punjab Agricultural University, Ludhiana.

Selection of Respondents: All the member-farmers of the different Farmers Committees of PAU constituted the population of this study and the farmers who sent back the duly filled-in questionnaires were taken as the respondents for this study.

Tool of Data Collection: A questionnaire was specially designed for the collection of data which included items to collect information regarding various agro-based enterprises taken up, sources of technical information and financial assistance and the mode of selling the produce for the convenience of the farmers. The language used in questionnaire was Punjabi.

Collection of Data: The questionnaire was distributed in the meetings and the farmers were requested to fill them and return through post. A self-addressed envelope was enclosed with the questionnaire for this purpose. The questionnaires from fifty-four respondents were received.

Analysis of Data: The data were analysed with the help of frequencies and percentages.

Results and Discussion

Enterprises taken up by the Respondents

Although the respondents were the members of the Farmer Committee Fruit and Vegetable Growers Committee and Animal Husbandry Farmers Committee yet it was felt necessary to find out different enterprises taken up by them. The information in this respect has been given in Table 12.1. The cultivation of field crops was taken up by majority (74.07%) of the respondents followed by vegetable production (59.74%), dairy and seed production (55.56% each). The fruit cultivation had been taken up by 40.74 per cent, flower cultivation by 25.93%, Bee-keeping (37.04%). The enterprises like poultry, mushroom growing, fish farming and agro-industry had been taken up by 11.11 per cent, 7.41 per cent, 3.70 per cent and 7.41 per cent respondents respectively.

TABLE 12.1: Distribution of Respondents according to their Enterprises (*multiple response*)

	Frequency	*Percentage*
Crop production	40	74.07
Vegetable production	32	59.74
Seed production	30	55.56
Horticulture	22	40.74
Flower cultivation	14	25.93
Bee-keeping	20	37.04
Dairy	30	55.56
Poultry	6	11.11
Mushroom growing	4	7.41
Fish farming	2	3.70
Agro-industry	4	7.41

It was further found that most of the farmers (92.59%) had taken up more than one enterprises and the maximum number of enterprises taken up by the respondents were seven (3.70%). It can be concluded that most of the farmers had taken up more than one enterprise so that income from different sources is generated and employment is available throughout the year. It not only helps in keeping the labour busy throughout the year, but also helps in bringing diversification in farming. Taking up enterprises like flower cultivation, seed production of various crops are some of indicators of progressiveness of the respondents which needs to be encouraged. This also helps maintaining the balanced use of natural resources and ecological balance.

Information Sources of the Respondents

It has been established that for the success of any enterprise latest technical know-how plays a very important role. The different sources of information used by the respondents were studied and the information has been given in Table 12.2. It has been noticed that for most (88.89%) of the respondents, PAU scientists were the source of information followed by newspapers (77.78%); Officers of the State departments and Television (70.37%).

On the whole different combinations of sources of information were used by the respondents from time to time which shows their anxiety to know about the latest advancements/recommendations made with respect to the enterprises they have taken up.

TABLE 12.2: Distribution of Respondents according to their Different Sources of Information

	Frequency	*Percentage*
PAU scientist	48	88.89
Officers of State Deptt. of Agriculture & other development departments	38	70.37
Radio	28	51.85
T.V.	38	70.37
Friends	36	66.67
Relatives	26	48.15
Newspapers	42	77.78
Changi Kheti	32	59.26
Package of Practices	32	59.26
PAU handbook	30	55.56

Sources of Financial Assistance

For smooth running of any enterprise the finances are needed from time to time. The information regarding various sources of finance used by the respondents have been given in Table 12.3.

TABLE 12.3: Distribution of Respondents according to their Source of Financial Assistance

	Frequency	*Percentage*
Nationalised banks	32	59.26
Co-operative societies	38	70.37
Commission agents	10	18.52
Relatives	6	18.52

The figures show that majority of respondents (70.37%) used co-operative societies for obtaining finances for their enterprises followed by Nationalised Banks (59.26%). It is very interesting to note that very few respondents took the financial assistance from commission agents (18.52%) and relatives (11.11%).

The reason for getting financial assistance from the recognised sources may be that most of the respondents are the regular payers of their instalments as per conditions laid down by these sources and are not the defaulters.

Role Played in Different Enterprises

From the figures given in Table 12.4, it has been found that majority of the respondents (66.67%) played the role of managers followed by 62.96 per cent as supervisors. However, 55.56 per cent of the respondents involved themselves for carrying out different operations. A little less than half (44.45%) of the respondents involved themselves as managers, supervisors and also carried different operations for running the enterprises. It can also be concluded that the progressive farmers involved themselves in different capacities for the success of the enterprises taken up by them.

It was very interesting to note that almost all the respondents (96.30%) kept the regular records related to their enterprises.

TABLE 12.4: Distribution of Respondents according to their Role Played in Different Enterprises

(*Multiple Response*)

	Frequency	*Percentage*
Manager	36	66.67
Supervisor	34	62.96
Self involvement in different operations	30	55.56
All of the above	24	44.45

Marketing of Commodities Produced by the Respondents

It has been found that all the respondents (100%) sell most of their produce raw. However, some part of the produce is sold after processing. The percentage of such respondents is only 22.22 per cent (Table 12.5). The processed items in addition to others include honey, *haldi, desi ghee*, paneer, chilli powder and pickles. The reason for not processing their produce at a large scale may be the non-availability of necessary infrastructure, lack of financial assistance and technical know-how and pressing circumstances for immediate returns. However, the farmers can be encouraged for value addition of their produce and the government should formulate some policies in this respect.

TABLE 12.5(a): Distribution of Respondents according to the Form in which they sell their Produce

	Frequency	*Percentage*
Raw	54	100.00
Processed	12	22.22

TABLE 12.5(b): Distribution of Respondents according to mode of sale of their produce

	Frequency	*Percentage*
Commission agent	42	78.19
Cooperatives	4	7.41
Shop-keepers	12	22.22
Direct to consumers	18	33.33

Regarding the mode of sale of their farm produce 78.19 per cent of the respondents informed that they sell it through commission agents, 33.33 per cent sold direct to the consumers, 22.22 per cent to shopkeepers and only 7.41 per cent sold their produce through co-operatives. It can be concluded that most of the respondents sold their agricultural produce through commission agents and very few sold direct to the consumers. It has been established that most of the profits are earned by the middlemen (commission agents) and the producers always remain at a loss. Therefore, efforts need to be made to eliminate the middlemen so as to get maximum profit.

REFERENCES

Kalra, Anjana and Kumar, K. (1995). "Farmers Profile and their Reactions about Farmers Service Centre of P.A.U.", *Journal of Research*, P.A.U., Ludhiana, 32(4): 474-78.

Kumar, K. and Kalra, Anjana (1995). "Opinion of Field Functionaries on Extension Programmes", *Agricultural Extension Review*, Sept.-Oct., pp. 18-21.

Kumar, Kanwal and Kalra, Anjana (2001). "Attitude of Farmers towards Veterinary Clinic of P.A.U.", *Journal of Research*, P.A.U., Ludhiana, 38(1-2): 104-11.

13

SEED BUSINESS IN PUNJAB AND HARYANA
Role of Private Sector

— *Arvind S. Malik* and *V.P. Chahal*

Seed, a living embryo, is a vital, basic and crucial input for attaining sustained growth in agricultural production in different agro-climatic conditions. Efforts are being made in the country since 1928 for production and distribution of good quality seeds to increase crop production. The organised seed industry in India, prior to independence, was largely confined to some ornamental plants and vegetables. The national level organisation on seeds, that is, the National Seed Corporation Ltd. (NSC) was set up in 1963 with an objective of developing a sound seed industry in the country. To meet the growing demand of quality seeds, the National Seed Project (NSP) was launched in 1976, which also covered the States of Punjab and Haryana.

Presently, India has around 500 companies engaged in production and marketing of seeds of different crops and 24 of them have their own R&D establishment operating in the country. The private sector seed share is more than 60 per cent while the public sector is contributing around 40 per cent in terms of value of seed distributed (Kapoor, 2000). Public sector is primarily concentrating on seed production and distribution of predominantly self-pollinated crops while the private sector is supplying primarily hybrids and high value and low volume seed in in-bred varieties and hybrids of sunflower, cotton, pearl millet, sorghum and vegetables like tomato, cabbage, brinjal, okra, chilli and melons. The area under hybrid seed in India is about 2[illegible] per cent whereas their contribution in production is more than 40 per cent.

The field survey was conducted to get an acquaintance of seed business in general and private seed sector in particular in the States of Punjab and Haryana covering nine districts from Haryana, *viz.*, Hissar, Sirsa, Fatehabad, Jind, Bhiwani, Rohtak, Mohindergarh, Rewari and Gurgaon, while from Punjab the districts, *viz.*, Bhatinda, Mansa, Muktsar, Ferozepur, Sangrur and Faridkot were covered for the purpose of data collection from farmers and dealers/sub-dealers of Agro-input companies on market share, brand awareness, promotion activities, purchase behaviour, packaging, price and demand of seed. A total of 200 farmers, 79 dealers and 145 sub-dealers were interviewed using schedule prepared for the purpose. The findings of the survey are presented and discussed below.

Results and Discussion

Market Share

Market share of different companies vary in different crops in both the States. Differences were also found in the same crops between the two States. In Haryana, Nandi was found to be market leader in pearlmillet (53%) followed by Mahyco in sunflower (35%) whereas in case of Punjab Proagro was found to have more market share (60%) in sunflower (Table 13.1). In case of paddy, wheat, and mustard maximum market share (70%) was shared by local company brands and rest (30%) market was covered by NSC and State Agricultural Universities (SAU) and State Seed Corporations (SSC). The companies *viz.*, Proagro, Mahyco and SPICPHI which were found to have more market share have also strong R&D units besides good marketing network at national level. Reverse is true in crops where hybrids are not available as local companies seed have more market share. All the local companies have no R&D Unit of their own, so they are not able to develop hybrids. These local companies simply multiply the breeder and foundation seed procured from government agencies and distribute among the farmers through their local network.

In Punjab there is very little area under pearlmillet as a foodgrain crop and it is mainly grown as fodder crop. As much as 60 per cent pearlmillet seed market was shared by local brands. Rest of the 40 per cent market was shared by Plantgene, Seed Tech., NSC and Proagro seed companies.

TABLE 13.1: Market Share of Different Companies in Punjab and Haryana

Crop	*Company*	*Percentage of Market Share*	
		Haryana	*Punjab*
Sorghum (SG)	Proagro	55	42
	Plantgene	15	13
	Mahyco	10	7
	SPICPHI	5	10
	Others	15	28
Sunflower	Mahyco	35	10
	Proagro	30	60
	SPICPHI	20	13
	Others	15	17
Pearlmillet	Seed Tech.	—	6
	Nandi	53	—
	Proagro	13	10
	Plantgene	10	8
	SPICPHI	7	9
	Mahyco	5	7
	Others/unbranded	12	60
Maize	NSC	15	10
	Ganga Cauvery	12	15
	Proagro	13	17
	Local brands	50	45
	Others	10	13
Vegetables	Mahyco	15	17
	Indo-American	10	20
	Proagro	8	10
	Sungro	10	5
	Local brands	45	40

Brand Awareness among Farmers

The farmers were asked to tell name of the seed company that first come to their mind (top of the mind). The farmers who could not name any company were given the aid mainly by showing brand marks and brand names of various companies. Nandi was the company which first came to the mind of most of the farmers in Haryana (26%), where as it was Mahyco in case of Punjab farmers (24%) (Table 13.2).

The findings of the survey also revealed that farmers of Punjab were more aware of brands as compared to farmers of Haryana. In general, the large farmers were found to be more aware of brands as compared to the small and marginal farmers.

TABLE 13.2: Brand Awareness/Top of the Mind Company among Farmers

Company	*Percentage of Farmers*			
	Haryana		*Punjab*	
	Number	*Percentage*	*Number*	*Percentage*
Nandi	52	26	29	3
Mahyco	24	12	36	24
Super Seed	28	14	16	2
Proagro	24	12	32	20
SPICPHI	20	10	22	12
Indo-American	10	5	15	10
Plant gene	22	11	19	8
ITC Geneca	4	2	8	6
Cargil seeds	4	2	9	7
Nath seeds	4	2	5	3
Others (Novaitis, Denocilete)	8	4	9	5
	200	100	200	100

Brand Promotion

Brand promotion in seed-business is basically considered as extension activities carried out to provide technical services to farmers and tips on use of modern package of practices with an objective to develop and sustain a strong brand image. Majority of the farmers (52%) in Haryana were of the opinion that Nandi was the most advertised seed company in their areas; whereas in case of Punjab it was Mahyco closely followed by Proagro. Also it was observed that Nandi seed company used maximum number of advertising media to promote its brands than by other companies (Table 13.3). The survey also revealed that Mahyco was found to be the only company, which had advertised through radio and TV also.

Majority of the farmers (44%) were of the view that Trolley-Painting was the most effective medium to influence their purchasing behaviour followed by wall paintings, pamphlets, exhibition stalls and demonstrations (Table 13.4).

Purchase Decision

There are number of factors which affect the decision of the farmers to buy a particular brand of seed available in the market. Table 13.5 reveals that farmers decision was mainly influenced by dealers and sub-

TABLE 13.3: Farmers Opinion on Most Advertised Seed Companies

Company	*Percentage of Farmers*		*Advertisement Media Used*
	Haryana	*Punjab*	
Nandi	52 (10)	— (3)	Banners, Posters, Stickers, Carry Bag, Trolley-Painting, Tin-Plates, Demonstrations, Pamphlets, Advertisement in Cinema Halls, Wall Paintings, Key Rings
Superseed	12 (4)	— (1)	Farm Magazines, Pamphlets, Trolley-Painting, Key Rings
Mahyco	8 (4)	32 (6)	Banners, Posters, Trolley-Painting, Radio & TV, Farm Magazines
Proagro	11 (3)	28 (4)	Trolley-Paintings, Farm Magazines, Demonstrations, Exhibition Stalls
SPICPHI	6 (3)	17 (2)	Demonstrations and Exhibitions, Trolley-Paintings
ITC Zeneca	2 (2)	12 (2)	Pamphlets, Trolley-Paintings
Plantgene	5 (3)	3 (3)	Advertisements in Newspapers, Trolley-Paintings, Farm Magazines
Others	4 (1)	8 (3)	Farm Magazines, Key Rings

Note: Figures in parentheses indicate number of types of media used for advertisement.

TABLE 13.4: Influence of Media of Advertisements on Farmers

Media	*Number*	*Percentage*
Trolley-painting	86	43
Wall painting in villages	24	12
Wall paintings in mandis/markets	14	7
Demonstrations	18	9
Exhibition stalls	16	8
Radio spots	10	5
Pamphlets	20	10
Farm magazines/newspapers	12	6
Total	200	100

dealers (54%) followed by extension personnel. Dealer's influence on purchase decision was found to be more when they provide seed on credit. So, farmers particularly small and marginal land holders purchase on credit whichever brand dealers offer. Only 10 per cent farmers purchased seed of a particular brand on the basis of their own decision followed by nine per cent farmers who purchased on the advice of scientists.

TABLE 13.5: Influencers in Purchase of Seed by Farmers

Influencers	*Number*	*Percentage*
Dealers/sub-dealers	108	54
Extension personnel	28	14
Advertisements	14	7
Opinion leader/fellow farmers	12	6
Own decision	20	19
Scientists	18	9
Total	200	100

Similarly dealers were asked to specify the factors for selling a particular brand. The survey revealed that profit margin given by the company was the foremost factor, which affected the dealers preference for selling a particular brand (Table 13.6).

TABLE 13.6: Factor affecting Dealer's Preference for Sale of a Particular Brand

Factors	*Number*	*Percentage*
Profit margin	68	46.90
Demand	33	22.76
Quality	27	16.62
Any other reason (Credit facility, variety etc.)	17	11.72

Marketing Channels

The private sector seed companies are selling their seeds through distributors/wholesalers/dealers and retailers. The channels of private seed marketing by which seed reaches to the farmers and are given below:

1. Channel-I Seed grower—Private processing plant—whole saler (dealer/distributor)—retailer—farmer
2. Channel-II Seed grower—Private processing plant—retailer—farmer
3. Channel-III Seed grower—Private processing plant—farmer

The survey revealed that major quantity of seed is being marketed by private companies through Channel I followed by Channel II and Channel III, *i.e.*, private processing plants sold seed direct to the farmer.

Barriers in Adoption of Quality Seeds

The farmers were also asked why they could not use the quality seeds. The data presented in Table 13.7 revealed that as many as 14 factors were considered by farmers as barrier in the adoption of quality seeds. These include: lack of technical guidance at sale points (57%), high price of seed (54%), non-availability of seed at government shops (46%), untimely supply (36.5%), fear of crop failure (35%), no need to purchase every year (33.5%), spurious quality (34%), desired quality seed not available (32.5%), lack of new varieties (28%), lack of knowledge (24.5%), lack of quality seed demonstrations (22.5%), low yield difference (19.5%) and non-availability of quality seed in the nearby local market (18%).

TABLE 13.7: Farmers' Perception of Barriers in Adoption of Quality Seeds

Sl.No.	*Barriers*	*Number*	*Percentage*
1.	Lack of technical guidance	114	57.0
2.	High price of seed	108	54.0
3.	Non-availability of government shops	93	46.0
4.	Low purchasing power	82	41.0
5.	Untimely supply	73	36.5
6.	Fear of crop failure	70	35.0
7.	No need to purchase every year	67	33.5
8.	Spurious quality	68	34.0
9.	Desired variety not available	65	32.5
10.	Lack of new varieties	56	28.0
11.	Lack of knowledge	49	24.5
12.	Lack of quality seed demonstrations	45	22.5
13.	Low yield difference	39	19.5
14.	Not available in local market	36	18.0

Packaging of Seeds

Packaging is not only a means to deliver the seeds in good condition but also a tool to differentiate from the competitors. Packaging design and material are major part of a seed sale policy. Findings revealed that tempering proof cloth packing was found to be the most preferred by the farmers (43%) in field crops (Table 13.8), whereas in case of vegetables tin packs were mostly preferred (40%) over cardboard packs and pouches for the simple reason that they could be utilised for daily use.

Conclusion

Of all the agro-inputs, seed holds the key position in India where

TABLE 13.8: Farmers Preference for Packaging Materials

Material	*Number*	*Percentage*
A. *Field Crops*		
Plastic packing	54	27
Cloth packing/jute, tempering proof	86	43
Quality packing only	40	20
No preference	20	10
B. *Vegetable Crops*		
Tin packing	80	40
Cardboard boxes	64	32
Pouches	26	13
Any quality material	10	5
No preference	20	15

majority of the farmers are small and marginal. Production of quality seed and making it available to farmers on reasonable rates through approachable sale outlets are of prime importance to agricultural development. Private companies having good R&D Units have more market shares in case of sorghum, pearlmillet and sunflower. Though the credit for popularising hybrid seeds in some vegetables and cash crops goes to the private sector, yet the vegetable growers of Punjab and Haryana still prefers local brands because of high cost of hybrids. On the other hand the public sector hybrid vegetable seeds barring few varieties have not reached to the farmers mainly due to inadequate attention of government to needed infrastructure and lack of trained manpower.

No single private company has been found to be popular among more than 24 and 26 per cent of the farmers in Punjab and Haryana, respectively. The companies found to have used as many as 10 media of advertisements but the trolley-painting followed by wall paintings and pamphlets distribution modes were most preferred by the farmers. Profit margin followed by demand of a variety were found to be the major motives of dealers in affecting the purchase decision of 54 per cent farmers who generally prefer to buy seeds on credit. The major quantity of seed is sold through the channel of marketing network, *i.e.*, seed grower-private processing plant-wholesaler (dealer/distributor)-retailer-farmer consumer. The lack of technical guidance at sale points, high price, non-availability of seed in government shops, untimely supply, fear of crop failure, spurious quality of seeds wee perceived as major barrier in the adoption of quality seed by farmers. The tempering proof cloth packing for the field crops and tin packs for vegetables seed were preferred by the farmers.

Suggestions for Promoting Private Sector Seed Business

Based on the findings of the survey the following suggestions could be of some value to government and private seed sector farmers as well.

- The private seed companies could only increase their market share and sustain brand image if they start providing pre- and post-sale technical advisory service to the user-farmers.
- Government and farmers organisations must recognise that efficiency in quality seed production and its supply to farmers could be enhanced with the active involvement of the private sector.
- Steps should be taken by the government and the private seed companies at policy level to work out coordination in the seed production and its distribution to farmers on reasonable cost.
- For this adequate public and private institutional linkages would need to be established and government may provide incentives and subsidies for private involvement in quality seed supply to atleast small and marginal farmers.
- The government should only give license for selling the seeds and agro-inputs to the farm graduates and the private companies should organise regular training programmes for dealers and sub-dealers so that they may be able to guide and advice the farmers whenever they come to buy seeds.
- The major quantity of seed is marketed through the dealers. They play an important role in boosting the sale of seeds but the companies should only fix a reasonable profit margin for them and adopt strict quality control measures to check the sale of spurious and low quality seeds.
- The promotional programme of seed companies should never mislead farmers but explain only the real benefits of their varieties. It must be targeted to build up the confidence of farmers in the use of quality seeds for effective advertisement of their brands. They must use a combination of media and methods including sponsoring farmers and dealers educational tours, meetings, fairs some special events like rural sports and medical check-up camps.
- Farmers are very receptive of new variety seeds. The private seed companies should estimate the quality seed demand of newly released varieties before starting seed production

programme so that seed of desired varieties can be supplied to the farmers.

- In marketing of private sector seed the consumer's price increases due to entrance of another intermediary, *i.e.*, wholesaler. So private seed companies should give more incentives to the dealer for promoting the sales.
- The SSCs should also appoint the contract dealers in private sector to promote their sale so that farmers may not have to pay more.
- The tempering proof packing materials should be used to maintain the purity and check the sale of the spurious seeds and avoid adulteration.
- Seed quality demonstrations should be conducted by private seed companies on farmers' fields to remove the fear of spurious seeds and to boost the credibility of the private seed sector companies.

REFERENCES

Dahiya, B.S. (2000). "Indian Seed Industry", *Agro-India*, September 2000, pp. 20-22.

Dhankhar, B.S. (1999). "Hybrid Seed Production of Vegetables" in *Krishi Vigyan Kendra—A Reality*, editors Das, P. and Hansra, B.S., ICAR, New Delhi.

Kapoor, Arvind (2000). "Private Sector Role Vital", *The Hindu Survey of Indian Agriculture*, 2000, pp. 211-12.

PART II

FARMERS' ORGANISATIONS AND FARMER-LED MARKETING IN PROMOTING AGRIBUSINESS

14

ROLE OF PUBLIC AND PRIVATE SECTORS IN PROMOTING AGRIBUSINESS

— P.N. Jha

Transfer of improved technology/practices for agricultural and allied development sectors has been, by and large, the monopoly domain of the government organisations, *i.e.*, public machineries. This system has worked now for the last fifty years ever since the Community Development Programme was started in 1952. Even with the launching of Intensive Agricultural District Programme and then Intensive Agricultural Area Programme throughout the country and later with coming up of high-yielding varieties programme, the transfer of technology responsibility and the support system, largely remained with the State machinery. The partnership/participation of private sector and non-government organisations remained negligible except a few sporadic cases here and there.

While the plus side of the government sector was that it had a well-knit organisational structure right from the State level down to the grass-root/village level, the negative aspects were that there was a lack of spirit of dedication, initiative, drive, sense of accomplishments and achievement motivation. This dampened the progress of speedy transfer of technology and no wonder, that even today it is said that not more than 30 per cent of improved technology has reached the farming community. Thus, the level of non-adoption of improved technology is alarmingly quite high.

When this was the situation in case of transfer and adoption of improved technology pertaining to mostly crop husbandry and partly animal husbandry, it is when we talk of commercial farming and agriculture as a profitable enterprise, the inseparable components are agro-processing, value-addition and agribusiness and thus making farming a business enterprise through promoting and ushering into agro-business. The experience gained in many of the States has revealed that agro-processing and agribusiness exercises undertaken by the government machinery had a poor fate. Many government sector fruit and vegetable processing units, Seed Corporations, input agencies etc., are languishing if not totally closed. Even the Sugar Industries which are purely based on Agricultural crop, *i.e.*, sugarcane are not running properly in many States and many of such Sugar Factories are closed for the last several years. On the other hand, the Agribusiness Industries and Input Agencies which are in the Private Sector are having a better efficiency and are running satisfactorily even if some of them not very well. This is mainly because there is a commitment from the employees, there is a sense of dedication, initiative and drive and the motive of the organisation is both to earn profit and to serve the cause of the beneficiaries. There are less of administrative channels, the decision-making is quick and things do not travel top-down but bottom-up.

During the last 50 years, the approach in farming research and extension has been shifting from mono-cropping to double cropping to multiple cropping and then gradually to cropping system leading to intensive agriculture and consequently green revolution that we achieved during seventies and eighties. However, the above shift did not guarantee stability, sustainability, equitability and gender neutrality and suitability for promoting agribusiness. It is now being realised that a technology developed by the research system should attempt to have these attributes for making agriculture commercial and a business enterprise. This is why the shift now is towards the commercial agriculture, complex, diverse and risk-prone agriculture and gradually to farming systems including crops of all kinds, horticulture, Animal Husbandry, Dairying, Poultry, Apiculture, Mushroom growing, Agro-Forestry etc., all these leading to a kind of rainbow revolution, *i.e.*, all around development in agriculture and allied sectors, promising rapid Rural Development through agribusiness and entrepreneurship development.

The Shifting Approach in Farming towards Promoting Agribusiness

In order to promote agribusiness and making agriculture a business enterprise rather considering agriculture as yet another industry, there has to be a shift in approach in farming conducive to promotion of Agribusiness not only in big cities and State capitals but also at the Blocks/Talukas and at Sub-division levels and if possible at grass-root level too. These requirements are as under.

- Shift from commodity-centred approach to farming system approach.
- Shifting to diversified and commercial farming.
- Research and Extension priorities to Post-harvest Technology and Management.
- Entrepreneurship Development both at micro and macro levels.
- Emphasis on value-addition and qualitative improvement.
- Developing organised market and processing centres.
- Gradual export orientation.
- Market decisions to precede production decisions.
- Market-driven and demand-driven approach rather than the supply-driven and produce-driven approach.
- Agro-based small Scale Processing unit at Block/Taluka level, Medium Scale unit at Regional level and Large Scale unit at State Level.
- Modern Technology to be wedded to Ancient Wisdom and I.T.K. for sustainability and for economisation of the investment in farming and processing.

For all this to happen we have now to examine as to what are the Agencies/Organisations which can participate in the task of bringing revolutionary changes in farming and promoting Agribusiness as explained in the beginning of this chapter. The Government Public Sector Organisations have been dealing with most of the activities and running the programmes for Agricultural Development. Even the research system though autonomous is the part of the government, as it is financed, guided and controlled by the government. Our experiences have revealed that the Public Extension system have been suffering from several weaknesses/discrepancies.

Weaknesses of the Public Extension System/Management Non-Conducive to Promotion of Agribusiness

- Difficulty in coordination and supervision.
- Top-down approach rather than bottom-up approach.
- Lack of initiative, drive and zeal.
- Delayed decisions, supplies and services.
- Lack of commitment and achievement motivation.
- Lacks accountability.
- Beneficiaries, stake holders and farmers at the receiving end, their involvement in decision-making is poor.

- Supply-driven.
- Lacks popular support.
- Lacks follow-up measures.
- Subsidy-based.
- Target-oriented and scheme-based.

Features of Private Extension Management

Through the government system and its management have several limitations as outlined above the Private Extension Management despite having several virtues does also suffer from certain limitations and weaknesses as under:

- There is initiative/drive/commitment.
- There is accountability.
- It is market-driven.
- Entrepreneur-oriented.
- Quick decision-making.
- Selective technology use.
- Neglect of Resource-poor, small and marginal farmers.
- More profit oriented.

Weaknesses of the NGOs/VOs

Of late, there has been a mushroom growth of non-government voluntary organisations in the country. In fact, in most of the districts of the country, some NGOs are functioning in different areas of agricultural/ rural development. Some of them have now gone for agribusiness also. While there are certain virtues of NGOs like initiative and drive and simplified procedure yet most of the NGOs are suffering from several handicaps and weaknesses and, therefore, the entire Agriculture/Rural Extension Work and Development could not be left to non-government organisations as they are not attuned to fulfil the gigantic task and are not well organised. Some of the glaring weaknesses are as under:

- Centered round a patriarch and an individual who has established the N.G.O. concerned.
- More and more energy and efforts directed towards getting as much fund as possible from the government and other national and international agencies rather than in effective implementation of the programme.
- Anxiety for achieving goals in the shortest possible time.

- Tendency to become publicity-oriented for cheap popularity.
- No proper accounting procedure.
- Hesitant to get the programme evaluated.
- Efforts are too much localised.
- Tendency to earn more and more rather than public service and benefit to the rural poor and farmers.
- Short of technical manpower and no staff want to stay with the NGOs for long and they are in search of lucrative jobs in other organisations hampering the continuity of work.

Apart from the Public and Private Extension System, we also now talk of participatory Extension System and its management, *i.e.*, my involving the clients, *i.e.*, farmers, the rural people, the functional leaders of the village, opinion builders, legitimisers, local influentials and self-help groups of men,women and youth.

The Features of Participatory Extension System could be summarised as

- Client oriented.
- Demand-driven.
- Client-initiated.
- Group support.
- Client participation from the very beginning: Planning to implementation to appraisal.
- Lasting effect.
- Promotes micro-enterprises at village and taluka level.
- Better price to the farmers/entrepreneurs for the produce.
- Helps in arranging inputs, credit and infrastructures.
- Number of middlemen and intermediaries is reduced, thus, increasing the share of producers in the consumers' money.

Looking to the above features as well as the plus and minus points of the Public, Private and Participatory Extension System, the role of these organisations could be delineated.

Delineation of Roles

Role of Public Sector Extension System

- Creating conducive environment for promoting agribusiness.
- Formulating broad public policies and directions in national interest.

- Encouraging the private sector, State holders and NGOs to participate in extension and social service and providing logistics and support particularly at the initial stage.
- Reaching the weaker sections and resource-poor farmers.

We have to identify and outline the roles and responsibilities of these agencies/organisations/groups which could be carried out by them on the expected line.

Roles of Private Sector Extension System

- Timely guidance, input supply and services.
- Agro-processing, value-addition.
- Agribusiness and entrepreneurship development.
- Faster transport of produce and finished products.
- To supplement the inadequate State funding of the extension system.
- Human resource development through training and in collaboration with the agriculture research system.
- Helping those farmers who want to go for specialised and commercial farming as well as those going for mechanised farming.

After summarising roles of Private Extension System, let us look at another set of organisations, *i.e.*, the Non-Government Organisation (NGOs/VOs) which have also to be actively involved in extension management for promoting agribusiness participation of non-government organisation could fruitfully be ensured in the following task.

Roles of Non-Government Organisations (NGOs)

- In ensuring the participation of grass-root workers and beneficiaries.
- In forming Self-Help Groups of men, women and rural youth.
- As secondary disseminators of technology.
- In motivating individual and group of farmers and late adopters and in increasing the number of adopter farmers.
- In mobilising the farmers and rural people to go for agro-processing and fetching higher return.
- As a link between average farmers and public sector extension agency.

- In ensuring timely availability of knowledge-based inputs at grass-root level synchronising with the season and time.

Having defined specific roles that could be assigned to and prescribed for above three types of organisations, let us identify the various components of agribusiness that could be taken up by these sectors with the active participation of farmers, rural people and self-help groups as well as stake holders. These components could be commercial farming, agro-processing, value-addition and qualitative improvement in the product and then standardisation and grading of the product, proper packaging and labelling. They have to be supported by logistics and infrastructures like adequate power supply for processing, adequate and proper transport system and organised market both within and outside the State and then again the agricultural exports. As of today, percentage share of agricultural exports in the national exports is hardly 14.21 per cent. Countries smaller in area and less production are attracting better world market than ours. India's share in world's agricultural exports has been hardly around one per cent and this has to be increased to 10 per cent in near future for making agriculture commercial and for earning more profit and income to farmers and foreign exchange. There is both the potentiality and the possibility of this and it can be achieved if we make sustained efforts.

Having prescribed the roles of the public and private sectors and the non-government organisations, it is to be made clear that there have to be a proper partnership mode and collaboration of these agencies coupled with the active participation of the clientele including farmers, rural people, farm women and farm youth as well as the self-help groups and for all this, proper motivation has to be created in the various extension agencies and a coordination mechanism is to be established to link and coordinate the programmes and activities of the agencies in order to avoid the confusion to prevent the wastage of men and material resources.

Knowledge-based Inputs and the Role of Public, Private and Non-Government Organisations

Most of the technology developed and recommended are knowledge-based. Farmers are required to first have the knowledge about the various improved practice/technology being recommended for use by them so that their application in field situation is not faulty. This is necessary for ensuring satisfactory results and performance of the technology otherwise the mere adoption of a technology and its application may not serve the purpose. For example, let us take the example of fertilisers. Farmers have to know the type of fertilisers available in different nutrient groups, their recommended dose, frequency of application, and method of application of fertiliser so that the farmers could procure the required fertiliser for application. He must have all the informations about fertilisers so that there is a correct application of the fertiliser in the field in the matter of type, dose, frequency and method of application. Similar is the case in insecticides, fungicides and weedicides in which the type, dose and method of application have to be well-known by the farmers. Even in case of varieties of different crops, farmers have to make choice depending upon properties of the varieties and the soil type, for example, low land rice variety cannot be grown in up-land situation and vice-versa. Similarly, the Diara composite variety cannot be grown in plains. Farmers have to have adequate knowledge of the suitability of the variety for different situations and for defined purpose. Even in the case of vegetable and fruit preservation for value-addition and processing sufficient knowledge-input is necessary.

In other words, most of the inputs and technologies developed are knowledge-based and their availability have to be coupled with suitable farm advisory and guidance service. In this case again three agencies, *i.e.*, the Public Sector Organisation, the Private Sector Organisation and the NGOs have to play a significant role. It is also a common experience that many a time inputs are not available and there are cases when inputs are available after the sowing season is over which is because of delayed decision-making for various reasons. Agriculture is highly season-bound and if the operations are not performed timely then the season is lost and there is a heavy loss to the yield. Even in case of agribusiness and agro-processing, knowledge-based inputs are to be provided in time. There could be several ways, a few of them as under, by the public, private and non-government organisations in a participatory mode for improving the availability and transfer of knowledge-based inputs at grass-root level.

Improving the Availability and Transfer of Knowledge based Inputs at Grass-Root Level

- Information shop and input sale centre with single door delivery system in every Panchayat alongwith Farm Advisory System.
- Agricultural Technology Information Centre at Block level.
- Formation of Self-Help Groups and Raitu Bazars in order to make knowledge-based inputs available to the farmers.

- Imparting sequential learning through lessons on All India Radio and Doordarshan.
- Conducting method demonstrations in villages like on seed treatment, soil treatment, top-dressing, insecticide use, seed drill operation and post-harvest technology.
- Training in soil and water testing, apiculture, mushroom culture at research stations, K.V.Ks for imparting knowledge in these areas.
- Stake-holders/input manufacturers while publicising their products could also give the requisite information on those products as to what are their properties and how best to use them.
- Primary Agricultural Credit Societies (PACs) should also handle inputs and literature on them as well as on micro-enterprises based on market-trend.

To conclude, it could be said that while the entire extension work cannot be left to the management of Public Extension System because of the obvious weaknesses but at the same time everything cannot be left in the hands of private sector and NGOs. In case of latter, self profit motivation and localised development will predominate and the real benefit to the average farmers will not accrue to the desired level. Thus, the best course will be to have a combination of the three and the partnership mode between them. Some functions like reaching the weaker section, down-trodden and resource poor farmers with the government support could be left to the public sector organisations while the private sector organisations could basically deal with the input supply and services and also agro-processing, value-addition and more of agribusiness. The government will have to provide necessary policy, guidance/directions and will also have a monitoring role from time to time in order to keep the development on the right tract in the desired and defined direction. The NGOs could help in ensuring the participation of grass-root workers and beneficiaries through self-help groups as secondary disseminators of technology.

There has to be a coordination mechanism to coordinate the activities, programmes and progress of these agencies so that they do not deviate in different directions. This kind of participatory mode will also help in improving the availability and transfer of knowledge-based inputs at grass-root level synchronising with the season and time constraint will also be helpful in promoting agribusiness.

15

METHODS OF ORGANISATION OF FARMERS THROUGH SELF HELP GROUP

— *V.G. Sunil* and *K.A. Ponnusamy*

Agriculture is a way of life, a tradition, which for centuries has shaped the thought, the outlook, the culture and the economic life of the people of India (ANP, 2001). Continued development in this sector is crucial to meet the demands of the burgeoning population, widening income disparity and problems of sustainability. The productivity, profitability, sustainability and equitability of our production system are in the declining trend. The rainfed areas are characterised by a number of complex bio-physical and socio-economic constraints affecting the basic sustenance of the farmers. Similarly, irrigated and the high value commercial areas needs modern technology and infrastructure to cope up with the changing demand. With the reduced donor contribution and mounting public expenditure, we cannot foresee any considerable government intervention to reduce these burdens.

In this context, the farmers are left with no options but to join themselves and form a collective strategy to address many of their problems. The common property resource nature of many of the present problems adds more weightage to a joint problem management team. Community organisations like Self Help Group (SHG) acquires greater significance in this context (MANAGE, 1997). Here, the experience of various government organisations who have already experimented in organising farmers through Self Help Groups (SHGs) acquires greater relevance. Kerala Horticultural Development Programme (KHDP) forms a successful case example in this direction.

Hence the present investigation was undertaken with the objective to identify the different methods of organisation of Self Help Groups.

RESEARCH METHODOLOGY

The study was conducted in Trichur district of Kerala State adopting an *ex post facto* design and case study as method. Six SHGs of three-year old were selected for the study purposively. The samples were selected in such a way to have maximum heterogeneity between SHGs. To bring variability, the SHGs of the district were first classified into six groups based on their geographic distribution. Hereafter, the SHGs of the district were classified based on their functioning into better, medium and poor performing ones by developing appropriate indicators. Random matching was carried out to have two SHGs each of high, medium and low performance, distributed to the six geographic groups. In this way every geographic region got a representation and also the SHGs vary among each other in their performance. The selected SHGs were: 'Arattupuzha' and 'Nedambadam' SGHs with high level, 'Kaynoor' and 'Melachira' SHGs with medium level and 'Nayarangadi' and 'Sarithapuram' SHGs with low level of performance. The total sample size of 100 respondents for the study was constituted by 90 farmers (@ 15 farmers per SHG) and 10 Change Agents associated with the groups.

Organisational process in the study referred to all those activities, initiatives and intervention in the village that had resulted in the establishment of the SHG. Data regarding this variable was collected using participatory tools like focussed group discussion, semi-structured interview schedule, transact walks, timeline analysis, and non-participatory observation, and were analysed qualitatively using process analysis methodology to identify different methods of organisation. The different methods thus identified in the course of investigation were named using different terminology.

Results and Discussion

Methods of Organisation

The results of the study revealed six different methods of organisation in SHGs. The different methods identified and named in the course of the investigation are discussed below:

Grass-root Organisation

Grass-root organisation was noticed in the SHGs of Arattupuzha panchayat. Here majority of the farmers was from very poor socio-economic status and does not have any early experience in working with groups. So the Change Agents had to start awareness creation activities from the very basic level, contacting individual farmers. His job became tough because of a local rift and the resultant reluctance from many of the large farmers to take over the stick for organisation. Through regular contact and informal interaction the change agent was able to carry the mobilisation function and also to organise the first SHG in 'Arattupuzha' ('Arattupuzha' SHG).

The process had taken a large amount of time and efforts from the Change Agent to mobilise and bring farmers together. However, the educative approach followed in the process had facilitated the group to get stabilised in a very short period. This method can be recommended on a general basis for organisation of Self Help Groups.

Split-Half Organisation

Splitting a parent group in to two equal halves on regional basis and admitting more farmers from neighbourhood to achieve the required group strength formed the characteristic feature of split half organisation. This type of organisation had resulted in the formation of the Kaynoor SHG. Before the organisation of Kaynoor SHG, 'Marotichal' SHG was functioning well in the village. The better functioning of 'Marotichal' SHG attracted more farmers to join in the group. However, admitting more members into the same group can create many managerial problems. Understanding this basic reality, the Change Agent of 'Marotichal' SHG split the group on regional basis into two equal halves. New members were admitted with the old members and 'Kaynoor' SHG was formed. The same procedure was adopted with the other SHG also.

The split-half method of organisation had resulted in early establishment of the SHG. The leaders from the parent group played the cognizance and motivation roles effectively. In this way it reduced the mobilisation role of Change Agent to a greater extent. Wherever further expansion of groups need to be carried out, this method can be adopted.

Make-shift Organisation

A large number of Farmer Organisations were functioning in 'Pananchery' Panchayat when the intervention of KHDP first occurred. These groups had proven their capability in mobilising farmers' interest. To have a SHG with a sound foundation, the change agent sought the help of some of the political leaders from the panchayat. With their consent, one of the Farmer Organisations in the panchayat was transferred into a SHG (Command Area Development Authority Group), with little change in the group structure. The genesis of 'Melachira' SHG thus occurred in a make-shift fashion.

Although the process had resulted in early formation of the SHG, it also brought many problems. The prominent one was the corrupt and preconceived leadership inherited during the process. Besides this, most of the members did not take the concept of SHG in its real sense. The ultimate result was the destabilisation of the SHG after a short period. Additional efforts from Change Agent to revitalise the group were thus found necessary. So before going for this method the Change Agents should undertake indepth study on the functioning of the parent group.

Tier-down Organisation

Tier-down organisation was characterised by definite stages in the organisation with different actors and orientation. The organisational process that had resulted in the establishment of Nedumbadam SHG formed a best example for this type. The initiation of the process had come from a meeting of KHDP officers with political leaders and prominent farmers of Chalakudy municipality. The strategies to organise a cluster of groups in the different panchayats of the Municipality were worked out at that meeting. Hereafter, the task to carry the mobilisation role in the various panchayats was allotted to appropriate persons. As a second stage, a panchayat level awareness meeting was arranged wherein officers of KHDP and those attended farmers formed various committees to do the job. It was the committees thus formed who had implemented the organisational role at grass-root level with Change Agents assistance. The remote location of the village was one of the factors that made the Change Agent to adopt for such a strategy.

The process of organisation was found successful mainly because of the high socio-economic condition of most of the members. The committee members themselves carried out most of the organisational activities with little change agent involvement. This method can be recommended for organising SHG with large and educated farmers.

Self-Induced Organisation

How local issues can influence the process of organisational was clearly demonstrated in the example of 'Nayarangadi' SHG. Here, a local rift between the farmers of 'Kunnapilly' and 'Nayarangadi' villages formed the impulse behind the organisation. The intervention of KHDP to 'Kunnapilly' to organise a SHG lighted this conflict once again. With no time left, the farmers of 'Nayarangadi' mobilised whoever they can to form a group. With nearly 15 members they approached KHDP and got their group registered. And the birth of 'Nayarangadi' SHG thus resulted out of a local issue in a very short period.

A local rift between two village groups acted as the root cause in the organisation of the SHG. Here the mobilisation work was done all in a hurry without any set criteria. And all the activities were done through the hands of a few village leaders. The process ultimately resulted in severe dropout of the members. Although the rift later transfer to a competitive mood. The Change Agents, therefore, should educate the members well on the concept of SHG before mobilising such group sentiments.

Name Sake Organisation

Name sake organisation was characterised by a target oriented and formal approach. Here the ultimate objective of the Change Agent was to meet a set target with little effort. This type of organisation was identified from 'Sarithapuram' SHG. Here, because of the remote location of the village and the highly conservative nature of farmers, the job of Change Agent became very difficult. Instead of adopting a grass-root type of organisation, the Change Agent had gone for a formal one. Further, in the initial days of group development, the Change Agent did not provide much nourishment. The consequence was that, the group functioned only when there was an officer attached to it.

Organising SHGs needs a deep sense of commitment from the part of Change Agents in selecting and orienting the members on the group concept. This assumes significant when we consider the poor socio-economic condition under which most of the members live. Hence to build a good foundation for a group, the target-based approach have little relevance.

Conclusion

The study revealed six basic processes of organisation of farmers into SHG. The documented processes are: grass-root organisation, split-half organisation, make-shift organisation, tier-down organisation, self-induced organisation and name sake organisation. The advantages and disadvantages of the different methods of organisation were also discussed in this chapter.

REFERENCES

MANAGE (1997). *Farmers' Participation in Agricultural Research and Extension Systems*, National Institute of Agricultural Extension Management, Hyderabad.

NAP (2001). *National Agricultural Policy Document*, Ministry of Agriculture, Government of India, New Delhi.

Sengupta, N. (1998). "Empowerment: A Socio-psychological Approach to Self-Help Group Formation," *Prajnan*, 26(4): 523-25.

16

DAIRY CO-OPERATIVES AND RURAL DEVELOPMENT IN INDIA
Problems and Prospects

— *V. Venkatasubramanian, A.K. Singh* and *S.V.N. Rao*

The pace of Dairy Development in India has picked up only after 1970 with the launching of Operation Flood. This period has also seen the emergence of dairy co-operatives as the dominant force. The milk producers have been assured with reasonable markets and dairy industry has become producer oriented. Presently, India has emerged as the worlds' largest milk producer and become a model for other countries with respect to its Anand Model milk co-operative system. The Operation Flood part of the dairy development has clearly emphasised the social aspects of developing through the importance given for the establishment and growth of Village Milk Producers Co-operative Societies (VMPCS) throughout the country followed by the initial success of Anand Milk Producers Union in uplifting the farmers.

The Anand model of co-operative structure is built on a vertically integrated single industry co-operative that links rural producers with urban consumers by introducing intensive milk production model, improved technology and marketing. The extension and transfer of Anand type Dairy co-operatives to other parts of the country did not take into account the political, economical and bureaucratic impediments which were not encountered at Anand. For a country of India's size and dimensions, any such effort of replicating the Anand type co-operative would prove very expensive in terms of time and resources. In 1970 there were only 1,588 member village milk producers co-operative societies and at present the number is 60,825 with an Annual Milk procurement of 3.58 million tonnes. The above achievement is viewed as insignificant, when it is viewed against the challenges faced by the sector. The researchers have identified a number of challenges which were summarised and discussed in this chapter under different sub-heads for the purpose of brevity and clarity.

The Problems

Socio-Economic Problems

The dairy development in the country is hamstrung by its characteristics socio-economic situations. In addition to the ever increasing human and animal population, a number of other factors are decelerating the pace of dairy development in the country. They are:

Small and Scattered Herds

Out of about 98 million farm holdings in the country over 75 million are owning land below 2 ha. This is in addition to about 30 per cent of the rural families who are landless (Patel, 1993). Dairy Production is largely taken as a subsidiary occupation to crop production in rural areas.

Investment in Dairying

The investment in dairying has decreased from 0.82 per cent of the total plan expenditure in First Plan to 0.58 per cent in the Seventh Plan, whereas the value of output from the dairying increased from Rs. 6,500 million in 1951 to Rs. 2 lakh 60 thousand million in 1990 (Economic Survey, 1990-91; National Accounts Statistics, 1991; Planning Commission; MOA; Annual Report and Dairy India, 1992). It is clear from these figures that the expenditure on AH and Dairying is very much out of proportion to their importance in the economy of the country. The paucity of funds is a serious constraint which impedes investment in manpower development, creation of proper communication, infrastructure, establishing reliable data based and strengthening the extension network in the villages.

Conflicting Goals

The conflicting goals of the policy-makers, researchers, dairy development personnel and farmers quite often create confusion. Policy-makers are usually interested in populist measures which may or may not contribute to the development of dairy co-operatives. The development professionals concentrate on large farmers to achieve the targets (e.g., number of co-operative societies constituted) in short duration by expending less energy. The co-operative input suppliers also aim at profit by supplying inputs to few established large co-operatives rather than spending time and other resources in covering co-operatives in remote areas.

The researchers are basically interested in developing technologies for satisfying the need for publications which are usually linked to their career advancement. These technologies were developed mainly because the researchers, extension workers and farmers operate in isolation and as such the researchers may not be aware of the needs as perceived by the farmers. Many a time the research is being conducted on research stations for the interest of large farmers. Hence, many of the technologies developed are more applicable to the resource rich farmers rather than to a large number of small and marginal farmers who need the research support more. In addition, doubts are also expressed about the adequacy of the present system of dairy development, ability of the extension workers in absorbing and communicating the modern technologies to the farmers in an understandable form and the capability of the research system to communicate the multi-crop, multi-animal and multi-disciplinary knowledge down the lane.

Milk Procurement and Processing

At present milk is being processed in the dairies under government, co-operative and private sectors. It is stated that not more than 20 per cent of the total milk produced is being handled by this organised sector. The private dairies are mostly involved in the manufacture of milk products with little or no attention to the supply of fluid milk to the consumers. Wherever, the private dairies are existing the government and co-operative dairies are facing stiff competition in milk procurement as well as sale of products such as ghee and skim milk powder. Incidentally, many of these private dairies are located in the north-western part of the country. With the delicensing policy of the government many more dairies are entering into the fray in private sector. These private dairies are concentrating on collecting buffaloe milk which required for manufacturing products. Very little attention is being paid to increase milk production, produce clean milk or procure hygienic milk. As a result, many dairies are incurring losses due to spoilages.

The absence of regular and remunerative market for milk and milk products is a major constraints encountered by the farmers. Lack of or inadequate credit facilities to the farmers for dairy purpose and exploitation by the middlemen hampers the farmer from deriving maximum profit from dairying. Similarly, there are no organised markets for fodders and animals in the country.

Organisational

At present the dairy development activities are being carried out by the national level organisation NDDB and its establishments and State level organisations like co-operative milk federations and Animal Husbandry departments. The success of these organisations in achieving their objectives depends upon the availability of funds, morale and capabilities of their employees and policies of their organisation.

It is not out of place to mention that there is nothing like dairy extension personnel solely responsible for the education of dairy farmers. Dairy Extension activity is always treated as secondary to many other activities such as milk procurement in case of milk union employees and treatment and insemination of animals by Animal Husbandry personnel. The net result is that majority of the farmers are not aware of the concepts of the dairy farming and also the facilities available to them through various organisations.

Some of the significant constraints perceived by the members and dairy co-operatives were, milk vendors and contractors, lower price of milk given by the co-operatives, low economic motivation of the member-farmers, groupism ad casteism prevailing in the villages, low risk bearing ability of the farmers and delayed veterinary aid by the district co-operatives.

Factors Responsible for Failure of Milk Co-operatives

The important factors responsible for the failure of milk co-operatives have been studied by Singh, 1995 and the salient finings were given hereunder.

The significant factors were, groupism and casteism in villages, ulterior motive of milk vendors and contractors, no serious efforts by District Co-operative Milk Producers Union for running the newly opened village level milk co-operatives, disloyalty of member-farmers, low economic motivation of villagers, groupism and politisation of the co-operatives and low commitment level of the member-farmers. All these factors were responsible for the poor performance of the milk co-operatives.

Policy Perspectives for improving the Performance of Milk Co-operatives

Improving the performance of milk co-operatives in our country is a multidimensional task and needs attention at various levels. At the village or farmers level, adequate steps are to be taken for educating the farmers towards the better utilisation of their local resources. The creation of co-operative values among the farmers will help them to improve their commitment level to their organisation, create the feeling of loyalty and increases their economic motivation. Efforts are need to be taken to depoliticise the milk co-operatives, and eliminate the local caste funds which are the main culprits for closing down of many milk co-operatives in the States like Haryana, Punjab, Bihar and U.P.

At the programme management level, improving the management process of our dairy development programmes need three levels of attention, i.e., an overall general/common strategies and specific techniques. The focus is to be made on human resources. First, the employees are to be offered participation in the management process. Secondly, the organisations philosophy concerning employee needs and the values of co-operation and team work are to be well articulated. The specific techniques to be adopted are right and transparent functioning of milk co-operatives, emphasis on work groups and team work, open communication and less red tapism, consensus in decision-making, relief from bureaucratic approach and concern for the member-farmers benefit at all levels of the organisation.

Change is difficult but it is a must. Bringing about a constructive change in the management policies of dairy development programmes will certainly bring about faster growth rate in this sector through farmers' participation.

Opportunities for Rural Development through Dairy Co-Operatives

There is no doubt that the milk Co-Operatives started in different parts of the country has brought forth the participation of rural producers and secured them from the clutches of traders but it has also made to realise the rural milk producers that the success and failure of the system lies in their hands only. The milk co-operatives though expected to evoke a uniform success story throughout the country, due to the prevailing socio-economic, cultural and political differences among the States, it could not be able to perform uniformly in all the States. In States like Haryana, Punjab, U.P. and North Eastern parts of the country, the co-operative structures have yielded to the local politics and become defunct. Therefore, a serious thought is to be given to find out where the fault is and corrective measures need to be carried out in the structure to function accordingly so that the desirable impact can be brought out in these States also for benefiting the real producers.

India being the country with a lot of diversified farming locations, dairying being practiced by the masses rather than mass production. Dairying is therefore largely viewed as an important tool for the poverty alleviation bringing self reliance to the rural poor and farmers. Agriculture in this country even though practiced as main occupation, due to its dependence on monsoon and uncertain price for the produce, the farmers are bound to depend on animal husbandry as a source of security and additional income. Therefore, the small size cattle herd need to be encouraged rather than recommending large size dairy sector simulating the western countries where the mass production of milk is practiced.

The recent liberalisation policy has paved the ways for multinational entry in the dairy sector which in turn threatens the indigenous dairy sector. Some may argue that such fear over the liberalisation issue is unwarranted and the entry of multinationals will no way harm the Indian Dairy industry. Entry of NDDB to organise the Sri Lankan Dairy Industry in line with that of Anand pattern and its subsequent failure due to the organised efforts of multinationals is a standing example before us to take every step for protecting our dairy industry from the multinationals. Any such politically driven policy of allowing the field open to multinationals will certainly negate the progress in this field which lies in the hands of millions of dairy farmers in the country.

Growing demand for the grains and high cost of concentrates prohibits the farmers from providing the required nutritional feed to their cattle. The cross-breeding programme has led to the requirement of compulsory feeding with high quality feed and rich plan of nutrition failing which the performance of crossbreeds comes nearer to the indigenous cattle. Efforts are, therefore, need to be made towards the solving of above problem for ensuring the sustainability to Indian Dairying.

Summary

The decision to transfer the Anand type Dairy Co-Operatives to other parts of India was a recognition of the importance of lessons learnt at Anand. The people involved and the situations cannot be recreated as each setting will be unique, but what remains is the concept of the dairy co-operatives and how it can be transplanted to other regions. In transferring the concept under OF-I and OF-II, a core group of people who have been trained at Anand have been formed a nucleus for transferring the Anand type milk producers co-operative society to the whole country.

The National Commission on Agriculture (1976) indicated that most of the co-operatives have failed to develop properly. Doubts have been expressed about the efficacy of co-operative structure for development of dairying in the country. Our experience revealed that "Anand Pattern" which was highly successful in Gujarat State, when introduced in Punjab, Haryana and Bihar had not proved to be very successful due to different socio-economic and political situation in these States. Their poor performance in these States could be attributed to various factors such as: irregulár payments, malpractices by secretary of societies, lack of proper technical persons, competition by vendors, political interference and unconducive socio-personal, economic, psychological, organisational and environmental factors etc. The good performance of milk co-operatives in States like Gujarat, Tamil Nadu and Andhra Pradesh needed to be recognised and suitable initiatives need to be taken up in the States like Haryana, Punjab and Bihar for educating and training the people involved in the co-operatives for a psychological change and capacity building required to run the milk co-operatives successfully.

REFERENCES

Acharya, K.T. and Huria, V.K. (1986). "Rural Poverty and Operation Flood", *Economic and Political Weekly*, 21: 37 (13 Sept. 1986): 1651-53.

Aneja, R.P. (1991). *Indian Dairying—Opportunities and Challenges*, 24th Dairy Industry Conference, New Delhi.

Baviskar, B.S. (1988). "Co-Operation and Politics", *Economic and Political Weekly*, 3(12): 490.

——— (1990). "Dairy Co-Operatives and Rural Development in Gujarat", in M. Doornbas and K.N. Nair (eds.), *Resources, Institutions and Strategies: Operation Flood and Indian Dairying*, Sage Publications, New Delhi.

Bhanja, S.K. (1991). *A Study of Socio-economic and Psychological Correlates in Organisation and Functioning of Dairy Co-operatives*, PhD thesis, Kurukshetra University, Kurukshetra.

Desai, H.N. (1978). *A Study of Leadership Characteristics in Selected Societies of the Kaira District Co-operative Milk Producers Union Ltd. (AMUL), Anand*, M.Sc. thesis, NDRI, Karnal.

George, S. (1985). "Operation Flood and Rural India: Vested and Divested Interests", *Economic and Political Weekly*, 20 (7th Dec. 1985): 2163-70.

Gopala, K. and Maraty, P. (1989). "Impact of Primary Milk Producers Co-operative Societies on Beneficiaries in Nalgonda District of A.P.", *Indian Co-op. Review*, 26(3): 278-82.

Gurcharan and Degra, B. (1984). "Operation Flood and Dairy Development in Villages around Delhi", *Economic and Political Weekly*, 19(20-27 Oct. 1984), 1816-17.

Jalal, R.S. (1983). "Problems and Progress of Dairy Co-operatives in Northern Hilly Region of U.P.", *Indian Co-operative Review*, 21(1).

Joshi, V.H. (1990). "Operation Flood: Constraints and Potentialities in Saurashtra", in Martin Doornbos and K.N. Nair (eds). *Resources, Institutions, and Strategies: Operation Flood and Indian Dairying*, Sage Publications, Delhi.

Krishnaraj, R. and Dubey, V.K. (1990). "Managerial Traits Associated with Organisational Efficiency of Milk Producers Co-operative Society", *Indian J. Dairy Science*, 41(1): 19-23.

Kulandaisamy, V. (1984). "Dairy Co-operatives and Rural Development: The Indian Experience", *Review of International Co-operation*, 77(1): 38-41.

Singh, A.H. (1995). *Performance of Milk Co-operatives in Haryana and Gujarat—A Comparative Study*, PhD. thesis, NDRI, Karnal.

Venkatasubramanian, V. and Fulzele, R.M. (1996). *Dairy Farming in Tropics*, MD Publishers, 11, Daryaganj, New Delhi

Venkatasubramanian, V.; Singh, A.K. and Rao, S.V.N. (2002). *Dairy Development in India*, Concept Publishers, New Delhi.

17

PROMOTION OF AGRIBUSINESS AMONG RURAL WOMEN
Role of NGOs

— *Santha Govind* and *K. Prameela*

Rural women play a key role in development, both in the context of the family and in society at large including its economy and social system. No country or society can achieve its potential without adequately investing in developing the capabilities of women and encouraging the empowerment of women. The Human Development Reports of UNDP indicate that the Human Development Index (HDI) is extremely gender-sensitive. The United Nations Development Report (1995) further states that 'investing in women's capabilities and empowering them to exercise their choices is not only valuable in itself but is also the surest way to contribute to economic growth and development'. Sustainable human development implies engendering the development paradigm.

All the Five-Year plans had given importance to the development of women. Women's development was recognised as one of the development sectors in the Sixth Five-Year Plan. In the Seventh Plan, the developmental programmes for women had a major objective of raising the economic and social status. In the Eighth Five-Year Plan, there was a definite shift from development to empowerment of women. The commitments of the Ninth Plan (1997-2002) envisages the empowering of women as the agents of social change and development.

One of the vital strategies for empowering rural women is to facilitate their involvement in economic activities through development of their entrepreneurial and income earning capabilities and access to credit. Several such initiatives have been mounted in developing countries largely through the efforts of NGOs to provide credit and related entrepreneurial and educational facilities to help women become income earners. The experience of the Self Employed Women's Association (SEWA) in Ahmedabad is known the world over. The Working Women's Forum in South India is also of similar nature.

The UNDP Human Development Report (1992) observes that 'non-governmental organisations in general can be of great assistance in developing countries'. The UNESCAP/UNDP (1991) document on NGOs clarifies four ways in which the government might make use of NGOs in any country.

1. NGOs may substitute for the government in the formal delivery of specific services.
2. NGOs may supplement the service delivery role of the government, either through delivering certain services for which government structures are not suited or by delivering the same types of services as the government but to population sectors to which the government structures find it difficult to reach.
3. NGOs may engage community development work that is often essential to ensure that normal delivery of services achieves its objectives.
4. NGOs can assist in ensuring that statutory development reflects grass-roots realities by contributing their experience directly to the planning process, acting as a pressure on the government to be responsive and facilitating the people's own access to all stages in the planning and delivery of services.

The NGOs became prominent after independence, especially after 1970s. For the first time, the Planning Commission included a separate sub-section on 'involvement of voluntary agencies' in the Seventh Five-Year Plan document (1985-90). Plan expenditure of Rs. 100-150 crores was earmarked towards the collaboration between the government and NGOs. To channel these resources for NGOs involved in poverty alleviation and rural development programmes, the Council for Advancement of People's Action and Rural Technology (CAPART) was established in 1986. This agency was also entrusted with the responsibility of providing support to the NGOs.

NGOs started to play a greater role in technology transfer activities by the end of 1980s. To increase the involvement of NGOs in efforts to strengthen the research-extension delivery system, a pilot Central Government scheme, agricultural extension through voluntary organisations was launched in 1994-95. The major objective of this scheme is to involve NGOs in agricultural extension with a view to integrate their efforts with those of the main extension systems. The impact of the scheme clearly points a greater involvement of the farm-women in the economic activities.

NGOs had the skills in group formation and promotion of joint action which government services are unlikely to develop. In this way, the government had involved a larger number of NGOs in many development programmes like Mahila Samakhya, Rajasthan Women's Development Project, IFAD assisted Tamil Nadu Women's Development Project, Women in Agriculture Project, UNFPA-assisted Integrated Women's Empowerment and Development Project in Haryana and Swarnajayanti Gram Swarozgar Yojana (SGSY) all aimed towards organisation of the rural women into self-help groups, training, credit, technology, infrastructure and marketing, etc.

NGOs had successfully tried to involve similar women's group in savings and loan activity. Results and many of these projects have conclusively shown that such women's groups could become cost-effective tool for poverty alleviation efforts.

India Rural Development Project to Empower Women

The world bank assisted the project to Government of India during 1997. Women's self-help group established by NGOs have successfully developed a system of revolving credit for the benefit of group members based on their own savings. NGOs further facilitated the self-help groups to maintain a high ratio of loan repayments and successfully generated additional income, jobs and small enterprises for their members. The NGOs of the projects have established support mechanisms for income-generating activities.

Tamil Nadu Women Development Project

The IFAD initiated Tamil Nadu Women's Development Project (TNWDP) in April 1989 and project became effective in January 1990. The main objective of the project was to bring about the economic and social betterment of women to improve the welfare of the families and their status both within the family and in the community. The project was designed to achieve its objectives through (a) income-generating activities in farm and crop development, animal husbandry and cottage and village industries, (b) Group development, and (c) NGOs involvement in beneficiary identification, group formation, training of animators, supervisors and project field staffs, selection of activities and provision of bank loans and the recovery thereof. The project relied on NGO involvement in all five districts to support the project's overall activities.

Training and Capacity Building

Training became a key factor in ensuring sustainability, *i.e.*, intensively trained group members became more aware of their capabilities and the benefits they might derive through participating in the project. Participation by the stockholders in training design was actively encouraged and multi-functional programmes (perspective building, technical and managerial aspects) were developed to suit the needs of individual stockholders. Classroom training was the main strategy for implementing this module. It would be worthwhile for the project to broaden its range of training approaches by including visits to well-functioning groups and federations, information/marketing fairs, etc.

NGOs played a central role in project implementation, particularly in the identification of beneficiaries, formation and supervision of groups, establishment of credit linkages and training. About 65 per cent of all bank loan beneficiaries claimed they had acquired new technical skills and another 20 per cent had strengthened their existing skills with training received under the project. More than half took up livestock enterprises. Apart from animal husbandry, other income generating activities included bee-keeping, cultivation of jasmine, mangoes, betel, grapes, coconut seedlings, rope making, mat weaving, etc.

Swarnajayanti Gram Swarozgar Yojana (SGSY)

SGSY was launched from April 1999. One of the salient features of SGSY is establishing a large number of potential micro-enterprises among rural women with emphasis on skill development through well-designed training courses by involving NGOs. The NGOs play an important role in the formation and nurturing of women self-help groups as well as in the monitoring of the progress of the beneficiaries. Their services are also utilised for provision of technology support and quality control of the products.

Lessons from Other Women Development NGOs

DHAN Foundation

DHAN Foundation has promoted Community Banking Programme

focusing on women. Dairy activity was scaled up in two federations in a significant manner. Tie-up with private agencies was made for marketing of milk. The Kalanjiams have extended credit to members to support the traditional business activities and for some members to initiate new income-generating activities.

MYRADA (Mysoor Resettlement and Development Agency)

MYRADA an NGO located in Bangalore formed women credit management groups. The design of the project included the delivery of credit to women through small savings and credit groups. Interest in joining groups exceeded their expectations. Moreover, women joined the groups and actively participated but did not necessarily take loans. MYRADA has been working effectively and empowering women. So far vocational activities were concentrated on self-employment but now the focus is shifting to entrepreneurship.

AWAKE (Association of Women Entrepreneurs of Karnataka)

AWAKE conducts entrepreneurship training programmes for rural women and they have been recognised in the country as effective and successful.

Entrepreneurship Development Institute of India (EDI)

EDI has developed a number of training programmes aimed at women. EDI is working with NGOs in village areas to develop entrepreneurship designed for both small and large enterprises for women.

Rural Development and Self-Employment Training Institute (RUDSET)

The objective of RUDSET is to identify, orient, trained, counsel and assist young people in rural areas to take up various self-employment and entrepreneurship both in agriculture and others.

Sumangali Sewa Ashrama

It promotes all round development of women with particular emphasis on economic independence.

Strategies for the Promotion of Agribusiness among Rural Women

Based on the lessons learned, it is suggested that rural women's agro-based income-generating activities could be enhanced through:

- Increase collaboration between government, research and extension organisation and non-governmental women development organisations.
- Identification of appropriate and location specific income generating technologies to promote agribusiness by the scientists.
- Use of NGOs as intermediate users as in the case of Eastern Bolivia where the package research information is delivered to intermediate users and mandated to transmit feed-back on farmers needs to the researchers.
- The agro-based enterprises for women should be tested by the NGOs before final adoption.
- The identified NGO officials should be adequately trained on the various dynamics of agribusiness, characteristics of women entrepreneur, achievement motivation training, steps in building up agribusiness, product selection, market survey, formulation of project, project appraisal, production and operation management, structuring training programmes for agribusiness and organisational support for the same.
- The liberalisation policies pursued since 1991 should be considered to provide viable agro-based business to rural women.
- NGOs need to form themselves into networks with women empowerment NGOs by overcoming their differences in influence government women development policies at various levels.
- NGOs need to build the capacity of the women target group to access the resources from the government as their right.
- NATP support for agro-eco system research should be utilised by NGOs to identify the appropriate location specific agribusiness for rural women.
- Each research centre should have a strong and viable agribusiness cell for women to study and plan the activities of NGOs in its jurisdiction.
- The functionaries of NGOs should participate in the monthly zonal workshops regularly and discuss the viable agribusiness activities of rural women.
- Proper follow-up for ensuring agribusiness activities among rural women especially in the marketing of products.

Thus the women empowerment NGOs should be treated as an important part of the overall transfer of technology mechanism to rural women. They should be linked with the existing research and extension systems. The Ministry of Agriculture, Government of India and State Department of Agriculture should involve NGOs in the process of ToT to rural women and to integrate their efforts with the main agricultural extension system in the States.

REFERENCES

Anonymous (1995). Entrepreneurship Development for Rural Women, Asian and Pacific Centre for Transfer of Technology of the Economic and Social Commission for Asia and Pacific, New Delhi, 69-130.

Annual Report (2000). DHAN Foundation, *DHAN Foundation*, Madurai.

Farrington, J. (1998). "Role of Non-Governmental Organisations in Extension", in *Improving Agricultural Extension—A Reference Manual*, Rome, pp. 217-18.

Rajasekhar, D. (2000). "Non-Governmental Organisations (NGOs) in India: Opportunity and Challenges", *Journal of Rural Development*, 271-72.

Rana, N. (1998). "Rural Small and Micro-Enterprises in Pakistan: Issues and Options", *Asia-Pacific Journal of Rural Development*, 89-90.

Rao, T.V. (1996). *Human Resources Development*, Sage Publications, New Delhi, 231-33.

Tapan, K.M. (2000). "Empowering Women—A Catalyst in Development", *Employment News*, 16-22 September 2000.

18

AGRIBUSINESS AND FARMERS' ORGANISATIONS IN INDIA

— Vijaykumar Chole

Agriculture was essentially subsistence farming rather than a business till the beginning of the planning era in India. Agriculture production was meant mainly for home consumption and very little or no marketable surplus was available with the farmers. The focus of agriculture is increasingly shifting from raising the production to market orientation and value addition. The approach to farming has undergone radical changes in the last few decades. Most of the farmers, even small land holder, now-a-days view farming as a business proposition, rather than a means of living.

The recently drafted National Agricultural Policy has accorded priority for promoting agribusiness activities at different levels. Faced with competition from globalising forces, the rural sector is unable to penetrate into new markets and even sustain the existing ones. Undoubtedly, marketing is key to growth of any enterprise, small or large, as it influences production, institutions and over all viability of a particular sector. Therefore, any imaginative initiatives in marketing techniques augurs well as bonus to any enterprise and farmers agribusiness organisations are no exception to it. Hence marketing aspect of agribusinesses is of immense importance for the farmers organisations in India.

Agricultural Marketing: Current Scenario and Need for Farmers Organisations

Agricultural marketing plays an important role not only in stimulating production and consumption, but in accelerating the pace of economic development in rural areas. Its dynamic functions are of primary importance to promote agribusiness and economic development in rural areas. For this reason it has been described as the most important multiplier of agricultural development. The recent far-reaching technological break through has led to a substantial increase in production on farms and to the larger marketable and marketed surplus. To maintain this tempo and pace of increased production through technological development, an assurance of remunerative prices to the farmer is a pre-requisite for developing modern agribusiness centres in rural areas. This assurance can be given to the farmer by developing a modern, dynamic/diversified and efficient marketing system led by farmers themselves.

TABLE 18.1: Marketed and Marketable Surplus of Some Agricultural Commodities

Commodity State/Particulars	*Farm Size*			
	Small	*Medium*	*Large*	*Overall*
Maize (Rajasthan)				
Marketable surplus	17.27	57.78	71.96	52.90
Marketed surplus	23.34	53.21	69.71	52.10
Bajra (Rajasthan)				
Marketable surplus	40.58	46.67	63.74	51.29
Paddy (Andhra Pradesh)				
Marketable surplus	47.10	58.20	68.10	—
Marketed surplus	46.30	56.40	63.70	—
Wheat (Rajasthan)				
Marketed surplus	33.20	44.80	57.40	50.30
Groundnut (Gujarat)				
Marketed surplus	70.33	78.47	80.01	78.56
Gram (Rajasthan)				
Marketable surplus	71.10	75.70	79.70	76.60
Marketed surplus	78.70	81.20	86.30	83.60
Mustard (Rajasthan)				
Marketed surplus	91.88	93.29	93.98	92.88

Source: Agricultural Marketing in India, 1999.

Note: In case of fruits and vegetables, almost 99 per cent is marketed surplus.

The great obstacles still encountered in the marketing of agricultural commodities have compounded the difficulties confronted in production. The marketing of agricultural commodities, especially perishables is associated with a unique set of conditions, which makes the task difficult and highly risky. Firstly, the nature of produce itself, due to its high degree of perishability, it is difficult to create time and place utilities for these commodities with small expenditure outlay. This problem is further compounded by the poor transport and storage facilities. Secondly, the marketing of agricultural commodities at present is mostly in the hands of a few proverbial middlemen who channelise the produce from the producer to the ultimate consumer. This has an adverse effect on the prices received by the cultivators. The middlemen dominate the market and exploit the producer/seller. The profits of these middlemen compared to the services rendered by them are outrageously high. They exploit not only the producer but also the consumer to a considerable extent because of which a middleman is rightly termed as "Double-Edged knife". Thus mostly the private agencies dominate the Indian agricultural trade with prevalence of malpractice.

The efforts of the government to improve the marketing system of agricultural commodities have been only partially successful in creating healthy conditions for scientific and efficient marketing. Moreover the progress of regulated markets is not uniform in all areas. Under such circumstances and in the wake of new liberalised market economy renewed interest needs to be taken in the farmer led co-operative or some sort of group marketing organisations. The need for strengthening co-operative organisation has therefore been recognised for the marketing of produce for farmers and for making inputs available for them at the right place and time. A paradigm shift towards farmers marketing organisations in times bound need to promote agribusiness in rural areas. These organisations should function as competitors of private traders/private business organisations. They should resemble private business organisations in the method of their operation, but different from the capitalistic system chiefly in their motives and organisations.

S.S. Acharya and N.L. Agarwal have defined, "A Co-operative sales association as a voluntary business organisation established by its member patrons to market farm products collectively for their direct benefit". It is governed by democratic principles and savings are apportioned to the members on the basis of their patronage, *i.e.*, the profit earned by the society is distributed among the members on the basis of the quantity of product marketed by him.

Given the dominance of small holders in India, agriculture and private traders in agricultural marketing, co-operative or some sort of group marketing alone can help in realising the advantages of scale of economy which is becoming more and more relevant in the light of emerging competition (Ramesh Chand, 1999). These farmer led organisations should emphasise the concept of commercialisation. In order to face the challenges as well as to take advantage of economic liberalisation, Indian agriculture would require co-operative organisations with high professional skills.

Success Stories

(a) *Amul: The Taste of India*

Whenever we talk of co-operative movement in India, one name is bound to be discussed and that is Amul, the brand promoted by GCMMF. GCMMF is taking on transnational across a clutch of product categories. Amul is brand worthy of the trust of 1,000 million Indians. It is a must that other co-operatives should take note of the success achieved by GCMMF, which can be well understood by the following market share monitor.

TABLE 18.2: Market Share Monitor: Amul *vs.* the Rest

Product	*Amul*	*Britannia*	*Nerka*	*Nestle*	*Mother Dairy*	*HLL*	*Others*
Processed Cheese	63	36	—	—	—	—	1
Butter	86	4	4	—	—	—	6
Infant Milk	68	—	—	24	—	—	8
Dairy Whiteners	45 (Amulya)	12	—	24	—	—	19
Curd	75	—	—	5	15	—	5
Ultra high treated Milk	36	—	—	14	—	—	50
Ice-cream	27 Recently Launched	—	—	—	11	40	22

Source: Business Today, September 2001.

TABLE 18.3: Pizza Market

(*Pizza sold per day*)

Amul	*Pizza Hut*	*Nirula's*	*Domino's*
19,000 (Within 15 days of launch. No Multinational has done this)	21,000	5,000	50,000

Source: Business Today, September 2001.

What has made Amul so successful brand? Its own whopping marketing network, diversified activities, price competitiveness, co-opting partners.

- Amul sources milk from over its two million dairy farmer members in Gujarat and is highly cost competitive. GCMMF is too price obsessive. It is a co-operative with a vision to give the farmer the best price for his milk and the consumer the lowest price for the products he buys.
- With 40 product category, 300 stock keeping units 1,00,000 retailers with refrigerators, a 18,000 strong cold chain and 5,00,000 non-refrigerated retail outlets, GCMMF is strong at its distribution system.
- Amul has moved its marketing efforts and diversified its activities. Amul has entered pizza and parotha market and aims at becoming major food brand in India.
- Co-opting partners.

B.M. Vyas, M.D., GCMMF says that, "The relationship Amul enjoys with other farmer based co-operatives all over the country is the biggest strength for their endeavour to become a major food brand".

Delhi and Bangalore based Mother Dairy or Akluj Dairy in Maharashtra help GCMMF in the critical functions of procurement and manufacturing. Amul has realised that if it is to succeed as a food company, it can be done so only by procuring and manufacturing through other co-operative federations.

Amul's game plan for the confectionery business is built around on alliance with CAMPCO, the Karnataka based cocoa farmers co-operative which produces 5,000 tonnes of cocoa products every year. Today GCMMF has nine such alliances with other co-operatives.

It is on the strength of just such an arrangement with Maharashtra Dairy and Pune Milk co-operative in Maharashtra that Amul is re-launching its brand in the 35 lakh litres a day milk market in Mumbai.

- Once the frozen pizza business takes off GCMMF has decided to tap Mother Dairy's Safal as a source of frozen vegetables.
- To equip its farmer members with latest technical know-how training programmes are organised.
- Advertisement and propaganda is done on large scale.

(b) *Mahagrapes: A Farmers' Cooperative*

- Mahagrapes is farmers' organisation catering to the needs of Maharashtrian farmers and has been responsible in a big way to increase the returns to the farmers.

- Mahagrapes, farmers' co-operative is one of the largest exporters of fresh grapes mainly seedless grapes from India. Keeping in view the international trend of fruit growers becoming exporters, the vineyard owners of Maharashtra entered international market with their own brand.
- The grape growers from areas such as Pune, Nasik, Solapur, Latur, Sangli farmed 18 Co-operative Societies with almost 2,500 farmers. The entire members' co-operatives (18) are partners of Mahagrapes.
- Mahagrapes has excellent infrastructure of its own for pre-cooling, cold storage and packaging (15 units), the technology for which has been imported from California.
- The share of Mahagrapes in grape exports is about 15 per cent with turnover being to the tune of more than Rs. 100 million. Inspite of the stringent quality parameters for export to Europe, Mahagrapes has succeeded as a reliable brand in super market chains of Europe.
- Mahagrapes is expanding its wings by diversifying its activities and extending exports to cover other fruits. It received an enthusiastic and successful response in export of fruits like pomegranates, mangoes, strawberries, limes, lemons etc. With more than 15 countries on its customer list Mahagrapes has not stagnated at all, it is continuously in process of exploration of new areas for marketing of fresh produce all over the globe. It has already initiated export of vegetables like okra, brinjal, capsicum, green peas and beans.
- Mahagrapes has a special team of quality control officers which monitors fruit quality at different stages of operation. To meet the quality parameters, Mahagrapes supplies inputs like verticilium, trichoderma, CPPU, tissue paper to its member farmers.
- A team of professional managers assist the executive partners, which constitutes of General Manager (Exports), Manager (Liaison & Logistics), Manager (Finance), Manager (Quality Control) assisted by a team of quality control officers. This team of professionals performs the delicate balancing act of locating internationally acceptable "Quality Product" on one hand and identifying the lucrative markets to increase net returns to its growers on the other.

(c) *Fruit and Vegetable Producing Farmers Societies in Maharashtra*

There are 32 farmer's procurement and marketing societies in Maharashtra. They have their own cold storages and transport systems. Few of these societies have tied up with hotel association in Pune for supply of farm fresh vegetables on fixed prices. For example one farmer's organisation Nandini fruit and vegetable growers association, District Kolhapur daily handles about 200 metric tonnes of fruits and vegetables. The case study shows that farmers organisation reduces the risks and spreads the benefits of farming business. The organisation has allowed the farmers to diversify the market arrangements. These organisations also supply inputs to the member-farmers.

(d) *Market Promoting Activities by Farmers' Organisations*

Maharashtra Rajya Draksha Bagaitdar Sangh, a public trust with more than 20,000 farmer-members is organising a Raisin Festival—Raisin exhibition and Raisin based recipe competition to promote raisin industry. This for the first time that such an activity is organised in India. Such promotions will definitely boost sale.

Thus from these case studies we can see that the success of these organisations lies in:

1. Farmers participation in major decision-making and farmers' faith in these organisations.
2. Co-operation with other co-operatives.
3. High professional skills/talents to manage the affairs.
4. Own infrastructure.
5. Market tapping/exploration exercises.
6. Market promotion activities.
7. Training and input supply activities.

Thus, a unique blend of co-operation, commercialisation and professionalism has led these farmers' organisations to promote and develop agribusiness centres for rural development. Such a paradigm shift towards farmers organisations is needed in order to comprehend and grab new opportunities and face the challenges thrown by new liberalised market economy in the wake of WTO and GATT for sustaining the growth achieved till now and expand new horizons success.

Strategies for Strengthening of Farmers' Agribusiness Organisations

(a) *Infrastructure Development and Training*

Farmers' organisations should establish their own procurement centres with facilities for grading, sorting, washing, packing and pre-cooling in the centralised locations at production areas. These organisations should have well maintained transport system for quick and effective procurement and delivery. This will help in effective post-harvest care and handling. They should develop a chain of their own warehouses and retail outlets at potential consuming areas.

E-commerce sites for agriculture should be launched by these organisations for input as well as output marketing. Farmers' needs vary with season, crop, weather and location. So most of agribusiness services need to be regional in scope. They should maintain updated market information. Many farmers do not have so much time or information access to make and implement informed marketing decisions because commodity prices are always changing. So, they can get information on comprehensive produce marketing and risk management programmes from any clicking on a link. This also gives the farmers peace of mind, and ability to concentrate more on producing their crop.

Member-farmers and small processors should be trained for use of pre-harvest treatments to regulate growth, delay maturity, reduce post-harvest disease and disorders, and to use small cost effective short-term storage structures.

(b) *Product Design and Diversification*

Along with marketing of raw agricultural produce, the farmers organisations should diversify and also concentrate on value addition to the produce by investing in agro-processing industries in rural areas. These agro-product organisations should get their act together on building their own brands. Newer and effective brand *mantras* should be utilised for marketing overdrive of the produce.

An agro-processing industry (API) is defined as "one which processes bio-mass, *i.e.*, agricultural raw materials, which include ground and tree crops as well as livestock and fisheries to create easily exportable forms, enhance nutritive value and extract chemicals for other uses". Because of the crucial farm industry linkages they generate, the APIs are considered to be an important component of the agriculture led growth strategy. They not only initiate industrial development in a region but are also capable of opening up new crop and livestock avenues to the farmers and motivating them for higher productivity and diversified crop-mix. By reducing the post-harvest losses and by adding value to the farmers' products they help in higher price realisation for farmers. Capable of being set up anywhere they also bring about decentralised industrial structure reducing regional inequalities and help in raising employment especially in the rural areas. Rural skills, technology, managerial abilities develop and help in emergence of new kind of activities further strengthening the linkages between different sectors and institutions.

New products with minimal processing operations needs to be developed to meet the domestic as well as global market as the export of these would be more profitable and less risky than export of fresh product. Full potential needs to be tapped. The product quality analysis, sensory evaluation and consumer acceptance studies have to be carried out for different processed products.

Large number of by-products can be manufactured from food processing wastes. Emphasis should be made on utilisation of food processing wastes for manufacturing and marketing of these wastes. Presently some methods are available for the use of food processing wastes, but the available technologies are under utilised. Hence the scope should be widened to meet the need of farmers for on farm use of wastes.

TABLE 18.4: By-products from Food Processing Wastes

Fruit	*% Waste*	*Nature of Waste*	*By-product*
Mango	27-50	Peels, stones, pulping waste	Starch, far, pectin, vinegar, syrup alcoholic
Pine-apple	50-60	Peels, cores, trimmings	Vinegar, syrup, citric acid, candied cores
Grapes	5-10	Stem, seeds, seed hulls	Cream of tartar, seed oil tannin vinegar, pectins, wines, stock seed
Apple	60-70	Pomace, Cores cull	Pectin cider, vinegar, soft drink jelly base
Citrus	50	Peels, pomace, seeds	Pectin, essential oil and seed oil
Peaches, apricots & cherries	—	Piths, kernels	Oil, stock feed

Source: *Yojana*, 1998.

(c) *Marketing Memo*

The business should be truly customer driven. For this Philip Kotler suggests following marketing memo for each department of the organisation.

R&D

— They should spend time meeting customers and listening to their problems.
— They should benchmark competitors' products and "seek best of class" solutions.
— They should solicit customer reactions and suggestions as the project progresses.
— They should continuously improve and refine the product on the basis of market feedback.

Manufacturing

— They should invite customers to visit and tour their plants.
— They should continuously search for ways to produce goods faster or at lower costs.
— They should continuously improve product quality, aiming for zero detects.
— They should meet customer requirements for "Customisation" where this can be done profitably.

Marketing

— They should study customers needs and wants in well-defined market segments.
— They should measure brand image and customer satisfaction on continuous basis.
— They should continuously gather and evaluate ideas for new products, product improvements and services to meet customer needs.
— They should influence at company departments, customer centered thinking and practices.

Sales

— They should have specialised knowledge of the customers.
— They should strive to give the customer "the best solution".

— They should make only promises that they can keep.
— They should feed back customers' needs and ideas to those charge of product development.

Logistics

— They should set a high standard for service delivery time and they meet the standard consistently.
— They should operate a knowledgeable and friendly customer service department that can answer questions, handle complaints and resolve problems in a satisfactory and timely manner.

Finance

— They should understand and support marketing expenditures (*e.g.*, image advertising) that represent marketing investments that produce long-term customer preference and loyalty.
— They should make quick decisions on customer credit worthiness.

(d) *Product or Brand Management*

In agricultural produce marketing, very few farmers' organisations hold a established brand. There is need to create and use effective brand *mantras*. But the job does not end after band creation. Product and brand management process should be a continuous and brand management process should be a continuous process and it should involve.

- Developing a long-range and competitive strategy for the product.
- Preparing an annual marketing plan and sales forecast.
- Working with advertising and merchandising agencies to develop copy, programmes, and campaigns.
- Stimulating support of the product among the sales force and distributors.
- Gathering continuous intelligence on the product's performance, customer and dealer attitudes, and new problems and opportunities.
- Initiating product improvements to meet changing market needs.

Remodelling of Farmers' Organisations

In the dawn of the new millennium, the issue of enhancing

competitiveness of farmers' co-operatives to face the challenges of liberalised market economy is becoming crucial and strategic. Critics of co-operatives argue that these co-operatives were not in good condition earlier and with this liberalised economic environment the situation will further worsen. There is no denying the fact that co-operatives are the foundation pillars of the rural economy and even today they influence significantly our socio-economic political structure. The planners and policy-makers of co-operative movement will have a tough task in reorienting and remodelling co-operatives during this transition phase of shift from controlled regime to open economy. Against this backdrop an attempt has been made to outline a few areas for remodelling co-operatives for enhancing professionalism and competitiveness in their working.

1. *Reinforcement of Meritocracy*

Co-operatives in India have never made any serious attempt to attract the best talents available in the country. In regulatory and projectionist regime, they have not influenced much. Now when they have to operate in a different environment where "Survival of the fittest" is the *mantra*, merit and competence cannot be ignored. Instead of looking at largely mediocre and security oriented individuals for managing the affairs of co-operatives, there is a need to identify people having traits like risk taking ability, resourcefulness vision and innovativeness. In addition to attracting talents, rewarding of competence and training for greater competence are the other issues addressed to this problem.

2. *Team Building*

It is being said that an Indian as an individual is the best. But when he is asked to work in Teams, the results are not encouraging. As co-operatives are the symbols of team spirit, critics contend that the members of co-operative rarely work as a team. Teams are built in an organisation to pool the talent energy and initiative of several persons with a view to achieve what individuals could not achieve. Efforts should be made to build teams for better results. The following points are to be kept in mind for building effective teams.

(a) Share common ideology for dedicated and devoted efforts.
(b) Access to resources for all.
(c) Pooling of wide spectrum of expertise, skills and abilities.

(d) Division of work, specialisation and delineation of responsibilities.

(e) Sub-ordinate oriented and tasks oriented leader.

3. *Self Supportive*

Co-operatives were groomed in our country with the help of state support and initiatives. This has affected their working adversely, slackness, poor initiatives, red tapism and bureaucratic culture are the hurdles to face the press and speed of liberalised market economy. If they have to stand on their own feet, good initiative and professionalism should be at primary consideration. State support and protection should be accepted only in case of exigencies.

4. *Downsizing*

Wherever possible, this needs to be carefully examined. The flatter and thinner organisation will enhance their competitiveness. Technological advancement should not be ignored. Automation, computerisation and training work force will enhance the speed and efficiency and this will strengthen their competitive abilities.

19

MARKETING OF POMEGRANATE IN WESTERN MAHARASHTRA

— *S.S. Bhosale, P.V. Kale* and *D.V. Kasar*

Cultivation of fruit crops for market is marked by commercialisation of agriculture. Successful fruit growing requires knowledge, skill and accuracy in production and marketing. Marketing of pomegranate involves a number of intermediaries for performing different activities like transport, packaging, grading and sale of fruits in different markets. These intermediaries in turn, take away high margins from the price paid by the consumers and to that extent producers are deprived of the legitimate share in consumers price. To overcome these problems the marketing system needs to be improved either by bringing about complete regulation of fruit trade or by resorting to an efficient alternative marketing system. Some attempts have been made by the producers of pomegranate in Maharashtra to form the producers' association on co-operative line to take up marketing activities. The present study *viz.*, relative economic efficiency of organised *vis-a-vis* unorganised marketing of pomegranate in Western Maharashtra was undertaken to examine the impact of producers' association in safeguarding the interests of the producers.

Methodology

Sangola area of the Solapur district has been emerged as the well-known pocket for pomegranate cultivation. The fruits of Ganesh variety from this locality are famous for their quality, which are disposed of by pomegranate growers as fresh fruits to distant markets in the country as well as in foreign markets. In view of this, Sangola area was selected purposively for the present study. Ten cluster villages having larger areas under pomegranate were selected. Pomegranate growers are broadly classified into two major groups. The growers who sale their pomegranate through pomegranate grower association in the different markets were included in organised group and the growers who sale their produce through Hundekari's in the different markets were included in unorganised group. A sample of 45 pomegranate growers each from organised and unorganised marketing groups were purposively selected for the study. The data pertaining to the year 1999-2000, were obtained from the sample pomegranate growers, with the help of a schedule designed for the purpose. The data so obtained were analysed by a simple tabular method and the results briefly explained as under.

Results

Marketing Practices followed in Organised Marketing

The harvesting of pomegranate commences from second fortnight of December and continues upto the end of March. About 85 to 90 per cent farmers harvest the fruits of early Hasta Bahar of Ganesh variety and they call it as 'August Bahar'. After picking the fruits are brought to the packing shed. The produce is then sorted into four grades on the basis of weight and size of a pomegranate fruit. The pomegranate growers make four grades, Grade-I weight of a fruit more than 650 grams. Grade-II weight of a fruit between 550 and 650 grams. Grade-III weight of a fruit between 400 and 550 grams. Grade-IV weight of a fruit less than 400 grams. After grading the produce is packed in various sized corrugated boxes. The details of boxes used are given in Table 19.1.

TABLE 19.1: Details of Boxes used for Packing

Sl. No.	*Grade*	*Trade Name of Box*	*Box Size (Inch)*	*No. of Fruits/box*	*Weight of a Fruit (gram)*	*Price/ Box (Rs.)*
1.	I	Maharaja	12 × 9 × 5	6	Above 700	5.50
2.	I	Ganesh	12 × 8 × 4.5	6	650 to 700	4.45
3.	I	Samrat	16 × 11 × 4.5	12	650 to 700	7.70
4.	II	Manik	15 × 11 × 4.5	12	600 to 650	6.70
5.	II	Vanraj	15 × 11 × 4	12	550 to 600	6.50
6.	III	Saraswati	14 × 10 × 4	12	500 to 550	5.70
7.	III	Kohinoor	14 × 10 × 3.75	12	450 to 500	5.60
8.	III	Ratna	13 × 9 × 5	12	400 to 450	5.50
9.	IV	Karandi	10 × 10 × 10	27	350 to 400	6.70
	IV	Karandi	10 × 10 × 10	36	300 to 350	6.70
	IV	Karandi (Loose)	10 × 10 × 10	Small	Below 300	6.70

While packing paper cutting are used as cushioning material in corrugated boxes. Then these boxes are stitched, tied with plastic tape. About 80-85 per cent of the total production of pomegranate is sold through the co-operative marketing societies, which provide various marketing facilities and organises training to the farmers. The producers who wish to sale their produce through co-operatives, bring their produce after suitable grading and packing, to the assembling centres, the societies arrange to take this produce to the main office at Sangola. After assembling the produce at Sangola, societies take this produce quickly through trucks to the Delhi and Kolkata markets.

Marketing Practices followed in Unorganised Marketing

The producers who are not interested in marketing of pomegranate fruits through co-operative marketing societies, they send their produce to Nagpur, Vijayawada, Hyderabad and Mumbai markets through private transport agencies. Hundekaries at Shetphal and Sangola because they get payment within 3 to 5 days while, the co-operative marketing societies make payment after 21 days of sale of the produce. Some of the producers have sold their fruits to the representatives of wholesalers of Delhi and Mumbai markets. They provide boxes and labour for picking and grading and the rate per box is finalised and payment is made to the producers on spot. Also the pre-harvest contractors purchase the orchards and the payment is made in instalments to the producers.

Cost of Marketing of Pomegranate in Distant Markets through Organised Marketing

The information on item-wise per quintal cost of marketing, gross and net price realised in distant markets is presented in Table 19.2.

The highest per quintal marketing cost was incurred for Kolkatta market (Rs. 555.77) and it was followed by Delhi market (Rs. 411.98). The lowest (Rs. 255.84) per quintal cost of marketing was incurred for the Mumbai market. The per quintal cost of marketing incurred in Nagpur, Hyderabad and Vijaywada markets was worked to Rs. 307.53, Rs. 275.17 and Rs. 272.56 respectively.

The major items of marketing of pomegranate were the commission charges, transport cost and packing material cost which constituted about 92.00 per cent of the total marketing cost. The transport cost of fruits sold in Kolkata and Delhi markets was 50 and 47 per cent of the total cost of marketing. The highest per quintal gross price was realised in Kolkata market (Rs. 1,310) and it was followed by Hyderabad market (Rs. 1,103.63). However, the per quintal net price realised was higher in Hyderabad (Rs. 888) and Vijayawada (Rs. 783) markets because of relatively low cost of marketing.

TABLE 19.2: Item-wise Marketing Cost, Gross and Net Price realised through Organised Marketing

Sl. No.	*Items of Marketing Cost*	*Markets*					
		Delhi	*Kolkata*	*Nagpur*	*Hyderabad*	*Vijayawada*	*Mumbai*
1.	Grading charges	5.88 (1.43)	7.17 (1.29)	5.17 (1.68)	5.51 (2.00)	4.93 (1.81)	6.22 (2.43)
2.	Packing charges	8.88 (2.16)	6.17 (1.65)	9.03 (2.94)	8.33 (3.03)	7.63 (2.80)	7.88 (3.88)
3.	Packing material cost	101.72 (24.70)	113.28 (20.43)	99.93 (32.49)	94.59 (34.38)	88.21 (32.36)	103.15 (40.32)
4.	Transport charges	193.94 (47.07)	276.80 (49.90)	80.39 (26.15)	66.55 (24.19)	76.74 (28.16)	52.26 (20.43)
5.	Commission charges	88.71 (21.53)	134.63 (24.26)	94.99 (30.87)	91.61 (33.29)	86.00 (31.55)	74.78 (29.23)
6.	Postage and telephone	0.32 (0.01)	0.15 (0.01)	0.14 (0.05)	0.09 (0.03)	0.14 (0.05)	0.05 (0.02)
7.	Hamali	8.45 (2.05)	13.63 (2.45)	11.92 (3.88)	8.49 (3.08)	7.86 (2.88)	11.50 (4.49)
8.	Other charges	4.10 (1.02)	0.00 (0.00)	5.96 (1.94)	0.00 (0.00)	1.05 (0.39)	0.00 (0.00)
9.	Total marketing cost	411.98 (100.00)	555.77 (100.00)	307.53 (100.00)	275.17 (100.00)	272.56 (100.00)	255.84 (100.00)
10.	Average price	1,043.0	1,310.2	1,092.99	1,103.63	1,055.23	1,031.65
11.	Net price realised	631.07	754.25	758.46	888.46	782.67	775.81

Note: Figures in the parentheses are the percentages to the total marketing cost.

Cost of Marketing of Pomegranate in Distant Markets through Unorganised Marketing

Per quintal marketing cost, gross and net price realised through unorganised marketing is presented in Table 19.3.

The pomegranate producers had sold their fruits in nearby markets like Nagpur, Hyderabad, Vijayawada and Mumbai for avoiding the delay in payment. The per quintal cost of marketing was Rs. 295.55, Rs. 283.43, Rs. 281.22 and Rs. 261.00 in the above said markets,

TABLE 19.3: Item-wise Marketing Cost, Gross and Net Price realised through Unorganised Marketing

(*Rs./quintal*)

Sl. No.	*Items of Marketing Cost*	*Markets*			
		Nagpur	*Hyderabad*	*Vijayawada*	*Mumbai*
1.	Grading charges	4.89	5.48	5.22	6.18
		(1.65)	(1.94)	(1.86)	(2.37)
2.	Packing charges	8.94	7.77	6.53	7.92
		(3.02)	(2.75)	(2.32)	(3.03)
3.	Packing material cost	96.72	97.68	88.83	104.17
		(32.72)	(34.59)	(31.59)	(39.91)
4.	Transport charges	78.19	67.71	83.49	58.52
		(26.45)	(23.97)	(29.69)	(22.42)
5.	Commission charges	89.37	95.19	88.14	72.96
		(30.24)	(33.70)	(31.34)	(27.96)
6.	Postage & telephone	0.11	0.08	0.14	0.06
		(0.04)	(0.03)	(0.05)	(0.02)
7.	Hamali	11.87	8.52	7.83	11.19
		(4.02)	(3.02)	(2.78)	(4.29)
8.	Other charges	5.46	0.00	1.04	0.00
		(1.86)	(0.00)	(0.37)	(0.00)
9.	Total marketing cost	295.55	282.43	281.22	261.00
		(100.00)	(100.00)	(100.00)	(100.00)
10.	Average price	977.80	1,057.46	881.44	934.57
11.	Net price realised	682.25	775.03	600.22	673.57

Note: Figures in the parentheses are the percentages to the total cost of marketing.

respectively. The highest marketing cost was recorded in Nagpur market and lowest in Mumbai market. The transport charges were highest in Vijayawada market and lowest in Mumbai market. The per quintal gross price realised was the highest in Hyderabad market (Rs. 1,057) and it was followed by Nagpur market (Rs. 978). The per quintal net price realised was higher in Hyderabad (Rs. 775) and Nagpur (Rs. 682) markets because of relatively low cost of marketing.

Conclusions and Suggestions

The big consuming markets like Delhi and Kolkata in which very less fluctuations in prices prevailing all the year round, the organised groups have sold the pomegranate of superior quality in large quantities. Also the unorganised and organised groups have sold their fruits in nearby markets and it shows that the per quintal marketing cost of organised groups was comparatively less and the gross prices received were much better than the unorganised group. The net prices realised by the organised group were more than unorganised group nearly by Rs. 100/quintal. However, efforts need to be made to minimise the marketing cost of organised group and for this purpose the following suggestions were made. The co-operative fruit marketing society arrange the transport through Railway by fixing the quota at Miraj and Daund Railway station. The transport cost may be charged on the truck load basis instead of per box basis. For getting better prices, the day-to-day information on arrivals and prices should be provided through internet facilities to the farmers. The co-operative fruit marketing society may appoint the representative in Delhi, Kolkata and Nagpur markets for providing day-to-day information on arrivals, prices and actual losses. The price of the produce may be decided on the weight basis instead of box basis. By considering the marketing cost and gross prices realised in organised marketing, the pomegranate growers from unorganised marketing group may follow the marketing practices adopted by the organised marketing group.

20

WOMEN SELF HELP GROUPS AND AGRIBUSINESS MANAGEMENT

— *B.P. Mishra, C. Satapathy* and *N. Nayak*

The new agricultural policy has given boost to the agriculture and allied sector products to provide an international market platform. However, the agricultural export has not given considerable share in the total export of the country. Though there exists a large potential to be harnessed, the managerial competence is very poor specifically in small farm production system. In view of growing economic and business importance there is an emerging need to build up efficient managers to handle the agribusiness in domestic as well as international markets.

In the rural livelihood analysis it is proved that the women constitute a major work force in small farm production and other disadvantaged sector of agri-based occupations. They play a significant and crucial role in farm and allied sector. The role of women in crop production, livestock production, horticulture, post-harvest operations, forestry, fishery and marketing of the produce is very well known. The realisation of the need to look at gender and development in a holistic manner has been modified in the recent initiatives for a National Policy for Women Empowerment—2001 and the State Policies. With this backdrop the present study was carried out on "Women Self Help Groups and Agribusiness Management" in Ganjam district of Orissa with the following objectives:

1. To study the socio-personal and economic profile of the member of SHGs.
2. To study the problems associated with managing agri-based enterprises.
3. To study the attitude of the members of SHGs towards the agri-based enterprises.

4. Preference of group members towards some agri-based enterprises.
5. Suggestions for improvement of agri-based enterprises.

Research Methodology

Ganjam district was selected purposefully as this district is one of the leading district in formation of Women Self Help Groups and the main author is working at Krishi Vigyan Kendra of Ganjam, Bhanjanagar at present as Training Associate in Agriculture Extension. The three blocks, namely Bhanjanagar, Aska and Belaguntha were also selected purposefully because these blocks are situated adjacent to KVK, Bhanjanagar, Ganjam and hence it was easy to carry out the study. The entire number of women SHGs who have taken bank loan for starting the agribusiness were collected and sort listed. From among the sort listed women SHGs, 24 SHGs were selected randomly following proportionate random sampling procedure. Altogether 200 respondents were selected whoever were present during the study period and were interviewed through a pre-tested and structured interview schedule. The data were then analysed and the findings have been reported.

Socio-Personal and Economic Profile

Age

Out of the total respondents 34 per cent, 42 per cent and 24 per cent of women were in the age group of upto 30 years, 31-40 years and above 41 years respectively (Table 20.1). Here it may be implied that the women with age group of 40 years are more eager to start the agri-based enterprise.

TABLE 20.1: Age of Women Members

Sl. No.	*Age*	*Respondents*	
		Number	*Percentage (%)*
1.	Upto 30 years	68	34
2.	31-40 years	84	42
3.	41 years and above	48	24

Martial Status

Table 20.2 reveals that out of total respondents 97 per cent were married and only 3 per cent were unmarried. It also may be a fact that married women have more responsibility towards the family economy and income than unmarried ones. So, this married women group were more interested to start the agri-based enterprises.

TABLE 20.2: Marital Status of Women Members

Sl. No.	*Marital Status*	*Respondents*	
		Number	*Percentage (%)*
1.	Married	194	97
2.	Unmarried	6	3

Family Background

Majority of women were from small and marginal land holding category (34% and 36% respectively) and 26 per cent were of landless agricultural labourers. Only 4 per cent were from medium land holding family category (Table 20.3). None was from large size land holding category. This may be because these majority respondents were financially not sound and hence needed to be productive through Self Help Group by agribusiness.

TABLE 20.3: Economic (Land holding) Background of Family of Respondents

Sl. No.	*Land holding Category*	*Respondents*	
		Number	*Percentage (%)*
1.	Landless labourers	52	26
2.	Marginal land holding	72	36
3.	Small land holding	68	34
4.	Medium land holding	8	4

Educational Level of Women Members

Out of the total respondents 10 per cent were illiterate, 8 per cent can read only, 60 per cent had education upto primary (lower and upper primary) level education, 14 per cent had education upto middle school level and only 8 per cent had education upto high school level. So it may be implied that the women with at least some education are willing to start any agri-enterprise than that of illiterate ones.

TABLE 20.4: Educational Qualification of Women Members

Sl. No.	*Educational Qualification*	*Respondents*	
		Number	*Percentage (%)*
1.	Illiterate	20	10
2.	Can only sign their name	16	8
3.	Primary school	120	60
4.	Middle school	28	14
5.	Upto high school	16	8

Occupational Status

Most of the husbands, fathers, fathers-in-law, brothers, brothers-in-law of the respondents were engaged in agriculture and allied activities.

Motivational Status of Members of Women SHGs

Psychological Motivation

Major psychological motivations were to use spare and leisure time followed by to achieve something new through agribusiness and to be own boss (Table 20.5).

TABLE 20.5: Psychological Motivation of Women Members

Sl. No.	*Items*	*Respondents*	
		Number	*Percentage (%)*
1.	To achieve something through agribusiness	170	85
2.	To be own boss	150	75
3.	Use spare and leisure time	190	95
4.	To know more about the world	72	36

Economic Motivation

Financial stability and additional income (76%), saving for future (68%) to be productive and lack of financial support from husband (62%) were stated to be major motivating factors for forming SHGs (Table 20.6).

Sociological Motivations

Major psychological motivations are "to generate employment", "to upgrade status of women", "to gain prestige/recognition and respect in the society". These are major factors for forming SHGs (Table 20.7).

TABLE 20.6: Economic Motivation of Women Members

Sl. No.	*Items*	*Respondents*	
		Number	*Percentage (%)*
1.	For financial stability and additional income	152	76.0
2.	To be productive in life	125	62.5
3.	Lack of financial support from husband	124	62.0
4.	Savings for future	136	68.0

TABLE 20.7: Sociological Motivation of Women Members

Sl. No.	*Items*	*Respondents*	
		Number	*Percentage (%)*
1.	To generate income	200	100
2.	To upgrade status of women	194	97
3.	To gain prestige/recognition and respect in society	168	84

Majority of respondents (59%) used face to face contact followed by group discussion (33%) and exhibition and fairs (28%) for communication. In mass media communication majority used radio followed by TV and audio cassettes. However very few women referred newspaper and magazines. This may be due to low literacy level (Table 20.8).

Professional Training

Majority of respondents did not receive any professional training (93%).

Most of them perceived that their enterprises were not much successful (89%) and 11 per cent said as fairly successful. No group member mentioned it to be highly successful.

Problems faced by the Members of SHGs

Technical Problem

The problem of knowledge and skill on budgeting and inadequate availability of land/shed/premises were presented by all the respondents, followed by non-availability of agencies to guide and lack of management skills (Table 20.9).

TABLE 20.8: Media of Communication used by Women Members

Sl. No.	*Items*	*Respondents*	
		Number	*Percentage (%)*
1.	Face to face contact	188	59
2.	Exhibition and fairs	56	28
3.	Group discussion	170	85
4.	Visit to established enterprises	42	21
5.	Mass media		
	(i) Magazines/Newspaper	22	11
	(ii) Public shows	—	—
	(iii) TV	28	14
	(iv) Radio	104	52
	(v) Audio-visual cassettes	42	21

TABLE 20.9: Technical Problems with Development/Production of Product/Produce

Sl. No.	*Items*	*Respondents*	
		Number	*Percentage (%)*
1.	Non-availability of agency to guide	172	86
2.	Inadequate availability of land/shed/premises	110	55
3.	Lack of management of experience	158	79
4.	Lack of knowledge and skills on budgeting	200	100
5.	Problem of transport	98	49
6.	Lack of training on agri-enterprises	200	100

Managerial Problems

Majority of respondents faced the problem of lack of skill of group members followed by low productivity of group members and defaulter in saving/repayment of loans (Table 20.10).

TABLE 20.10: Managerial Problems

Sl. No.	*Items*	*Respondents*	
		Number	*Percentage (%)*
1.	Lack of skill with group members	177	88.5
2.	Defaulter in savings and repayment of loan	92	46.0
3.	Low productivity of group members	95	47.5

Socio-Personal Problems

Major social-personal problems were, their family life as affected adversely, they had to overcome the belief that women were not as serious as men in business, lack of respect for women, lack of confidence to start own business and lack of support by family (Table 20.11).

TABLE 20.11: Socio-Personal Problems

Sl. No.	*Items*	*Respondents*	
		Number	*Percentage (%)*
1.	Affect family life adversely	162	81
2.	To overcome belief that women are not as serious as men in business	152	76
3.	Lack of respect for women	146	73
4.	Lack of confidence to start the agribusiness	128	64
5.	Lack of support by family members	96	48

Input Problems

Majority of women members had difficulties in production process, pre-assessment for demand for produce/product, non-availability of raw material, poor knowledge on market and competition and inadequate sales promotional expertise (Table 20.12).

TABLE 20.12: Input (Raw Material Inventory and Material) Problems

Sl. No.	*Items*	*Respondents*	
		Number	*Percentage (%)*
1.	General difficulties in production process	144	72
2.	Pre-assessment of demand for product not done	140	70
3.	Non-availability of raw material	132	66
4.	Poor knowledge of market and competition	126	63
5.	Inadequate sales promotional expertise	120	60

Financial Problems

Majority of women members faced the problem of price fluctuation of produce/product, lack of working capital, difficulty in financing the enterprise and non-ability of adequate loan facility (Table 20.13).

Problem of Government Assistance

Majority of respondents had poor government assistance. Government policy did not help much to procure the produce/product and marketing.

TABLE 20.13: Problems with Financial Implications

Sl. No.	*Items*	*Respondents*	
		Number	*Percentage (%)*
1.	Price fluctuation—variations in raw material	44	22
2.	Difficulty in financing of enterprise	104	52
3.	Lack of working capital	108	54
4.	Inadequate loan facility	102	51
5.	Price fluctuation of produce/product	130	65

Preference of SHG Members towards some Important Agri-based Enterprises

Most of the women members put first preference to vegetable garden, dairy management, flower cultivation, second preference to mushroom cultivation, cultivation of banana and papaya, poultry farming, pisciculture, apiculture and flour mill establishments and third preference to preparation of marketable milk products, bamboo/sabai/jute crafts, rabbit rearing, khali stitching by machine, preparation of dehydrated products (badi, papad muduki etc.), bakery and tailoring (Table 20.14).

TABLE 20.14: Preference of Women Members towards Agri-based Enterprises

Sl. No.	*Items*	*Respondents*	
		Mean Score	*Rank*
1.	Vegetable gardening	2.40	I
2.	Mushroom production	2.14	II
3.	Dairy management	2.40	I
4.	Cultivation of banana and papaya	2.14	II
5.	Preparation of marketable milk products	1.92	III
6.	Bamboo/Sabai/Jute craft	1.92	III
7.	Poultry farming	2.14	II
8.	Pisciculture	2.14	II
9.	Rabbit rearing	1.92	III
10.	Flower cultivation	2.40	I
11.	Apiculture	2.14	II
12.	Khali stitching by machine	1.92	III
13.	Preparation of dehydrated products (Badi, papad, muduki etc.)	1.92	III
14.	Flour mill establishment	2.14	II

Attitude of the SHG Members towards the Agri-based Enterprises

The respondents have the opinion that the "Agri-based enterprises are more profitable, use family labour and thus save labour expenditure, do not have complex production process, best use of locally available resources and can change the socio-economic status of their family with good income". However the majority of respondents opined that the agri-based enterprises need more supervision, very time specific, more risky due to climatic hazards/disease and pest problems and faced the problem of marketing (Table 20.15).

TABLE 20.15: Attitude of Women Members towards Agri-based Enterprises

Sl. No.	*Items*	*Respondents*	
		Mean Score	*+/– than average*
1.	More profitable	2.76	+
2.	Use of family labour and thus saves labour expenditure	2.45	+
3.	Needs more engagement	1.86	–
4.	Best use of resources available	2.32	+
5.	Not more complex	2.27	+
6.	Needs more supervision	2.08	+
7.	Do not need more technical training	1.78	–
8.	Changes socio-economic status	2.12	+
9.	More risky due to climatic hazard/pest and disease problem	1.84	–
10.	Time specific	2.08	+
11.	Face the problem of marketing	2.04	+
	Mean Average	1.88	

Suggestions of Respondents for Improvement of Agri-based Enterprises

Rural Agri-based Business are to be encouraged

It has been suggested by 82 per cent of the respondents that the major business activities take place in big cities with full facility. But there is a need to make extra effort to encourage the rural men and women in villages providing good infrastructure and professional competence to harvest the benefit of agri-based enterprises at village level.

Education and Training

The respondents have low level of education. So they need more skill training facility for them to acquire skill in agribusiness management at their own level.

Guidance and Periodical Counselling

It is suggested that there must be women guide and counsellors at panchayat level with agriculture background to guide them on budgeting, financial management and exposure to market opportunity (Table 20.16).

TABLE 20.16: Suggestions for Improvement of Agri-based Enterprises

Sl. No.	*Items*	*Respondents*	
		Number	*Percentage (%)*
1.	Rural agri-based enterprise should be encouraged	164	82
2.	Education and training	132	66
3.	Guidance and periodical counselling	156	78
4.	Govt. should fix price for produce/ product	148	74
5.	Adequate loan facility	172	86

Fixation of Market Price and Opening of Sale Centres

As suggested by the women members government should fix market price for the agricultural produce and open sale centres at village level or local market for easy marketing of the produce/product as because most of agricultural produce are perishable.

Conclusion

The study reveals that there is an urgent need to involve the rural women in agribusiness activities. It requires the correct identification of enterprise, ensuring their participation and taking their attitude as key factors of success. Exposure of rural women to extension agencies and developmental organisations is of prime importance to bring them to the mainstream of development through agribusiness. Special banks and co-operatives should be established at village level for easy access of women. The government departments and NGOs who are engaged in development and empowerment of women need to review their past experiences and reframe rules and policies for women specific agro-enterprises and essentials of agribusiness management for rural women Self Help Groups.

REFERENCES

Shalini, N. (1994). "Women Entrepreneurship—Profile, Problems in India: A Review Analysis", *Asian Journal of Entrepreneurship Management*, 4(1&2): 29-41.

Satapathy, C. (1996). "Farm Women and Land Based Enterprises: Some Case Studies", in *Development of Women—Present Status and Future Strategy*, ed., B. Mohanty, pp. 120-31.

21

ORGANISATION OF FARMERS GROUPS FOR SUSTAINABLE AGRICULTURE DEVELOPMENT

— *Bhaskar Gaikwad*

Agriculture extension is playing very important role in the rural development process as well as in the overall development of the country. However, the conventional agriculture system in India remains primary hurdle for the further development amidst the changing scenario of the WTO era. The farmers are more involved in the production of the conventional crops. However, there is very less adoption of the technologies and crops as per the demand of the changing population particularly in such a situation when the customers are becoming more cautious about the quality of the agriculture produce. Similarly there is great demand of the new agriculture products, which are not commonly grown by the farmers. The agro-processing sector is much lacking behind where only 2 per cent of the produce is being processed. Though, there are many new avenues for agriculture development in India, due to many reasons farmers are not getting proper technical support besides poor infrastructure and poor backward and forward linkages. Similarly individual farmers are involved in production of new commodities as well as moving towards diversification. But the individual efforts lead to failure due to lack of group actions and coordination and due to unavailability of the technical information, skills required to do the new enterprises, marketing of the produce. Many times, the research, development and financing agencies are also not providing the proper support to the individual farmers to start new enterprises or any new agribusiness. Now-a-days a qualified, highly intellectual young generation with very strong enthusiasm is entering in the field of agriculture. They require good support from the various agencies and organisation so that they can go in the right direction. For the last 50 years, we have concentrated on grow more concept. However, in future we have to give more emphasis on quality production with minimum cost of production by effective use of natural resources besides improving productivity and the net returns. Similarly, total agriculture should be treated as agribusiness where all the inputs, resources and produce should be properly managed to get the maximum returns.

Assessment of Present Agriculture Situation

Assessment of the present agriculture situation reveals that the technical information, knowledge and skills of the new enterprises and agribusiness is lacking in the rural areas. For providing various information and skills, every State government has their own agriculture extension agencies. However, with the changing scenario, it is getting difficult to upgrade the extension workers as per needs of the farming community. Similarly, they are mostly overloaded with the technologies evolved by the public sector. However, they are many good technologies generated by private sector, which are being neglected by these extension workers. Similarly, the technologies generated from the research stations are not situation specific and not fine-tuned. These are the main reasons for less adoption of technologies. The private institutions, which are providing the information, are much commercialised and are providing the technologies at unaffordable rates to the rural community particularly unemployed youths who want to enter this business. Similarly the market information is the major problem. This has resulted in the poor sustainability and ultimate failure of enterprises.

The financing agencies and banks are declaring various loan facilities for the enterprises, however, when the well trained individual contacts these financing agencies, they are getting very poor response. The State as well as Central governments are implementing various schemes and projects for the motivation of farmers, but the concerned officers do not know many times these informations. There is very wide gap in the knowledge between the technology generator, provider and receiver. The guidelines for preparation of projects and proposals are many times not known to the rural population due to which they are not getting the benefit of the schemes, which are generally availed by people who are not involved in the agriculture.

There are many opportunities in future for agribusiness. For this purpose there is a need for solving the problem and preparing action plan as per the needs of the rural community and for concentrating more on cultivation of non-conventional crops as well as development of non-conventional enterprises as per demand of domestic and international market. Post-harvest management particularly harvesting, grading, packing, pre-cooling, transport, storage, marketing, processing for value addition are the important issues which can be more strengthened. Besides providing backward and forward services like inputs, analytical services, market information, consultancies, technical informations, the social mobilisation of the community should be the focus area for plan of action. Research organisation can work on the problems and needs of the community for providing the required technologies to the developmental departments, which should be more location specific. The financing agencies should involve the research and developmental agencies besides entrepreneurs while deciding their policies and plan of action of funding for the agribusiness. Individual efforts many times lead to failure due to one or many reasons. Therefore more emphasis should be given on formulation of groups particularly commodity or enterprise-wise groups. These groups can work better cohesively for technology and skill adoption, seeking credit and for marketing in better way than the individual. Through group formation the diversification in the agriculture can be accomplished at faster pace.

Out of the national female population of 48 per cent, 77 per cent is in rural areas. Large-scale involvement of women is in various unorganised agricultural and allied sectors. However, male dominance in the rural family structure has restricted the women from utilising their resource capabilities.

Model Developed by Krishi Vigyan Kendra (PIRENS)

Krishi Vigyan Kendra (PIRENS) has given major focus on group based activities rather than individual approach. During 1996, Innovative Farmers Clubs have been established where the innovative farmers and entrepreneurs from different commodities are the members of this club. This club has been specially formulated for assessment and refinement of the technologies suitable for various agro-ecological situations. Through these clubs various technologies and practices were refined and fine-tuned as per the needs, problems, available resources and agro-ecological situation. After the refinement of the technologies, adoption has been improved to a considerable extent. With the help of these Innovative Farmers Clubs, KVK has made assessment and refinement of various production technologies of the crops particularly grape, pomegranate, drumstick, sugarcane, cotton and zero energy cool chambers. Establishment of farmers clubs was planned by KVK in the maximum villages for faster dissemination of the technologies.

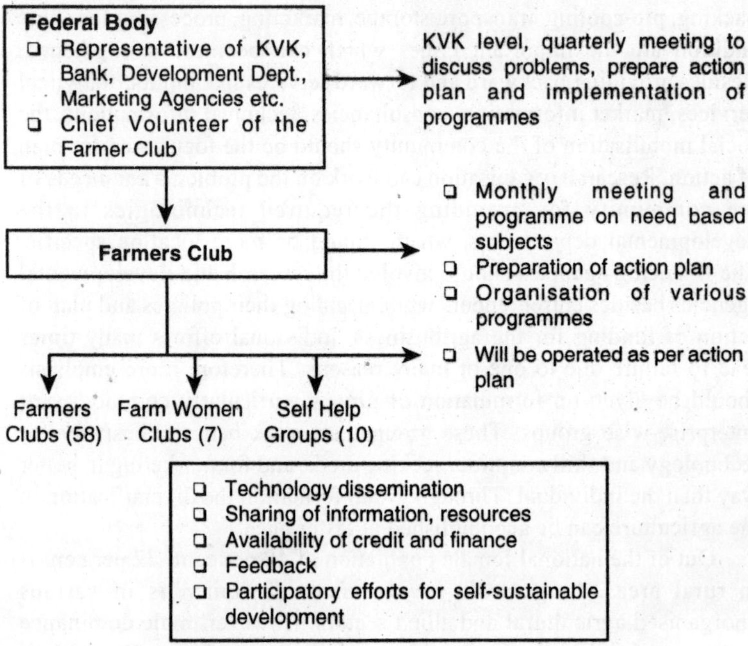

FIG. 21.1: Federal Structure of Farmers Club

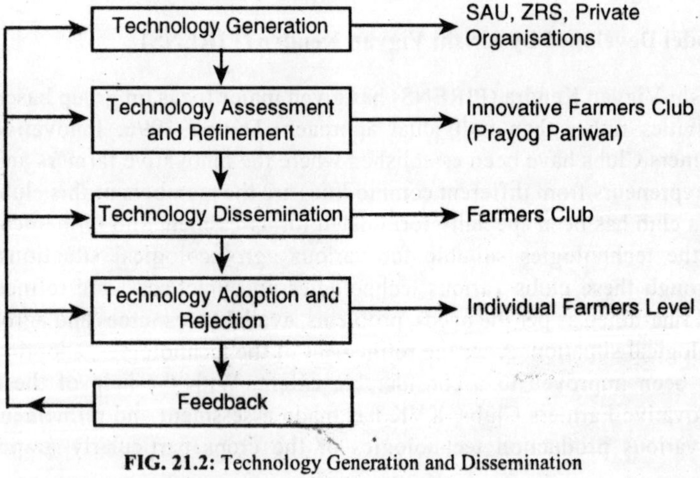

FIG. 21.2: Technology Generation and Dissemination

The Farmers Clubs are informally organised forum that included volunteer, self-realised and self-motivated progressive farmers of the villages. Each club has more than 15-20 volunteer farmers from various enterprises. The clubs are sponsored by National Bank for Agriculture and Rural Development (NABARD) and affiliated with the village level Nationalised or Co-operative bank. This is mainly due to requirement of credit or capital for adoption of new technologies or agriculture business. Objectives of these clubs are:

1. Proper utilisation of credit for improving the yield through application of technologies for intensification and diversification of enterprises thereby rise in income level of farming community.
2. Development of rural entrepreneurship through harvesting of local resources and increase the capital investment in the agro-based industries.
3. To create the suitable climate for thrift and saving.
4. To promote group based approach in the developmental activities.
5. To improve the socio-economic condition of the rural population.

These clubs have linkages with the various development departments, research institutions, private agencies and the institutions/persons involved in the various activities related to agriculture. All the members of the clubs have positive attitude and aspiration towards application of technologies in the field. One active member is working as Chief Volunteer and spares his time for coordinating various activities of the club. Due to formation of farmers clubs, there are different benefits like:

1. KVK can easily disseminate various technologies through these clubs.
2. Group purchase of inputs, group processing, marketing, etc.
3. Utilisation of credit facilities as well as saving and thrift programme through bank.
4. Development of linkage with other institutes, development of library or other infrastructure.
5. Improve productivity and net returns from various enterprises, which will improve socio-economic status in these villages.

Working Pattern of Farmers Club

Krishi Vigyan Kendra established 67 farmers clubs in 5 tehsils, *i.e.*,

Sangamner, Kopargaon, Rahata, Shrirampur and Rahuri. Total planning, implementation and evaluation of the activities has to be done at two level, *i.e.*, apex level (KVK) and village level (Farmers club). At KVK level federation of farmers clubs is established in which representatives of KVK, bank, development department, marketing agencies and farmers clubs are the members whereas at village level interested farmers from the various enterprises are the members. At farmers club level, initially action plan is prepared for the year. This action plan is based on their needs and problems for providing sustainable solutions of the members as well as farming community of the villages. The action plan is being collected by the KVK and accordingly separate action plan is being prepared based on the action plan of the farmers club. In this action plan, all the activities are being included based on the needs and problems of farmers club. KVK has identified 55 farmers from innovative farmers clubs from the various enterprises. List of these farmers is being circulated and the same farmers are being used for off campus training programmes besides subject matter specialist of the KVK. This system is being well accepted and has been found to be effective for conducting off campus training programmes. On campus training programmes are being conducted at KVK with the help of KVK SMSs, progressive farmers as well as expert from the various organisations.

At village level, farmers club members get together at monthly interval to discuss the various issues as well as to conduct the programme. At KVK level all the members get together at quarterly basis where they discuss their problems, issues, needs and opportunities. The feedback is simultaneously noted. At apex level various programmes especially skill based on campus training programmes are being finalised. In this way, the members of federation body work as a link between the farmers clubs and the various departments. For conducting the various programmes, National Bank for Agriculture and Rural Development is providing some financial support.

What KVK wants to achieve from This?

Individual contact method of technology dissemination is very laborious and time consuming. Similarly the demand of the technology from the farmers is increasing where KVK cannot satisfy the demand and needs of the farming community with existing staff. Therefore, KVK has planned to work on group-based transfer of technology system as well as second stage (distance learning) education through these clubs. For this purpose some expert farmers from different enterprises and technologies selected from clubs are being promoted by KVK. Through these clubs, Internet Service Centre (Information Technology Centre) will be started in the villages. These centres will provide all the information required by the farmers like market information, weather forecasting, projects and schemes of government, availability of inputs, besides the technologies required by the farmers. These Internet centres will be supported by KVK by providing required information. This will help KVK to disseminate the technology at faster rate besides providing all the sources and inputs available with the KVK.

These clubs will be specially developed for promotion of the enterprises in the villages. To start the new enterprises technology availability, financial support, availability of inputs, processing and market information will be given to these groups. Due to this group approach, diversification in agriculture will be possible. Similarly through these clubs, self-help groups will be established which will be helpful them to make available financial support up to some extent. Educated, unemployed youths will get self-employment opportunities. Through these clubs, the KVK will develop self-motivated, self-reliable and self-sustainable rural community for sustainable development of the villages.

Conclusion

The conventional agriculture extension system cannot fulfils the needs and demands of the farming community in the changing scenario of agriculture due to globalisation. Technical information, knowledge and skills of the new enterprises and agribusiness is lacking in the rural areas. Similarly the farming communities are facing problems of availability of credit, marketing of the farm produce, availability of inputs for adoption of new technologies and processing of their farm produce. The individual farmers cannot solve these problems. Similarly, all these facilities cannot be made available through the public sector. Therefore, there is urgent need to modify the present system of communication or develop some system, which can solve the problem or serve as the intermediate agency to implement the various programmes.

KVK has made special efforts to establish some models for effective and faster transfer of technology which is be more location specific, problem solving, participatory and easily adoptable to the farming community. KVK has interestingly worked on assessment and refinement of the technologies suitable at various locations through the Innovative Farmers Clubs (Prayog Pariwar). However, for dissemination of these technologies, KVK has now established 67 farmers clubs in the different villages. Progressive farmers from the different enterprises are the members of these clubs. These clubs are linked with KVK, local banks, and development departments for various purposes. Through these clubs, besides dissemination of the technical information, group purchasing of the inputs, intensification and diversification in agriculture, marketing, processing, availability of credit facilities will be possible. These clubs will be well connected to various agencies through Internet. Similarly, the club members prepared their action plan as per their needs and demand. At KVK level, there is federal body, which meet once in three months to discuss the progress and action plan of the farmers club. The federal body will prepare the yearly action plan based on the action plan of the Farmers Clubs. This type of structure and model will improve the socio-economic status of the farming community with the participatory mode of development.

22

FARMER OWNED HORTICULTURAL MARKETS
A Case Study of Kerala

— *Mothilal Nehru* and *N.K. Jayachithra*

It is often lamented that the traditional agricultural markets involve high marketing costs due to the presence of intermediaries in the marketing channels. The producer's share in the consumer price on an average is 45-50 per cent. Individual farmers are not able to bargain better for their produce because of distress/forced sales of the produce at available prices to the village merchants/traders. The scenario is again complicated by the farmer's indebtedness to the private money lenders at exorbitant interest rate. The traditional agricultural extension system was unable to give the attention to the marketing problems of small and marginal cultivators especially the farmers of perishable commodities like fruits and vegetables.

It was in this context Kerala Horticulture Development Programme (KHDP), a joint venture of European Union and Government of Kerala was launched in 1993 and started its field level activities. This is the precursor organisation of the new legally registered body called Vegetable and Fruit Promotion Council Keralam (VFPCK). The objective of VFPCK is to improve the livelihood security of dependent farmers in Kerala by empowering them through access to technology, skills (material) inputs, credit and infrastructure. The vision of VFPCK is "Empowered farmer groups leading the nations in reaping the fruits of Self-Help". Neither does it give any subsidy nor it procures any produce, but guides and support the fruit and vegetable growers to form "Self-Help Groups" (SHGs) and encourage them to undertake activities in Production and Technology, and to avail the innovative credit package. It provides technical consultancy service in various aspects of cultivation through training, campaigns, demonstrations, study tours/field visits etc. Over and above it has established network of adjacent SHGs to form a group marketing centre called Field Centre (FC). The organisation was appreciated for its initiatives in introducing many eco-friendly and low cost cultural and plant protection techniques for banana and vegetable cultivators in Kerala.

SHG is operationally defined as a group of 15-25 farmer members, voluntarily joined together on neighbourhood principle and have some marketable surplus. The member-farmers must be genuine cultivators who are really interested in cultivation. SHG members are empowered through many farmer participatory interventions to take up activities on their own to attain self-help stage of SHG. About 10-15 SHGs join together to form a FC, which addresses the marketing needs of about 200-300 farmers from the affiliated SHGs.

All the FCs are governed by a set of FC bye-law framed by the organisation and registered under Indian Societies Act, 1955, Section 12. Each FC is managed by a committee of Marketing Master Farmers (MMFs) of the affiliated SHGs. As a tool to ensure successful operation and transparency in the accounting system of the agribusiness, regular monthly auditing is implemented in all the FCs. The audit reports are discussed at FC level and the efficiency is enhanced. As the FC management team gets prior information on previous day's market prices from Market Information Centre (MIC) of VFPCK, Cochin, the participating farmers in FCs will be in a better position to bargain with the traders.

The markets are owned and managed by farmers themselves. They need to pay only 5-7 per cent of sale value as a nominal commission for the functional and over-head expenses of FCs, but, can realise about 70-80 per cent of consumer price, which would not have been achieved through the traditional marketing channels involving middle men and huge marketing cost. The transportation cost is also minimal as it is marketed nearby the production points and loading and unloading will be carried out by farmers themselves. As on 31 August 2001, there are 69 FCs which were able to sell 56,570 metric tonnes of fruits and vegetables with the participation of 40,277 farmers belonging to 7 district level Project Areas.

In this backdrop this study was conducted with the following objectives:

Objective

- To analyse the functional efficiency of Field Centres in terms of parameters such as participation in trading (PT), quantity traded per day (QT), percentage of outstanding (OS), liquidity ratio (LR), turn over ratio (TOR), gross profit margin ratio (GPMR) and net profit margin ratio (NPMR).
- To analyse the comparative performance of field centres in each district level Project Area.
- To suggest strategy to enhance the functional efficiency of FCs.

Methodology

A simple random sampling technique was adopted for selecting the study areas. Three district level Project Areas *viz.*, Trivandrum, Ernakulam and Palakkad out of seven Project Areas were selected. In the second stage all the FCs from these three Project Areas were selected which comes to a sample size of 34.

The operational definition, formulae, standards and the rating scale for each of the seven parameters are given in Table 22.1. A three-point continuum rating scale was adopted for each parameter. The maximum score for a parameter is 3 and the minimum is 1, so the maximum total score that each FC can score is 21 and minimum total score that can be obtained is 7. Based on the total score for all the seven parameters, the FCs are rated for their functional efficiency into three categories, those FCs with an overall score of 17-21 are rated as good, those having a score of 12-16 are rated as average and those ones having a score below 12 are rated as poor.

Results and Discussion

Table 22.2 indicates that out of 10 FCs in Trivandrum Project Areas, Marenelloor FC is the leader with a very good liquidity ratio of 18. GPMR and NPMR are also up to the standard level. Percentage of outstanding is almost nil in this FC, but it has to improve in terms of participation in trading and there is scope for higher volume of trade per day. Pappanchani and Balaramapuram FCs are in the second position and Parassala in the third rank position. On average, the percentage of participation in trading in Trivandrum is only moderate one, *i.e.*, 53 per cent. Only in Kunnathukal FC the participation level is exceptionally

TABLE 22.1: Operational Definition, Formula, Standards and Rating Scale of Efficiency Parameters

Sl.No.	Parameter	Operational Definition	Formula	Standard	Rating Scale Level	Rating Scale Score
1.	Participation in Trading (PT)	% of total number of farmers participating in trading to that of total member farmers in a FC	$\frac{\text{No. of part members in trading}}{\text{No. of total member farmers is FC}} \times 100$	At least 50%	< 50% 50-60% > 60%	1 2 3
2.	Quantity Traded per day (QT) (2 market days/ week, hence total of 104 market days/year	The quantity of fruits and vegetables traded per day (ton)	Quantity traded in the financial year 2000-01 (ton) divided by 104 days	On average t/day	<4 t 4-6 t > 6 t	1 2 3
3.	% of outstanding (OS)	Amount of Rs. receivable from traders in a financial year to the total sale value of produce in Rs. for the same year	$\frac{\text{Amount receivable from traders in the financial year 2000-01 (Rs.)}}{\text{Total sale value of 2000-01 (Rs.)}} \times 100$	'zero'	>/5% 2.1-4.9% ≤ 2%	1 2 3
4.	Liquidity Ratio (LR)	The ability of the FC to pay back liabilities in case of a closure	$\frac{\text{Value of total assets (Rs.)}}{\text{Value of total liabilities (Rs.)}}$	2 : 1	< 1 1-1.9 ≥ 2	1 2 3
5.	Turn Over Ratio (TOR) in days	Indicates how many days it takes to collect debts (outstandings) from traders	$\frac{\text{Amount receivable from traders (Rs.)}}{\text{Total sale value per day (Rs./day)}}$	Max. 7 days	> 7 3-7 < 3	1 2 3
6.	Gross Profit Margin Ratio (GPMR)	Ratio of gross profit of FC to total sale value in a financial year	$\frac{\text{Gross Profit of FC (Rs.)}}{\text{Total sale value for 2000-01 (Rs.)}}$	5% (market commission) 0.05	0.010-0.020 0.021-0.049 ≥ 0.005	1 2 3
7.	Net Profit Margin Ratio (NPMR)	Ratio of profit after deducting the expenses of FC (net profit) to the total sales value in a financial year	$\frac{\text{Gross profit – Total expenses for FC (Rs.)}}{\text{Total sale value for 2000-01 (Rs.)}}$	3% *i.e.*, 0.03	< 0.001 0.010-0.029 ≥ 3	1 2 3

high, *i.e.*, 98 per cent. Similarly the average quantity traded per day in Trivandrum is only 1.65 ton, but Parassala and Balaramapuram FCs have appreciably higher volume of daily trade, *i.e.*, 4.5 ton and 5.44 ton respectively. Extent of outstanding is with in a safe limit except in Neyyattinkara FC where it is 5.48 per cent and it needs to be reduced. Kovil Nada FC has to take concerted efforts to lower its higher TOR value of 4.58 to improve its financial efficiency. In Trivandrum Project Area, though GPMR is nearby the satisfactory level of 5 per cent (0.05), NPMR still has to go up to keep the standard of 3 per cent, *i.e.*, 0.03. Liquidity and Turn Over Ratio in the Project Area in general are up to the expected limit, *i.e.*, LR is 4.65 and TOR is 2.36.

From Table 22.3, it is clear that Ernakulam Project Area in general has a higher percentage of participation in trading, *i.e.*, 79 per cent. Koovappady and Thiruvaniyoor FCs have exceptionally higher participation in trading, *i.e.*, 90 per cent each, whereas Ayyampuzha FC is the only one exception in Ernakulam with less than 50 per cent for participation in trading, *i.e.*, 47 per cent. Liquidity ratio is also higher than the one in Trivandrum, *i.e.*, 5.2. GPMR is at the standard level, *i.e.*, 0.051. There is a higher scope for reducing the outstanding and TOR for the Project Area as a whole. It is also seen that quantity traded per day also has to move up from the present position of 2.16 ton/day to enhance the physical efficiency of FCs. Elanji and Malayattoor FCs are the only exceptional cases in this regard with a volume of daily trade of more than 4 ton. In the case of outstanding all FCs are with in safe limit, *i.e.*, < 7, but Kurumassery FC has a higher value of 7.2 which is to be addressed immediately. Karumalloor and Ayyampuzha FCs are with very good LR of 22 and 10 respectively. At the same time, TOR is high in Karumalloor and Kurumassery FCs, *i.e.*, 7 and 6.90 respectively. Koovappady, Kunnukara, Manjapra and Thiruvaniyoor FCs out of 14 FCs in Ernakulam are in the first position, with an appreciable level of all physical and financial parameters. Puthenvelikkara and Thiruvaniyoor FCs are in the second position. Though they are at the same rank, the former (Puthenvelikkara) has better LR, GPMR and NPMR than the latter (Thiruvaniyoor). Other parameters are at similar level for both FCs. Third position is attained by Elanji, Kadungalloor, Kakkad and Nedungapra FCs.

It is evident from Table 22.4 that the participation level in trading in Palakkad is quite similar to that of Trivandrum and it is below the standard level of 50 per cent. The percentage of participation in Thachanattukara FC is much below the standard, *i.e.*, 23 per cent.

TABLE 22.2: Efficiency of Farmer Markets of Self-Help Groups in VFPCK Project Area: Trivandrum

Sl.No.	*Name of FC*	*Participation Level*			*QT**	*OS**	*LR*	*TOR*	*GPMR*	*NPMR*	*Overall Score*	*Rank*
		No. of SHGs	*No. of Member Farmer*	*PT*								
	(1)	(2)	(3)	(4)	(5)	(6)	(7)	(8)	(9)	(10)	(11)	(12)
1.	**Balaramapuram**	22	170	48	5.44	1.57	3.26	2.46	0.040	0.020	16	2
2.	Kovilnada	16	140	57	—	—	1.08	4.58	0.050	0.030	12	6
3.	Kunnathukal	17	102	98	0.27	0.67	1.06	0.00	0.040	loss	14	4
4.	**Maranelloor**	20	205	44	2.50	0.33	18.04	< 1	0.050	0.030	17	1
5.	Mylachal	19	203	46	2.50	1.74	1.86	1.70	0.030	0.020	14	4
6.	Neyyattinkara	17	136	45	1.46	5.48	2.27	2.00	0.050	0.024	14	4
7.	**Pappanchani**	10	117	46	0.60	0.00	16.40	0.00	0.050	0.010	16	2
8.	**Parassala**	15	167	60	4.50	2.47	1.30	2.40	0.030	0.013	15	3
9.	Perumkadavila	14	128	41	0.65	0.00	5.39	0.00	0.030	loss	13	5
10.	Thiruvaniyoor	14	146	40	0.60	0.00	0.58	0.00	0.050	0.020	14	4
	Average	18	151	53	1.85	1.11	4.65	2.36	0.42	0.017	16	

Note: * . Period from 01.04.2000 to 31.03.2001.

PT = % of Participation in Trading.
QT = Quantity Traded/day.
OS = % of Outstanding.
LR = Liquidity Ratio.
TOR = Turn Over Ratio.
GPMR = Gross Profit Margin Ratio.
NPMR = Net Profit Margin Ratio.

TABLE 22.3: Efficiency of Farmer Markets of Self-Help Groups in VFPCK Project Area: Ernakulam

Sl.No.	*Name of FC*	*Participation Level*			*QT**	*OS**	*LR*	*TOR*	*GPMR*	*NPMR*	*Overall Score*	*Rank*
		No. of SHGs	*No. of Member Farmer*	*PT*								
	(1)	(2)	(3)	(4)	(5)	(6)	(7)	(8)	(9)	(10)	(11)	(12)
1.	Ayyampuzha	8	356	47	1.07	3.5	10.0	1.75	0.049	0.025	14	3
2.	**Elanji**	13	174	84	4.04	4.9	2.1	4.70	0.049	0.017	16	3
3.	**Kadungalloor**	10	91	88	2.88	4.1	2.3	3.90	0.050	0.019	16	3
4.	**Kakkad**	10	76	66	2.37	2.5	3.5	1.25	0.050	0.021	16	3
5.	Karumalloor	10	208	57	0.84	1.4	22.0	7.00	0.050	0.013	15	4
6.	**Koovappady**	11	153	90	2.00	2.0	5.1	1.92	0.050	0.028	18	1
7.	**Kunnukara**	13	105	74	1.87	0.9	2.2	0.86	0.049	0.039	18	1
8.	Kurumassery	15	300	83	3.50	7.2	6.3	6.90	0.052	0.024	15	4
9.	Malayattoor	12	163	85	4.18	4.4	3.3	4.20	0.049	0.990	15	4
10.	**Manjapra**	13	128	86	0.83	1.2	2.5	1.15	0.050	0.019	18	1
11.	**Nedungapra**	8	128	78	1.80	3.9	4.3	3.70	0.051	0.019	16	3
12.	**Puthenvelikkara**	8	131	85	1.92	2.5	3.7	2.40	0.052	0.026	17	2
13.	**Thiruvaniyoor**	15	135	90	1.02	1.7	2.6	1.60	0.056	0.028	18	1
14.	**Thuravoor**	8	113	90	1.97	2.4	2.8	2.30	0.051	0.011	17	2
	Average	11	162	79	2.16	3.0	5.2	3.10	0.051	0.022	16	

Note: * Period from 01.04.2000 to 31.03.2001.
PT = % of Participation in Trading.
QT = Quantity Traded/day.
OS = % of Outstanding.
LR = Liquidity Ratio.
TOR = Turn Over Ratio.
GPMR = Gross Profit Margin Ratio.
NPMR = Net Profit Margin Ratio.

TABLE 22.4: Efficiency of Farmer Markets of Self-Help Groups in VFPCK Project Area: Palakkad

Sl.No.	Name of FC	Participation Level			QT*	OS*	LR	TOR	GPMR	NPMR	Overall Score	Rank
		No. of SHGs	No. of Member Farmer	PT								
	(1)	(2)	(3)	(4)	(5)	(6)	(7)	(8)	(9)	(10)	(11)	(12)
1.	Alanelloor	9	92	57	0.60	0.25	0.32	0.27	0.050	0.004	14	2
2.	Ayloor/Kairady	9	95	68	2.90	0.30	NCL	0.00	0.040	0.021	14	2
3.	Elevenchery	13	129	60	0.83	0.34	3.16	0.51	0.050	-0.110	15	3
4.	Kottayi	12	180	36	2.19	1.92	8.45	1.50	0.057	0.005	15	3
5.	Kumarumputhur	12	125	48	2.39	0.50	1.76	0.53	0.050	0.010	15	3
6.	Machamthode	15	269	29	6.70	1.23	3.78	0.80	0.050	0.010	18	1
7.	Punchacode	15	169	37	0.92	4.00	NCL	1.90	0.047	0.018	11	6
8.	Thachanattukara	8	111	23	0.50	56.30	3.50	6.10	0.054	-0.009	11	6
9.	Vadakarapathy	11	135	58	0.66	—	0.89	—	0.045	0.005	13	5
10.	Viyyakurussi	15	190	41	7.80	4.17	3.60	3.28	0.050	0.019	16	3
	Average	12	136	45.7	2.48	6.90	2.55	1.49	0.049	-0.009	12	

Note: * Period from 01.04.2000 to 31.03.2001.

PT = % of Participation in Trading.
QT = Quantity Traded/day.
OS = % of Outstanding.
LR = Liquidity Ratio.
NCL = No Current Liability
TOR = Turn Over Ratio.
GPMR = Gross Profit Margin Ratio.
NPMR = Net Profit Margin Ratio.

Similarly there is much scope for improvement in the volume of daily trade. Percentage of outstanding has to be carefully tackled to bring down the level of the set standard. Similarly NPMR shows negative trends, which needs urgent attention. In general, it requires a close monitoring of the financial parameters of FCs in Palakkad. We can also see that Machamthode is ahead of all the ten FCs in terms of efficiency, but it needs betterment in terms of participation in trading (29%) as well as NPMR. Viyyakurissi FC in the second position has an appreciable volume of daily trade, *i.e.*, 6.7 ton, but it needs betterment in terms of participation in trading and NPMR. It has to reduce the outstanding and TOR. Exceptionally very bad financial situation is visualised in Thachanattukara FC due to very high outstanding (56.30%) and TOR accompanied by a low volume of trade, leading to a negative NPMR. The third position is attained by Elevenchery, Kottayi and Kumarumputhur FCs, out of which Elevenchery FC has a negative NPMR and all the three FCs have to go up further to keep the standard volume of daily trade and NPMR.

Strategy to Enhance the Functional Efficiency of Field Centres (FCs)

To Improve the Level of Participation in Trading

(a) Coordinate the activities of FC convenors at district level Project Areas to stress the point of improving the participation.
(b) Product diversification will attract more traders to FCs. For this, financial feasibility study to start a small scale processing unit at FC level should be conducted.
(c) The FCs should take up small scale campaigns and advertising/publicity using locally available resources to create awareness about the merits of group marketing through farmer owned horticultural markets and FC management. This will encourage more number of non-member farmers to become members in FC and will increase the participation level in trading.

To Increase the Quantity of Daily Trade

Take measures to ensure higher quantity of fruits and vegetables get collected from all the member farmers which eventually will create necessity for traders to come to FC to buy the produce. Even when sufficient traders are not available to dispose the produce, wholesale and processing units in the locality and nearby open markets can be approached. Never try to reach the outlets through village level merchants, commission agents or brokers.

To Reduce the Outstanding and TOR

(a) Identity of traders participating in FC businesses must be known to FC management. Registration for traders must be made a rule. A directory of traders with name, address, bank A/c No. and phone numbers (Residential and official), contact points, working hours etc. will be highly beneficial and this will serve the purpose to collect outstanding amount from traders within a stipulated period.

(b) Credit payments must be discouraged to bring down the outstanding. Insist cash/cheque payments. If the traders pay ready cash, introduce the price incentive for those traders in the form of 2-5 per cent reduction in sale price to boost up the morale of traders. Defaulting traders with outstanding amount of more than Rs. 500 must be prevented from participating in FC agribusiness.

(c) Convince the Marketing Master Farmers of all affiliated SHGs in a FC, about the importance of financial prudence required in FC management during management committee meetings.

Conclusion

The study points to the fact that the farmer owned markets are great success and farmers can manage their markets as effective as corporate people. The whole FC management is carried out by farmers themselves, ensuring a better producer's share in consumer's price. FCs are not only the easily accessible local markets, but they perform as a check also, against the financial exploitation of farmers from middlemen and traders in markets. Another visible impact is that even non-KHDP/FC member farmers get benefited by bringing their produce to FCs. FCs also get a higher commission of 7 per cent of sale value from non-members (5% from members).

Enhancement of participation in trading and the volume of daily trade, reduction in overdue and Turn Over Ratio (TOR) would enable a lot to strengthen the financial status of the FCs in Kerala. This will also serve the purpose of catering to the marketing problems of small and marginal horticultural farmers of Kerala.

23

ECONOMIC CONTRIBUTIONS OF SUGARCANE PROCESSING INDUSTRY
A Case Study of Maharashtra

— *B.N. Pawar, P.M. Kapase* and *V.S. Vikhe*

India is the largest producer of the sugarcane and sugar in the world. Sugar is the second largest agro-based industry in India next only to the cotton based industry. The sugar industry carries about 2.5 per cent weight in industrial production. This industry provides substance to about 35 million cane farmers and their families constituting 7.5 per cent of India's rural population. It also provides direct employment to over 0.4 million workers mostly from the rural areas. The sugar industry in India has a major contribution in raising the standard of living of rural population. The sugarcane system is a complex one, which involves the multiple stages of crop production, processing, marketing as well as consumption.

The sugarcane production system and sugar industry of Maharashtra represents a good example of interface between agriculture and industry with strong complementary relationship of development of both the sectors during the past four decades. At present there are 133 sugar factories (129 from co-operative sector and 4 from private sector) in working stage, while 67 sugar factories (63 from co-operative sector and 4 from private sector) are installing at various stages. A large number of natural, physical, technological, financial, social and economic factors have contributed to the establishment of this interface on a sound footing. In effect, both the sectors are highly dependent on each other and the inter-dependence is so strong that the analysis of progress of a particular sector cannot be accomplished without having reference to the growth of the other sector. Besides contributing to the growth of each other, these two sectors together have initiated and accelerated the process of development of agriculture and the State economy to a great extent.

The analysis contemplated in this chapter is based on time-series as well as cross-sectional data obtained from the secondary sources. The yearly data on acreage and production of sugarcane, number of sugar factories and their crushing capacity, quantities of sugarcane crushed and sugar produced and their value, labour employment in harvesting and transport of sugarcane and manufacturing of sugar, revenue contributions of sugar factories, etc., have been obtained from the published sources for the Maharashtra's sugar industry as a whole, for the period from 1962-63 to 1999-2000. At the same time, the farm level cross-sectional data on input use structure and production cost of sugarcane have been obtained from the Government of Maharashtra sponsored Scheme for Creating Permanent Machinery for Cost of Cultivation of Crops. The same have been used for working out estimates of total labour employment, total fertiliser consumption and total non-labour operating production cost of sugarcane production in Maharashtra as a whole. The estimates of family labour income and farm investment income of the farms involved in the sugarcane production system have been derived on the basis of the standardised crop production cost concepts of the farm management studies.

Growth of Sugarcane Production System and Sugar Industry

Maharashtra has been bestowed with certain favourable natural resource endowments especially, in terms of sunlight, temperature and humidity, which are quite conducive for the growth of sugarcane with high sucrose content. The dynamic leadership provided by the renowned social reformers, co-operators and planners has resulted into transformation of agriculture and rural economy in many respects in the State since the beginning of the plan era. Based on the successful performance of the sugar factory at Pravaranagar (Ahmednagar) in co-operative sector, increasingly large numbers of co-operative sugar factories have been established in different parts of the State. The massive investments in different types of irrigation works, *i.e.*, canal irrigation systems, open dug wells and irrigation lifts have resulted into increased availability of irrigation water resource useful for irrigated agriculture. The farming community in the State at large has always responded well to new technologies and new production systems subject to availability of complementary resource services and institutional supports. The growth of sugarcane production system is in effect an outcome of such a type of dynamic farmer response aiming at some positive change.

TABLE 23.1: Growth Indicators of Sugarcane Production System in Maharashtra

	Growth Indicators	*1962-63*	*1972-73*	*1982-83*	*1992-93*	*1999-2000*
1.	Area under sugarcane ('000 ha.)	135	182 (34.81)	326 (141.46)	404 (199.25)	590 (337.03)
2.	Proportion to gross cropped area (%)	0.72	1.70	1.07	1.92	2.66
3.	Proportion of net irrigated area allocation to sugar-cane (%)	11.99	14.26	16.93	15.69	16.25
4.	Total number of sugar factories in operation	29	46 (58.62)	78 (168.97)	99 (241.38)	133 (358.62)
5.	Installed crushing capacity of sugar industry ('000 tons/day)	32.71	66.56 (103.48)	138.29 (322.78)	230.16 (603.70)	428.16 (1,209.26)

Note: Figures in parentheses are the percentage change over the base year 1962-63.

As indicated in Table 23.1, the area allocation to sugarcane crop increased from 135 thousand hectares in 1962 to 590 thousand hectares in 1999-2000 in response to the expansion of net irrigated area and establishment of additional sugar factories in the co-operative sector in the State. The number of sugar factories in operation increased from 29 to 133 during the period of 36 years beginning from 1962-63. The growth process of sugar industry was accelerated during the period under reference. Besides establishment of new sugar factories, the existing sugar factories also tried to expand their crushing capacities by adding new machineries and enlarging factory sheds. The total installed capacity of the sugar industry increased by more than twelve times during the period from 1962-63 to 1999-2000.

Performance of Sugarcane Production System and Sugar Industry

Maharashtra has witnessed a spectacular growth of sugar industry in the co-operative sector during the past four decades. It possesses a unique position in the country in regard to the sugarcane production system and sugar industry not only because of its largest share in sugar production (nearly 30 to 35%), but of its maximum contribution to the national sugar production. The State could attain this position because of its dynamic development planning efforts and commendable success on the fronts of sugarcane production and sugar industry.

Table 23.2 provides empirical evidence that the annual total production of sugarcane almost trebled during the period under reference. During the early period, the production of sugarcane increased both due to area expansion as well as productivity improvement. However, the increase in production of sugarcane during the eighties and the nineties depended exclusively on area expansion because of decline in its productivity. With the establishment of additional sugar factories and rewarding prices for sugarcane, the sugarcane growers were attracted more to divert relatively a larger proportion of their sugarcane output for sugar manufacture. As a result, the total quantity of sugarcane crushed for sugar manufacture increased more than proportionately in comparison to the increase in production of sugarcane during the period under reference.

TABLE 23.2: Performance of Sugar Production System

Performance	*1962-63*	*1972-73*	*1982-83*	*1992-93*	*1999-2000*
1. Total sugarcane production ('000 tons)	10,667	11,918 (11.72)	31,360 (193.99)	30,853 (189.24)	53,140 (398.17)
2. Sugarcane productivity (tons/ha.)	79.01	81.63 (3.32)	96.20 (21.76)	76.40 (–3.30)	92.10 (16.57)
3. Quantity of sugarcane crushed ('000 tons)	4,348	10,057 (131.30)	27,625 (535.35)	29,739 (583.97)	57,099 (1,213.22)
4. Quantity of sugar produced ('000 tons)	530	1,074 (102.64)	3,025 (470.75)	3,360 (533.96)	6,511 (1,128.49)
5. Sugar recovery (%)	8.20	9.36	9.13	8.85	11.42
6. Value of sugar at current price (Rs. in crores)	58.11	181.59 (212.49)	1,173.80 (1,919.96)	2,936.40 (4,953.17)	7,320.51 (12,497.68)

Note: Figures in parentheses are the percentage change over the base year 1962-63.

The performance of sugar industry also paralleled the performance of sugarcane production system in the State. During the year 1962-63, the sugar industry of Maharashtra could manufacture 530 thousand tons of sugar by crushing 4,348 thousand tons of sugarcane. After 36 years, by way of crushing 57,099 thousand tons of sugarcane, the industry could manufacture 6,511 thousand tons of sugar, while the increase in the quantities of sugarcane crushed and sugar produced was almost by 1,213 per cent and 1,128 per cent, respectively. The value of sugar produced at current prices increased from Rs. 58.11 crores to Rs. 7,320.51 crores from 1962-63 to 1999-2000.

Inter-dependence of Sugarcane Production System and Sugar Industry

The sugarcane production system and sugar industry of Maharashtra has remained interdependent on each other throughout the time period. With the establishment of new sugar manufacturing units, the demand for sugarcane as the raw material increased at a fast rate. The purchase price of sugarcane also increased simultaneously, attracting thereby many more farmers towards sugarcane cultivation. The increased production of sugarcane necessitated further expansion in the total installed crushing capacity of the sugar industry over a period of time both through establishment of new sugar factories and expansion of existing crushing capacities of the successful factories. The information given in Table 23.3 gives us a broad idea of the type of inter-dependence between the sugarcane production system and sugar industry in Maharashtra in terms of per sugar factor acreage under sugarcane, sugarcane crushed, sugar manufactured and crushing capacity utilisation at different point in time.

TABLE 23.3: Indicators of Interdependence of Sugarcane Production System and Sugar Industry in Maharashtra

	Inter-dependence Indicators	*1962-63*	*1972-73*	*1982-83*	*1992-93*	*1999-2000*
1.	Per factory sugarcane acreage (ha.)	4,655	3,960	4,180	4,080	4,436
2.	Per factory sugarcane crushed (tons)	3,678	2,591	4,021	3,117	4,286
3.	Per factory installed crushing capacity (tons/day)	1,128	1,447	1,773	2,325	3,220
4.	Per factory sugar crushed (tons/day)	1,000	1,420	1,874	2,209	3,222
5.	Capacity utilisation (%)	88.61	98.11	105.69	95.00	100.06

The inter-dependence of the sugarcane production system and sugar industry of Maharashtra was of high order and both of them derived benefits from each other. Moreover, the performance of the individual sugar factories was dependent on the availability of sugarcane from the factory command area. The average factory could get sufficient raw material supplies only because of increased sugarcane production through larger area allocations for the crop. On the other hand, the guaranteed market for sugarcane motivated farmers to go for increased production of sugarcane with additional quantities of required resource input. The necessary research and extension efforts of the agricultural universities and credit supply from the co-operative and nationalised banks in the State proved to be quite effective for the prosperity of sugarcane farming.

Economic Contribution of Sugarcane Production System and Sugar Industry

The sugarcane production system and sugar industry of Maharashtra has proved to be the catalytic agents for development of agriculture and transformation of rural society to some extent. Both of them have resulted into generation of additional employment and income opportunities in the rural areas as well as contributed to revenue collection of the State and Central governments. The estimates of contributions made by the sugarcane production system and sugar industry in this regard at different points of time are presented in Table 23.4. The estimates of the total labour employment and total farm business and farm investment incomes derived from sugarcane farming system have been worked out on the basis of the farm level per hectare input use and production cost structure of sugarcane crop. As is evident from the table, the sugarcane production system and sugar industry together have made significant contributions towards employment and income generation in rural Maharashtra during the period of time. The employment in sugarcane production system and sugar industry has increased several times, and so is the case in regard to income generation. The government exchequer could get additional revenue receipts in the form of educational tax, sugarcane purchase tax, excise duty, etc. Additionally, the sugar industry has made direct financial contributions to strengthen various social, educational and co-operative institutions and towards other developmental activities such as irrigation water resource development, road construction, etc.

Besides, resulting into increased sugar production and increased employment, income and revenue generation, the sugar industry-sugarcane production system interface has provided a big boost to increased production of the industrially valuable products such as alcohol and alcohol based products (*viz.*, liquor, acetic acid and its esters, acetone, butanol, ethyl hexanol, acetates, vinyl acetate, monomer, PVC, polyethylene, styrene monomer, oxalic acid, synthetic rubber and of other chemicals), cattle feed, paper, particle boards, wax, etc., from the by-products of sugar industry (*viz.*, molasses and bagasse). Many of the industrial units using these products have made commendable progress in the manufacture of a wide range of chemicals which were otherwise being imported earlier by spending valuable foreign exchange. Presently, some of the innovative and progressive sugar factories have successfully tried to generate electricity at a low cost using sugarcane bagasse.

TABLE 23.4: Indicators of Economic Contribution of Sugarcane Production System and Sugar Industry in Maharashtra

(Employment in lakh days, Quantity in lakh quintals and value/income/revenue in Rs. in crores)

Economic Contribution Indicators	*1962-63*	*1972-73*	*1982-83*	*1992-93*	*1999-2000*
1. Labour employment in sugarcane production	473	637	1,223	1,317	1,457
2. Labour employment in sugarcane harvesting and transport	106	121	332	357	459
3. Labour employment in sugar manufacture	212	336	569	723	1,335
4. Total labour income generation	278	1,216	5,885	22,214	1,62,550
5. Value of fertiliser demanded	3.40	10.30	77.90	168.60	291.10
6. Family labour income sugarcane producers	27.90	75.20	151.50	356.00	463.88
7. Farm investment income	36.80	99.20	212.60	442.50	1,242.71
8. Revenue contribution of sugarcane industry					
(a) Education tax	0.50	0.90	3.30	6.10	9.44
(b) Sugarcane purchase tax	1.90	6.10	34.00	105.30	115.62
(c) Excise duty on sugar	15.60	33.10	118.00	233.60	452.70

Changing Economic Environment in Rural Maharashtra

As discussed earlier, the sugarcane production system and sugar industry together have made significant contributions to the development of agricultural sector and rural areas of Maharashtra. In real sense, the individual sugar factories have become the growth centres in the rural areas. By way of creating educational, transport and health facilities in the rural areas, they have opened new vistas for positive changes in the socio-economic systems of the rural society. The strength of dynamic political system of Maharashtra has always been derived from the complementary support of the people related directly or indirectly to the sugarcane production system and sugar industry of the State.

Another major contribution of these two systems of Maharashtra could be evidenced from the changing economic environment in the rural areas. The rural markets have been expanding at a fast rate to absorb various good and services produced from both within and outside the agricultural sector. With the increased purchasing power, changing consumer's preference and changing social norms and value systems, the rural society has been demanding a wide range of products for consumption and betterment of living conditions. The new economic environment created in the rural areas has become a great boon to a large number of products manufacturing and services supplying industrial groups in production and distribution of their products.

Conclusions

The chapter concludes that the interface of the sugarcane production system and sugar industry of Maharashtra have had significant influence on creating a new social order and economic environment for transformation of traditional poverty ridden rural society into a dynamic and progressive one with better living conditions. The complementary relationship between the two sectors has proved to be quite efficient and effective not only for their own development but also for the development of all other related organisations/institutions. The sugarcane production system-sugar industry interface of Maharashtra has created conducive economic environment for changing rural society. Besides, the infrastructural and institutional development support extended by the sugar industry has resulted into the transformation of the rural areas to a great extent.

24

MARKETING EXTENSION IN AGRIBUSINESS MANAGEMENT
Need for Reorientation

— *N. Srinivasan*

The vocabulary of 'Agribusiness' is uttered in the recent times by many scientists, policy-makers and the real world executives involved in the field of agricultural development, particularly in the present context of globalising Indian agriculture. It means different things to different people. To have conceptual clarity and avoid the perceptional bias, the concept is well defined first, followed by the description of the status of marketing extension with particular reference to Agribusinesses in India. Then, the need for reorienting the marketing extension strategy in Agribusiness management has been focused in the rest of the chapter.

Agribusiness Conceptual Dichotomy

Davis and Goldberg, the originators of the concept of 'Agribusiness', perceived agribusiness "as the sum total of all operations involved in the manufacture and distribution of farm supplies, production operations involved in the farm and storage, processing and distribution of farm commodities and items made therefrom". On the other hand; Wills, Duft, Snodgrass and others referred agribusiness to the firms that buy from and/or sell to the farms. The difference in the conceptualisation between these two schools of thought is that in the latter, Wills *et al*, excluded the on-farm activities from the definition of agribusiness, as the subject of farm management has its origin since long back. Thus, there existed dichotomy in conceptualisation. Nevertheless, this chapter goes with the conceptualisation of the originators.

Agribusiness System

The paradigm of agribusiness under Indian situation is depicted in Fig. 24.1. Considering agribusiness as a system, the various components in the system are grouped under five major sub-systems, *viz.*, (i) inputs supply sub-system, (ii) on-farm production sub-system, (iii) output marketing sub-system, (iv) agricultural processing sub-system, and (v) facilitative services sub-system. Among the five sub-systems, farm production sub-system is the fulcrum and it acts as the focal point around which the other four sub-systems revolve. Activities of all these sub-systems are inter-dependent in such a way that a stimulus given in one part of any sub-system would result in the reaction of the whole system. The whole agribusiness system is influenced by the macro forces like political, cultural, natural, economic and technological forces over which the agribusiness system has little control. So, in tune with the change in the macro forces, the agribusiness system has to adapt itself. Hence, the growth of the agribusiness depends on how effectively and efficiently the various agribusinesses function and also how promptly they adapt to the changing environment.

Agribusiness Opportunities

Economic growth is the *sin-quo-non* for change and better living in any country. The primary objective of developing countries like India is to achieve rapid, balanced and sustained economic growth. Agribusiness plays a key role in the economic growth and development of the country. Several research findings do suggest that entrepreneurship is the dominant factor in the growth process of any country. Therefore, the opportunities for new agribusinesses in India are listed in Table 24.1

Agribusiness Marketing Extension

The term 'Marketing Extension' has been very infrequently used; but has got rich meaning. Marketing extension, as defined by Thomson, means to extend the knowledge of marketing and solution to the problems to those who are in a position to apply this knowledge in improving marketing. While agricultural extension education involves the transfer of production technologies from lab to land, the marketing extension involves the transfer of marketing technologies from the marketing institutions to the stakeholders, namely primary producers (farmers), the intermediaries and the ultimate consumers. In other words, marketing

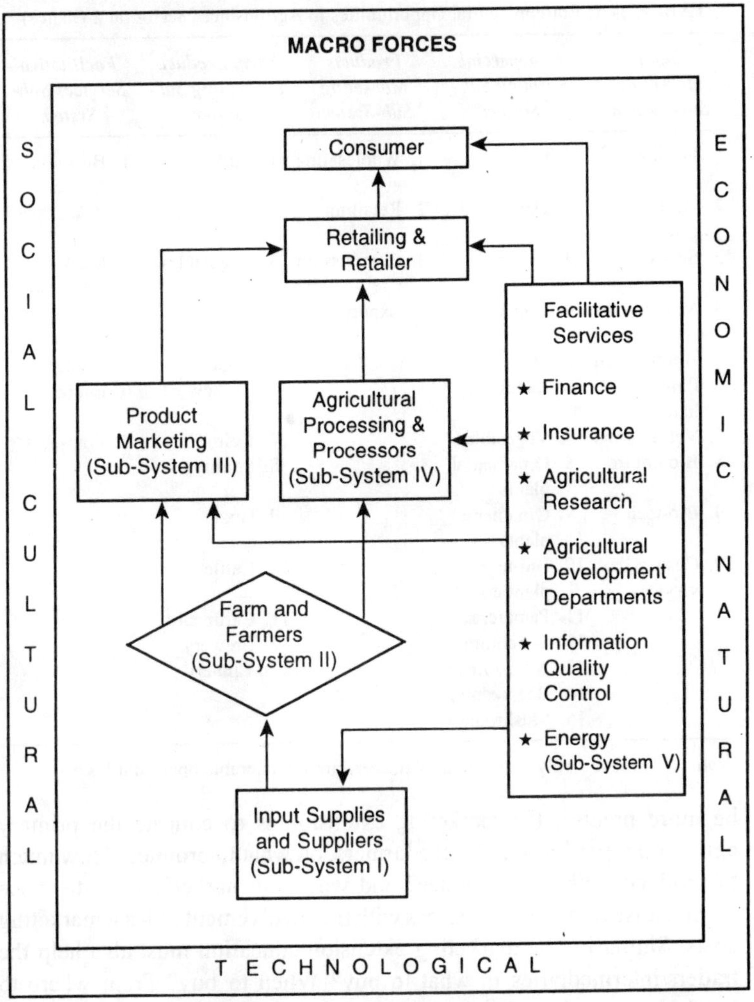

FIG. 24.1: Agribusiness System Model

extension refers to the dissemination of information/ideas/knowledge on the improved facilities available in the organised marketing institutions, market arrivals and prices to the farmers, merchants and the consumers and making them to utilise these facilities for their advantages. Therefore, the objective of marketing extension is to apprise the producers, traders and consumers of various voluntary and regulatory measures brought into force for the improvement of marketing system, so that they can all derive the benefits flowing from these measures. To

TABLE 24.1: Entrepreneurial Opportunities in Agribusiness Sector (at a Glance)

Input Marketing Sub-System	*Commercial Farming Sub-System*	*Products Marketing Sub-System*	*Farm Products Processing Sub-System*	*Facilitative Services Sub-System*
1. Fertiliser	1. Crop	1. Wholesaling	1. Milk	1. Business finance
2. Agrl. Chemicals	2. Dairy	2. Retailing	2. Fruits	2. R & D
3. Seeds	3. Poultry	3. Commission	3. Vegetables	3. Marketing Information
4. Machineries	4. Goat	4. Export	4. Paddy	4. Quality Control
5. Animal feed	5. Fish		5. Sugar	5. Insurance
6. Poultry hatchery	6. Rabbit		6. Cashew	6. Energy
7. Vet. Pharma	7. Vegetables		7. Coir	7. Transport
8. Bio-control	8. Ornamental plants		8. Poultry	
9. Bio-tech	9. Ornamental plants		9. Goat	
10. Custom services	10. Bio-tech plantlets		10. Cattle	
	11. Pamarcsa		11. Cardboard	
	12. Sericulture		12. Brewery	
	13. Agro-forestry		13. Tannery	
	14. Bee-keeping			
	15. Mushroom			

Note: These are only indications. Yet, there are innumerable opportunities.

be more precise, the marketing extension is to educate the primary agricultural producers, *viz.*, the farmers, on what to produce? How much to produce? When to produce? and where to market?, so as to make them realise remunerative prices with the involvement of least marketing costs. Similarly, the marketing extension education must also help the traders/intermediaries in what to buy? When to buy? From where to buy? How much to buy? And where to sell? so as to help them in buying, selling and handling the produce with least marketing cost and reasonable profit margin and to keep them going in their businesses. Yet another target group of marketing extension is the ultimate consumers. The marketing extension again must help the ultimate consumers in buying right type of goods in right quantity and quality, in right place and time and in right price, with least costs involved, so as to derive the maximum satisfaction. Thus, the marketing extension education helps in improving the marketing efficiency, which is a single ratio of output to input, by minimising the marketing costs and maximising the satisfaction at various stages of the vertically value adding agricultural marketing system (Fig. 24.2). Moreover, the acceptance of any newly evolved marketing technique relating to preparation of the produce, packaging, grading, storage, handling etc., by the farmers or traders invariably requires a great deal of convincing them about the advantages of adopting that technique. Therefore, marketing extension may also be regarded as marketing communication/promotion.

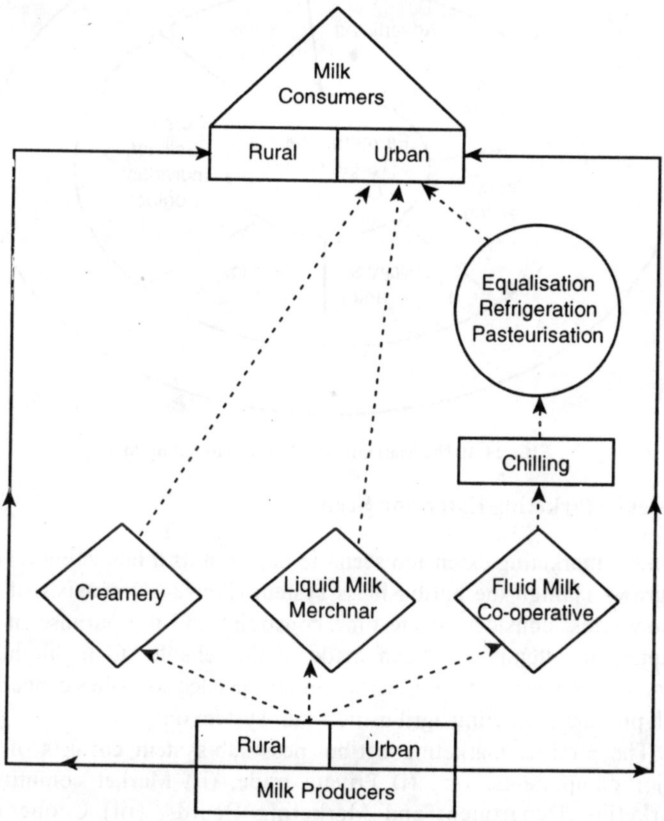

FIG. 24.2: The Liquid Milk Marketing System of the Dairy Sub-sector in Tiruchirapalli District

Any firm in the agribusiness system, to survive and grow, has to not only improve its internal efficiency; but also to adapt itself to the changing environment. For efficiency, among many strategies followed by the agribusiness firms, the most crucial one is the marketing strategy and to be more specific, the 'marketing mix', which consists of four P's, *viz.*, product, price, place and promotion. The fourth 'P', namely, the promotion in the context of agribusiness is equated with the marketing extension. Therefore, the 'Marketing extension/promotion mix' is detailed in Fig. 24.3.

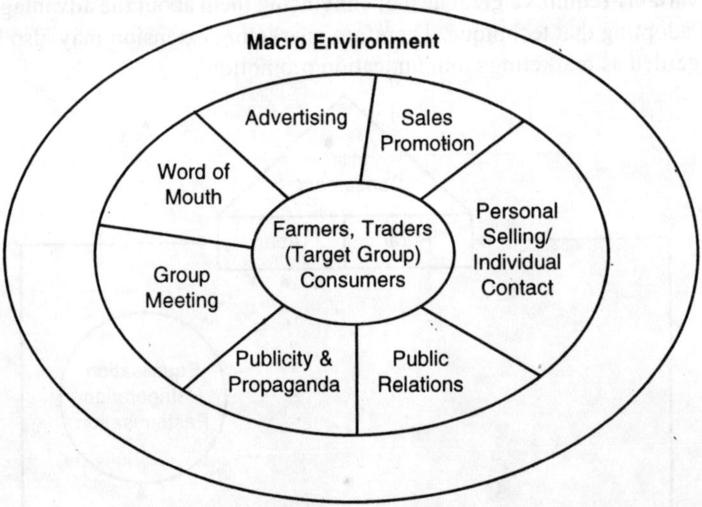

FIG. 24.3: The Marketing Extension/Promotion Mix

Present Marketing Extension Scenario

In India, marketing extension scenario is still in its infancy; but it is yet to grow. Though the agribusiness system (Fig. 24.1) consists of five sub-systems, considering the time constraint and the purpose of this chapter, an attempt has been made in this chapter to highlight the marketing extension statuses, issues and the needed strategies concerning with product marketing agribusiness sub-system only.

The product marketing agribusiness sub-system consists of five major components, *viz.*, (i) Private trade, (ii) Market committees, Marketing Department and Marketing Boards, (iii) Cooperative Marketing Institutions, (iv) Public undertaking, and (v) Exports. The nature and extent of marketing extension activities carried out in these various agribusinesses are assessed and issues identified.

Private Product Marketing Agribusinesses

The private product marketing agribusinesses play a very little role in marketing extension activities particularly in attracting arrivals from the farmers. Through personal contact and advancing credit to farmers, even from the stage of raising the crops, a better rapport is established. In promoting their product among the consumers, the private product marketing agribusinesses resort to advertisement through mass media and personalised services, particularly in the wake of rising competitiveness among themselves in the present liberalised economy.

Cooperative Marketing Agribusinesses

There are more than 7,600 Primary Agricultural Marketing Cooperative societies in the country. The marketing cooperatives in general carry out the marketing extension activities in a modest way to attract arrivals. They mainly depend upon extension education through advertisements in mass media like T.V., Radio, Daily Newspapers and distribution of pamphlets and booklets and through educating the member patrons by individual contact. During the celebration of cooperative week each year, intensive publicity and propaganda activities are carried out through mass media, group meeting, individual contact, etc. In this way, the milk marketing cooperatives are carrying out better marketing extension education activities among milk producers, particularly under the Operation Flood Programme, implemented by the NDDB. Among consumers, their products are promoted through advertisement in mass media. In the recent years many training programmes on "Marketing extension/Promotion" are conducted to the personnel of the marketing co-operatives. However, they are being constrained by lack of aptitude and experience for extension work, lack of better ability for effective extension education activities in rural areas and meagre budget provision.

Regulated Marketing Agribusinesses

Above 6,600 regulated markets are functioning in the country. The regulated markets and market committees in Tamil Nadu State do involve in marketing extension through publicity and propaganda activities to educate the farmers about the benefits accrued by selling their farm produce through regulated markets. For example, in Tamil Nadu State the superintendents and supervisors, through the group meetings of farmers and individual contacts, propagate the benefits of the regulated markets. The market committee, under which the regulated markets are functioning in the district, has one superintendent or supervisor position exclusively meant for attending the publicity and propaganda activities.

He coordinates the publicity and propaganda activities carried out by the various regulated markets in his jurisdiction, *viz.*, one district. The Secretary and Special Officer of the market committee are also involved in marketing extension activities along with the regulated market personnel at the village level. They organise exhibitions also in village fairs and festivals.

The State Agricultural Marketing Board is the Apex Body at the State level. The overall responsibility of developing agricultural marketing system in the State is vested with the Board, to which the various district market committees are affiliated. One of the important activities vested with the Board is publicity and propaganda, *viz.*, the marketing extension education activities. The Board has its own publicity and propaganda units located in four regions in the State *viz.*, Chennai, Coimbatore, Tiruchirapalli and Madurai. Each of these units has infrastructural facilities like publicity vehicle, cine projector, video van, public address system, etc. In coordination with the personnel of the market committees and regulated markets, the publicity and propaganda activities are carried out by these units. These centres were formed in 1982 with the positions of one Assistant Director of Agriculture, one cine operator and one Jeep driver. In the Tamil Nadu State, Agricultural Marketing Board is headquartered at Chennai. Propaganda materials like posters, stickers, pamphlets, etc., are prepared and distributed through the regional publicity units and market committees. The various extension education activities carried out by the Tamil Nadu State Agricultural Marketing Board through its regional centres are indicated in Table 24.2.

The Tamil Nadu State Agricultural Marketing Board derives its finance from the different market committees in the State, which contribute 15 per cent of their receipts. Out of this, fifty per cent is allocated as 'Market Development Fund' to meet the requirements of market development activities and the balance of 50 per cent is made available to meet the administrative expenditure of the Board and other activities, including marketing extension activities like training and publicity and propaganda.

The number of regulated markets and districts covered by each regional propaganda unit is given in Table 24.3.

It could be seen from Table 24.3 that each unit/centre has to cover 58 to 76 regulated markets spread over a notified area of about five to seven districts. Hence, each publicity and propaganda centre/unit manned by one extension personnel, *viz.*, the Assistant Director of Agriculture,

TABLE 24.2: Publicity and Propaganda Activities carried out by Tamil Nadu State Agricultural Marketing Board

Sl. No.	*Name of the Work*
1.	Identification of Villages in each regulated market area from where the arrivals are very less and concentrating publicity work in these villages.
2.	Village level intensive campaigns which include village meetings, film shows, small exhibitions, etc.
3.	Total number of villages to be visited are finalised.
4.	Arranging exhibitions during local festivals, Department meetings, etc.
5.	Participation in zonal workshops conducted by District Joint Directors of Agriculture.
6.	District level seminars.
7.	State level seminars.
8.	Flashing messages in All India Radio.
9.	Flashing messages in television.
10.	Flashing news through newspapers.
11.	Taking video films to be shown to farmers in villages (Long duration films—15-25 minutes).
12.	Wall paintings and hoardings in important places.
13.	Distribution of leaflets, pamphlets, hand bills, etc.
14.	Distribution of strickers.
15.	Preparation of Cinema Slides.
16.	Match box advertisements.

TABLE 24.3: The Number of Districts and Regulated Markets covered by Publicity and Propaganda Units in Tamil Nadu State

Sl. No.	*Centre*	*No. of Districts*	*No. of Regulated Markets*
1.	Chennai	6	70
2.	Coimbatore	5	76
3.	Tiruchirapalli	7	58
4.	Madurai	9	66
	Total	27	270

finds it very difficult to cover the jurisdiction during cropping seasons. Moreover, the personnel for publicity and propaganda activities, either the superintendent or supervisor at Market Committee level, seems to have not been assigned with any specific plan and budgetary provision. So the present arrangements for publicity and propaganda indicates the scope for strengthening with specific extension programmes and budgetary provisions at the levels of market committees and regional units.

At the regulated market level, the Superintendent or Supervisor who heads the Regulated Market has to take care of various enforcement activities like collection of market fee, inspection of licensed premises, issue of trader licenses, checking the movement of notified commodities etc., in addition to carrying out the publicity and propaganda activities through individual contact with farmers during his visits to the villages. In places where supervisors are positioned (in addition to superintendents), the supervisors carry out the publicity and propaganda activities in addition to other works. The Agricultural Officers positioned in the Post-Harvest Information (Commercial Grading) Centres by the State Agricultural Marketing Department, also attend the publicity and propaganda activities, particularly in the nearby villages and in the market premises.

Government Marketing Agencies

The government marketing agencies, such as, Civil Supplies Corporation, Food/Cotton Corporation of India and others also purchase from the farmers. Though their main aim is to procure from farmers to feed the Public Distribution System and to build buffer stock, they also have obligation of enforcing the minimum Support Price policy of the government. Wherever monopoly procurement was in operation, the farmers had to sell their entire surpluses through the Direct Procurement Centres of the Corporation only. Because of the changed government policy, the monopoly procurement was dispensed with and the parallel procurement policy has been enforced recently. Now the farmers have options to market their produce through any agency and hence the experience shows that the farmers get better prices. However, very little marketing extension activities are resorted to. The government agencies fail to attract enough arrivals because of keen competition from private trade and bureaucratic hurdles. Therefore, the procurement (and extension) strategy adopted by the government needs reorientation to improve the competition in the agricultural product markets.

Directorate of Marketing and Inspection

The Directorate of Marketing and Inspection, Government of India is vested with the full responsibility of developing the agricultural marketing system in the country. It is more concerned with the development of regulated marketing system, grading and standardisation, marketing news and intelligence and market survey. An Extension Cell functions since 1962 in the Directorate of Marketing and Inspection and its major activity is to prepare and supply to the various States the necessary publicity and extension materials in the field of agricultural marketing. The cell also advises the State marketing boards and departments on agricultural marketing extension, *viz.*, publicity and propaganda activities, relating to regulated markets, grading and market information.

The Needed Reorientation

As could be seen in the foregoing discussion, a modest attempt is being made as far as marketing extension activities are concerned in the product marketing agribusinesses in India. Therefore, the reorientation of marketing extension strategy may be aimed at on the following lines in the product marketing agribusinesses in India.

- Judicious combination of the levels of various variables in the marketing extension/promotion mix may be aimed at and for which sufficient funds may be earmarked in the budget.
- Attending to marketing extension education activities on the part of the marketing personnel of the regulated markets and cooperative marketing must be made as mandatory and it must find place in their job charts as a planned activity. The marketing extension activities must be planned, implemented and controlled by the concerned supervising personnel, for which, sufficient budgetary provision must be made.
- The Krishi Vigyan Kendras and the NGOs may play an active role in transferring marketing technologies particularly to farmers by positioning appropriately trained persons in marketing extension.
- In this era of information technology, marketing information system must be developed at each market committee level and at the State level in the Agricultural Marketing Board and Agricultural Marketing Department. The collection of marketing information and intelligence activities including forecasting must be systematically carried out on scientific lines.
- Large number of training programmes to farmers may be organised to educate them on what to produce?, when to produce?, how to produce quality products?, how much to produce?, and where to market? In order to realise better prices for their produce. The personnel in the marketing agribusinesses may also be trained on latest marketing technologies and marketing extension techniques.

- The "Marketing Extension Cell" in the Directorate of Marketing and Inspection, Government of India must play a very active role in marketing extension activities, in collaboration with the State Marketing Brands and Agricultural Marketing departments.
- To collect and disseminate information and advise farmers and traders, on export of farm products on a single window basis, a separate Agricultural Products Export Promotion Council may be set up in each State and at the Centre, the APEDA, may be entrusted with the task of effecting coordination among the States. This needs strengthening APEDA with many more branches all over the country.

25

TRANSFER OF KNOWLEDGE BASED INPUTS TO FARM WOMEN

— *Parminder K. Ghuman, B.S. Hansra* and *B.S. Boparai*

The role of agricultural extension in the coming years would be quite different from what it was in early ninety or earlier. In the changing agricultural scenario, major emphasis is laid on increasing the productivity, improving the quality standards, judicious use of farm inputs, processing, value addition and marketing of farm products. The progress so far achieved is not to the desired level of satisfaction due to intra-regional differences in the extent of progress achieved so far. The efforts have mostly been confined to the head of the family, though outcome of many of the farm decisions is influenced by other members of his family specially the housewife.

In fact, Indian rural women have always been an important and prominent partner in agriculture sector either as cultivators, effective supervisors, decision markers or as agriculture helpers. According to recent studies, 50-90 per cent of all agricultural activities are carried by farm women and also contribute towards 44 per cent of the world food today (Country Report, 1995). Sangwan, *et al.* 1990 reported that there is a distinct sphere of participation between males and females in farm and dairy sectors; the men were the planners and women implemented the activity. The pattern seems to suggest that information flow has to be effective for task completion. Secondly, if women are implementers of plans related to farm and dairy sector; they should also receive direct information.

The priority of women's programmes, on the contrary, is mostly on health, education and non-agricultural home enterprises. They include very little of agribusiness and failed to recognise the need for the specific women target group. As agriculture is changing fast, need based efficient extension management is essential to cope up with the new environment. Women farmers although in minority as compared to men, but a significant group, should be taken care of while planning the strategies of extension management of agricultural business and other related activities.

Hence, an attempt has been made to identify the strategies for technology dissemination with the following objectives:

1. To identify the tasks carried out by the farm women in agribusiness.
2. To identify the needs of the farm women for knowledge and skills of technologies of agribusiness.
3. To identify the extension strategies for technology dissemination to farm women.

Methodology

The study was conducted in the sub-mountain undulating zone of Punjab, which is predominantly a rainfed area having resource poor and illiterate farming families. Due to uncertainty of returns, able bodied men migrate to areas of more employment opportunities. Farm related operations are left for aged men and women. One hundred and seventy two farm women from 4 villages from two blocks of this zone were selected for the study. In a preliminary survey conducted, activities of crop production, grain storage and cattle management were identified in which farm women found to be participating. An interview schedule was used to collect the data. The data was analysed with the help of percentages and mean score.

Findings

Tasks carried out and Managed by Farm Women

It is evident from Table 25.1 that more than 20 per cent of the farm women were participating manually in majority of the operations of crop production, all operations of grain storage and cattle management. Hence, these operations were selected to develop subject matter areas in which to measure the farm women's knowledge and skills.

Level of Knowledge and Skills of Farm Women

The data enfolded in Table 25.2, revealed that majority of the farm women had low to medium level of knowledge and skills of technologies of crop production, grain storage and cattle management. Their mean knowledge scores in most of the subject matter areas were less than 50 out of 100. The reasons for low level of knowledge and skills as reported by farm women are numerous and are of socio-economic in nature. Hence, an integrated approach is needed to understand women specific problems and to disseminate improved technologies of agriculture and allied fields.

TABLE 25.1: Tasks carried out and managed by the Farm Women in Agribusiness

Sl. No.	*Operations*	*Participation (Percentage)*		
		Manual	*Supervision*	*Total*
A.	*Crop Production*			
1.	Land preparation	78.48	—	78.48
2.	Manure application	78.48	—	78.48
3.	Fertiliser application	21.50	—	21.50
4.	Buying of seed	50.00	—	50.00
5.	Seed preparation	93.60	—	93.60
6.	Sowing	65.70	—	65.70
7.	Hoeing & weeding	79.65	—	79.65
8.	Plant protection measures	—	—	—
9.	Irrigation	—	—	—
10.	Harvesting & threshing	78.49	—	78.49
11.	Unsheathing	79.06	—	79.06
12.	Storage of agril. wastes	72.68	—	72.68
B.	*Grain Storage*			
1.	Drying of grains	100.00	—	100.00
2.	Cleaning of grain	97.67	2.33	100.00
3.	Storage of grains	97.67	2.33	100.00
4.	Treatment of grains	—	—	—
C.	*Cattle Management*			
1.	Harvesting of fodder	93.02	—	93.02
2.	Chaffing of fodder	88.37	—	88.37
3.	Feeding of animals	100.00	—	100.00
4.	Watering of animals	50.50	—	50.50
5.	Milking of animals	100.00	—	100.00
6.	Sale of milk	8.72	—	8.72
7.	Cleaning cattle	97.67	2.33	100.00
8.	Cleaning cattle shed, collection of cow dung and making cow dung cakes	97.67	2.33	100.00
9.	Care of sick animals	100.00	—	100.00
10.	Care of neonatal calf and calved animals	100.00	—	100.00

TABLE 25.2: Level of Knowledge and Skill Competency of the Farm Women in Agribusiness

Sl. No.	*Operations*	*Percentage of Level of Knowledge & Skills*			
		Low	*Medium*	*High*	*Mean Score*
Level of Knowledge					
A.	*Crop Production*				
1.	Land preparation	25.19	45.19	29.62	79.62
2.	Buying of seed	45.35	32.56	22.09	48.06
3.	Seed preparation	39.12	37.88	23.00	47.21
4.	Sowing	37.17	37.17	23.00	47.21
5.	Hoeing and weeding	76.64	—	23.36	53.65
6.	Fertiliser application	54.05	29.73	16.22	41.66
7.	Manure application	100.00	—	—	41.22
8.	Harvesting and threshing	63.70	36.30	—	55.76
	Overall	55.15	36.47	22.88	51.86
B.	*Grain Storage*				
1.	Restoring practices	25.00	35.47	39.53	43.57
2.	Surface storing practices	27.32	37.21	35.47	47.87
3.	Metal bin storing practices	37.79	73.95	13.95	45.47
	Overall	32.56	29.07	38.37	45.63
C.	*Cattle Management*				
1.	Feed management	61.05	38.95	—	47.21
2.	Diseases management	56.98	28.49	14.53	33.84
3.	Pregnancy and post-calving management	21.90	30.81	41.28	64.41
4.	Milk production management	45.93	32.56	—	—
5.	Cattle shed/yard management	41.28	—	58.12	69.19
	Overall	45.35	26.16	28.40	52.25
Level of Skill Competency					
1.	Crop production	64.96	24.82	10.22	38.97
2.	Grain storage	55.23	36.63	8.14	51.56
3.	Cattle management	58.72	36.05	5.23	44.34

Strategies for Disseminating Technologies to Farm Women

Recruitment of More Women Extension Agents

In India most of the extension agents are men, who concentrate mainly on male farmers. Moreover due to certain socio-cultural barriers, they face difficulty to communicate to farm women. Hence, involving more women extension agents will facilitate the easy access of technology to farm women.

Technology Dissemination through Women Farmers Groups

Group approach alleviates many socio-cultural difficulties of male extension workers working with farm women. They can either work already existing women groups such as *mahila mandals* etc. or special groups of farm women can be made.

Selection of Contact Women Farmers

Identify a few local women in whom you find some interest and enthusiasm. These women can act as contact farm women to facilitate easy access and technical know-how to women farmers.

Access to Agricultural Inputs

Agricultural inputs should be made available nearby rural areas such as through village co-operatives or through other input agencies.

Ensure Access to Markets and Market Information

Market infrastructure should be developed for easy disposal of agricultural produces in rural areas itself. Market information should be provided through extension workers.

Access to Agriculture Credit

Attempts should be made to reduce cumbersome procedures needed in getting loans and to ensure financial help on priority through cooperatives and nationalised banks.

Exclusive Training Programmes for Farm Women in Agriculture and Allied Fields

Exclusive training programmes based on the location specific needs of the farm women will encourage them to contribute more competently.

A Diagnostic Approach for Overall Process of Technology Transfer

Diagnostic approach is required for technology transfer in order to understand the set of condition in which farm women live and work.

Total Development through Extension

The new concept of extension has become broad based which emphasises the total development of environment in a family resides and works. This mean development of the family as a whole: its value system, psychological and mental set-up, literacy level, etc. It will definitely help in fast technologies dissemination.

Dissemination Centres

Technology dissemination centres specifically for farm women in rural areas need to be set up to transfer technologies related to home agriculture an allied sector. Women programmes are mostly related to health, education and non-agricultural home enterprises. The training in home enterprises is mostly confined in tailoring and embroidery only.

Use of Mass/Electronic Media

Mass media such as T.V., Radio, Newspaper and farm journals should focus more on farm women. Farm women's experiences, success stories of farm women in agriculture sector through publication and telecast. Specific literature should be published on technology dissemination among farm women.

Conclusions

In India, women have a major place in agrarian scenario. Most of the women in rural areas participate and manage many agriculture related activities besides fulfilling their responsibilities as home makers. Their level of knowledge and skill competency in agricultural technologies and allied fields is low to medium level. Hence, it becomes necessary for the planners, policy-makers and extension personnel to consider the available potential of women in agriculture and develop appropriate strategies for improving the competency and capabilities of farm women and orient them towards the right perspective to achieve the sustainability in agriculture and allied fields.

REFERENCES

Choudhary, M.A. (1996). *Empowering Strategies for Rural Women,* Beijing, 1995, Country Report, Deptt. of Women and Child Development, Ministry of HRD, Government of India, New Delhi.

Sangwan, V., Munjal, S. and Punia, P.K. (1990). "Participation of Women in Farm Activities", *Indian Journal of Extension Education*, 26(1&2): 112.

PART III

EXPORT ORIENTED AGRIBUSINESS

26

EXPORT ORIENTED AGRIBUSINESS AND GLOBALISATION OF INDIAN ECONOMY

— *D.S. Navadkar, B.K. Mali* and *D.V. Kasar*

General

Indian agriculture has experienced spectacular changes in the recent period manifesting large scale commercialisation and diversification. They broadly include cultivation of new crops and varieties, increase in the share of area under cash crops, large scale spread of livestock activities and fisheries, bio-technology, horticulture, processing, etc. In response to the new economic environment ushered in by the process of liberalisation and globalisation, many new agribusiness activities are cropping up.

The critical role of agricultural marketing has been heightened in recent years because of the rise in arrivals of agricultural produce relative to total agricultural output. This proportion will increase as subsistence farming declines and prospects of exports of agricultural products are expected to expand rapidly. Also, the export markets have shown expanding trends in response to provisions under the new world trade order brought about by the WTO.

Performance of Indian Agriculture

The role of agriculture in the economic development continued to be of great importance. The country has achieved the self sufficiency in foodgrains and milk production. Presently agricultural sector in India contributes 29.4 per cent of GDP, employing 64 per cent of country's work force. During the Eighth Plan, 1992-97 agriculture registered an annual growth rate of 3.5 per cent, with foodgrain output grown at 3 per cent. In the Ninth Plan agricultural growth was aimed at 4.5 per cent.

In this chapter it is attempted to go into the details of issues concerned with export marketing, role played by APEDA, use of information technology and strategies of agribusiness in the context of globalisation of Indian economy.

Globalisation of Indian Agriculture

The agricultural scenario all over the world particularly in the post-GATT era, is undergoing a rapid change. Agribusiness in rural areas and also agro-industrialisation assumes significance in view of this changed scenario. In the perspective of liberalisation, Indian agriculture is now to meet double challenge, that of growing demand for food and industrialisation within the country and that of external demand and exports. Agriculture occupies a central position in the Indian economy. After achieving the technological change in the production process India could become not only self-sufficient but also attain a position of net exporter of agricultural commodities. Concentrated efforts are needed for boosting of farm export, especially in the context of globalisation of agriculture. Under the new world trade, liberalisation policy of our government started in 1991. Formation of WTO (1 January 1995) exploit the lucrative world market for the benefit of rural people and therefore agricultural marketing has opened into the world. We need to reshape our production and trade practices to reap the maximum benefit out of it and to put the economy to new heights.

India's Scenario with World Agriculture

As could be seen from Table 26.1, that India ranks first in the globe in terms of irrigated area, production of pulses, tea and milk and possesses highest number of cattle and buffaloes. The nation is the second largest producer of sugarcane and vegetables and occupies third position in the world in terms of production of cereals, cotton lint and fruits.

Growth in Crop Production

The production of major cereals, pulses and commercial crops during 1996-97 to 2000-2001 (Table 26.2) indicated negative percentage change excepting rice, wheat, sugarcane, tea and coffee. There was substantial increase of production of rice, sugarcane and coffee. The production of wheat and tea had positive change, however, this change was very meagre. The latest estimate for 2000-2001 points to an improvement in production of sugarcane and cotton whereas, it predicts fall in oilseeds

TABLE 26.1: India's Position in World Agriculture

Sl. No.	*Items*	*India*	*World*	*India's*	
				% Share	*Rank*
1.	Area (Mi. ha.)				
	Total area	329	13,387	2.4	7
	Land area	297	13,048	2.3	7
	Arable land	163	1,382	11.8	2
	Irrigated area	57	242	23.6	1
2.	Population (million)				
	Total	960	5,849	16.4	2
	Agriculture	541	2,564	21.1	2
3.	Economically active population (million)				
	Total	960	5,849	16.4	2
	Agriculture	256	1,302	19.3	2
4.	Crop production (MT)				
	Cereals	223	2,096	10.6	3
	Pulses	15	58	25.9	1
	Sugarcane	265	1,241	21.4	2
	Cotton (Lint)	2.86	19.7	14.5	3
	Tea	0.79	2.73	28.9	1
5.	Livestock position (million head)				
	Cattle	209	1,334	15.7	1
	Buffaloes	92	167	55.1	1
6.	Tractor in use	1,400	26,345	5.3	4
7.	Vegetables	54.97	595.56	9.2	2
8.	Fruits	37.13	429.45	8.6	3
9.	Milk	72	546.64	13.2	1
10.	Eggs	1.61	51.76	3.1	5

Source: Agricultural Statistics at a Glance, Government of India, 2000, New Delhi.

and jute/mesta. Coffee cultivation is spread over three southern States *viz.*, Karnataka, Kerala and Tamil Nadu. Coffee is freely exported commodity. About 80 per cent of coffee produced is exported. Domestic prices of coffee are higher than the international prices. The recent sharp fall in international coffee prices has adversely affected the Indian coffee exporters. India is the largest producer and consumer of tea in the world. There is no restriction on export of tea. Although import of tea is under the restricted list, it is freely importable from SAARC countries with effect from 1.8.1998.

Pre-liberalisation and Post-liberalisation Growth in Agriculture

The average annual growth recorded in agriculture and allied sector

TABLE 26.2: Crop Production

(million tonnes)

Crop	*1996-97*	*2000-2001*	*Percentage Change over 1996-97*
Rice	81.7	86.8	6.24
Wheat	69.4	70.0	0.86
Coarse cereals	34.1	29.9	–12.32
Pulses	14.2	12.3	–13.38
Foodgrains	199.4	199.0	–0.20
Groundnut	8.6	6.2	–27.91
Rapeseed/Mustard	6.7	4.3	–35.82
Soybean	5.4	5.2	–3.70
Other six oilseeds	3.7	2.9	–21.62
Total oilseeds	24.4	18.6	–23.77
Jute & Mesta	11.2	9.9	–11.61
Cotton	14.2	13.2	–7.04
Sugarcane	277.6	300.6	8.29
Coffee (lakh ton)	2.05	2.95	43.90
Tea (million kg)	780	784	0.51

Source: Economic Survey, 2000-2001.

during the post-reform years 1992-93 to 1999-2000 was 3.6 per cent compared to 3.9 per cent in pre-liberalisation period 1980-81 to 1991-92 (Table 26.3). As regards the growth in foodgrains, the most dominant segment of crop agriculture, decelerated from 2.9 per cent to 2.00 per cent in the post-reform period. Decline in coarse cereals (–1.6%) conformed to the trend of no growth even before the reforms, indicating preference for higher value foodgrains. Still there has been high growth maintained in wheat (3.6%) and rice (2.2%). Crop agriculture including 46 crops have registered 2.2 per cent annual growth in post-reform period compared to 3.4 per cent in eighties. This deceleration is obviously due to decline in coarse cereals production. Annual growth in rice and wheat has been fairly robust (Table 26.4).

TABLE 26.3: Pre-reform and Post-reform Growth in Agriculture (per cent) at 1993-94 Prices

Crop	*1980-81 to 1991-92*	*1992-93 to 1999-2000*
Agriculture and Allied	3.9	3.6
(i) Agriculture (all crops, animal husbandry, dairying)	4.2	3.7
(ii) Forestry	–0.1	0.8
(iii) Fishing	5.4	5.5

Source: Economic Survey, 2000-2001.

TABLE 26.4: Annual Compound Growth of Crop Agriculture (Base: Triennium Ending 1981-82)

Crop	*1980-81 to 1991-92*	*1992-93 to 1999-2000*
Crop agriculture	3.4	2.2
Foodgrains	2.9	2.0
(i) Cereals	3.1	2.1
(ii) Rice	3.7	2.2
(iii) Wheat	3.6	3.6
(iv) Coarse cereals	0.0	–1.6
(v) Pulses	1.4	0.8
Non-foodgrains	4.3	2.4

Source: *Economic Survey*, 2000-2001.

Agricultural Exports

The various commodities identified by the Commerce Ministry for sustainable agricultural export with their percentage changes are depicted in Table 26.5.

TABLE 26.5: India's Export of Agricultural Products

(*million dollars*)

Particulars	*1996-97*	*1999-2000*	*% Change over 1996-97*
1. Tea	292	407	39.38
2. Coffee	402	315	–21.64
3. Cereals	1,104	718	–34.96
4. Tobacco	213	229	7.51
5. Spices	339	393	15.93
6. Cashew	362	566	56.35
7. Sesame and Niger seeds	78	86	10.26
8. Guargum meal	100	189	89.00
9. Oil meals	985	370	–62.44
10. Fruits & vegetables	208	202	–2.88
11. Processed fruits & juices	59	113	91.53
12. Meat & preparations	200	180	–10.00
13. Marine products	1,129	1,180	4.52
14. Others	1,359	527	–61.22
15. Total agri-exports	6,828	5,475	–19.82
16. Total country exports	33,470	37,599	12.34
17. Percentage of agri. to total export	20.40	14.60	

Source: *Economic Survey*, 2000-2001.

The major agri-exports of India are cereals, spices, cashew, oilcakes, tobacco, tea, coffee and marine products. Value of agri-exports to total exports of the country has been ranging in between 15 to 20 per cent. The exported commodities *viz.*, tea, tobacco, spices, cashew, sesame and niger seeds, guargum, processed fruits and marine products have exhibited some up trend. The rest of the commodities listed in Table 26.5, the export of which substantially decelerated over the time period. This resulted in 19.82 per cent decline in the agri-exports during 1999-2000 over 1996-97. India's agri-exports face certain constraints that arises from conflicting domestic policies relating to production, storage, distribution, food security, pricing concerns, etc. Higher domestic prices of bulk exports like sugar, wheat, rice, etc. make out export commercially less competitive. Market intelligence and creating awareness in international market about quality of products need to be strengthened to boost agricultural export.

India's Position in World Export

Table 26.6 gives commodity wise shares of exports. According to the average shares in the total export bill, fish and preparations, rice, tea and coffee, spices, feeding stuff for animals and tobacco are at the top of our export list. India is having only 0.6 per cent share in the world export of agricultural products.

TABLE 26.6: India's Share in World Exports by Commodity Divisions and Groups

(US $ million)

Commodity	*1998*		
	World	*India*	*Share (%)*
Meat and meat preparation	46,088	203	0.4
Fish and preparation	45,532	1,140	2.5
Cereals and preparation	55,690	874	1.6
Rice	8,289	863	10.4
Vegetables and Fruits	71,852	712	1.0
Sugar and preparation	17,485	707	0.4
Coffee, tea, cocoa, spices and mfg.	35,334	1,176	3.3
Coffee and substitute	15,975	433	2.7
Tea and make	2,881	472	16.4
Spices	2,424	271	11.2
Feeding staff for animals	21,044	898	4.3
Tobacco and mfg.	24,407	235	1.0
Unmfg tobacco and refuse	6,439	235	3.6
Mfg. tobacco	17,968	0	0
Oilseeds and oleaginous fruit	15,082	241	1.6
Other exports	46,88,557	8,460	0.2
Total exports	50,75,047	32,039	0.6

Source: Economic Survey, 2000-2001.

Role Played by APEDA

Agricultural and Processed Food Products Development Authority (APEDA) was established in 1986 with an aim to boost Indian agro-based products through optimum utilisation of the enormous resource base of our country in terms of agro-climatic regions, fertile soil, abundant light and two monsoons. Therefore, APEDA has been referred as facilitator for developing conducive environment for trade. For increasing the value of added exports of farm produce by making them globally competitive APEDA conducted experiments and developed controlled atmosphere containers to be carried by sea unlike earlier times it was carried by air. APEDA covered almost all the production categories which include floriculture and seeds, fresh and processed fruits and vegetables, animal products, other processed foods or cereals in order to strengthen the agricultural exports. The development of data base on products markets and services by identifying new markets, providing better support system to exporters and manufacturers and finally by introducing new products to the international market.

Under its various schemes APEDA offers following services:

1. Financial support to tap vast export potential in agricultural sector.
2. Undertake market surveys and feasibility studies for identification and penetration of new markets.
3. Improving infrastructural facilities (*e.g.*, cargo handling at Mumbai, Delhi, Bangalore, Trivandrum and Chennai exit points).
4. In collaboration with Indian Institute of Packaging, Mumbai, APEDA is involved in developing packaging standards.
5. It provides assistance in the cost, subject to the ceiling of Rs. 5 lakh and 50 per cent of the cost subject to the ceiling of Rs. 2 lakh is provided for installing quality assurance and quality control system.
6. It is in constant touch with many research institutions like ICAR to bridge the research gap and bring out need oriented research.

Use of Information Technology

India has certainly come a long way socially and economically with the advertisement of colour TV's, paging services, cellular phones, etc. The scenario has changed drastically. Compared to other developing countries, India is much ahead in all-round development attaining proficiency even in having our satellites in orbit.

Connecting the domestic and foreign markets by Internet, E-mail will serve in strengthening the market research and market intelligence system, one can know the price change, demand and supply conditions, price trend in any markets within few minutes. Therefore, there is tremendous scope for it.

Strategies of Agribusiness

1. There is an urgent need for planned development of horticulture in the country.
2. Adequate marketing support should be provided.
3. The export markets demand high quality products. Hence there is a need to undertake research and development.
4. Packaging requires considerable attention. Proper packaging will reduce transit losses and help to maintain the quality of the product. This will also help to cut down costs.
5. Branded products are preferred in foreign markets. Thus, proper branding of products needs to be encouraged.
6. The efforts may be made to tap a large number of international markets in the bid to expand and diversify India's markets.

Summing Up

India has made lot of progress since independence in terms of growth in output, yields and area under many crops. Presently, agricultural sector in India contributes 29.4 per cent of GDP, employing 64 per cent of country's work force. The post-liberalisation period indicated decelerated picture in the production of coarse cereals, pulses, major oilseeds, commercial crops like cotton and jute. There is substantial increase in production of rice, sugarcane and coffee. The export scenario indicated that tea, tobacco, spices, cashew, sesame and niger seed, guargum, processed fruits and marine products exhibited some upward trend of export. However, other agricultural commodities showed decelerated picture. The decline in foreign trade is not only due to the trade policies of the developed countries but also due to our own policies. Under the new global order India has wide scope to develop her agricultural export market. APEDA has been referred as facilitator for developing conducive environment of trade and playing a significant role in the agricultural export of the nation. The information technology has much scope in the agribusiness activities. Connecting the domestic and foreign markets by Internet, E-mail will serve in the strengthening the market research and intelligence system. India with its rich genetic resource base has large potential to place the country in the forefront of the global agricultural trade in the years to come.

REFERENCES

Economic Survey, Government of India, 2000-2001.

Emerging Issues in Agricultural Export Marketing in India, Shivaji University, Sept., 1997.

Implications of GATT/WTO on Agriculture and Rural Development, NIRD, Hyderabad, March, 1996.

Indian Agriculture, 1999.

Market Oriented Agribusiness Management, M.P.K.V., Rahuri, 1999.

27

INDIA'S CASHEW EXPORT
Issues and Strategies

— *K.D. Kokate* and *S.S. Wadkar*

The Cashew (*Anacardium occidentale* Linn.) which belongs to the family Anacardiaceae, is an economically important tropical tree crop. It ranks second only to almond, among the nine tree nuts of importance in the world trade. A native of Brazil, where it flourished as a wild growth in the jungles and cashew was brought to east Africa and India by the early Portuguese settlers more than 400 years ago. In India, it soon established itself all along the west coast and later in the east coast. For several centuries cashew was merely regarded as a sturdy perennial tree yielding good soft wood and producing a rather delicious juicy apple. Nuts were thrown as its hard shell contained corrosive liquid. India is the largest producer of raw cashewnut in the world and accounts for 43 per cent of world production. As years rolled the interest in the nut slowly developed and extraction of kernel was discovered. More and more people world over have been consuming this unique nut, making the cashew an economic product.

Now, cashew is one of the major export earning crops which accounts nearly Rs. 2,453 crores constituted about 1.5 per cent of total Indian export. India is largest producer, processor and exporter of cashew in the world. Attempt is made in this chapter to acquire recent information on cashew export and some issues/policies which can enhance the cashew industry in India.

Recently, the structuring process in Indian economy has begun. Our economy is gradually opening up to face the challenges of liberalisation in the global market. In this context, an attempt is made to examine and review export performance of cashew in pre-liberalisation and post-liberalisation period in India.

Methodology

The study involved secondary data. The time series data were collected for the period from 1980-81 to 1999-2000 from the Institute of Cashew Export Promotion Council of India, Cochin. The collected data were analysed for two different periods, namely, period-I, pre-liberalisation of economy, *i.e.*, 1980-81 to 1989-90 and period-II, post-liberalisation of economy, *i.e.*, from 1990-91 to 1999-2000 to observe the effect of liberalisation policy on export of cashew. External value of any currency is fluctuating over a period of time. Hence, by considering the year 1980-81 as base year, constant prices were worked out to give the real picture of export scenario.

Trend Analysis

The growth in quantity exported, export value were worked out by using linear trend equation. The compound growth rates were worked out by using the exponential form of growth function.

Results and Discussion

Recent available information of major countries on production and export was gathered and compared to know the share of production of raw cashew and export. The information as export of cashew kernels by major countries and world production of raw cashewnuts is presented in Table 27.1 and Table 27.2, respectively. In the year 1991, India exported 46,841 MT of cashew kernels, which shared about 59.1 per cent of total world export. However, the India's share in the world production was 40.7 per cent with the production of 2,95,000 MT. The second position occupied by Brazil with export of 23,429 MT (29.5%) and annual production of 1,86,000 MT (25.7%). At third place, Mozambique ranked with the export of 3,334 MT (4.2%) and production of 4,11,100 MT (5.7%).

In the year 1995, India exported 72,640 MT, which share 65.9 per cent of total world export. The share of India in world production in the same year was 38.5 per cent with the production of 3,22,000 MT. This was followed by Brazil. In the export, the percentage increase over base year was maximum in respect of India, *i.e.*, 55.07 per cent. However, in respect of production the percentage increase was 9.15 per cent.

TABLE 27.1: Export of Cashew Kernels by Major Countries

(Quantity in MT)

Countries	*1991*	*1995*	*% Change over Base Year*	*1996*	*% Change over Base Year*
India	46,841 (59.1)	72,640 (65.9)	55.07	70,308 (61.9)	50.09
Brazil	23,429 (29.5)	31,752 (28.8)	35.52	36,855 (32.4)	57.30
Mozambique	3,334 (4.2)	1,134 (1.00)	–66.00	1,361 (1.2)	–59.17
Tanzania	1,043 (1.3)	182 (0.82)	–82.55	227 (0.2)	–78.23
Kenya	1,202 (1.5)	272 (0.2)	–77.37	340 (0.3)	–71.71
Vietnam	544 (0.7)	1,315 (1.2)	141.72	1,361 (1.2)	150.18
Indonesia	1,701 (2.1)	1,043 (0.9)	–38.68	1,134 (1.0)	33.33
China	680 (0.9)	1,021 (0.9)	50.14	1,134 (1.0)	66.76
Others	499 (0.7)	839 (0.8)	68.13	907 (0.8)	81.76
Total	79,273 (100.00)	1,10,198 (100.00)	39.01	1,13,627 (100.00)	43.33

Note: Figures in parentheses are percentages to the total.

TABLE 27.2: World Production of Raw Cashewnuts

(Quantity in MT)

Countries	*1991*	*1999*	*% Change over Base Year*
India	2,95,000 (40.7)	4,60,000 (51.11)	55.93
Brazil	1,86,000 (25.7)	1,60,000 (17.78)	–13.98
Mozambique	41,100 (5.7)	40,000 (4.44)	–2.76
Tanzania	48,000 (6.6)	80,000 (8.89)	66.67
Kenya	19,000 (2.6)	—	—
Vietnam	35,000 (4.8)	70,000 (7.78)	100.00
Indonesia	31,000 (4.3)	20,000 (2.22)	–35.48
Others	70,000 (9.6)	70,000 (7.78)	0.00
Total	7,25,000 (100.00)	9,00,000 (100.00)	24.14

Note: Figures in the parentheses indicate percentage to the respective total.

In the year 1996 with the export of 70,308 MT India shared 61.09 per cent of total world export, followed by Brazil. The percentage increase over the base year in the respective year was 50.09 per cent. It was concluded from the Table 27.2 and 27.3 that, India had efficiently used the production for export as it was indicated from the fact that as compared to other countries, India's export in regard to production was very well in boom and made best use of production for export and earned valuable foreign exchange.

TABLE 27.3: India's Share in Total World Export

Year	*World Export (in tonnes)*	*India's Export (in tonnes)*	*% Share*
1980	67,181	31,836	47.4
1985	66,043	40,553	61.4
1990	79,245	47,753	60.3
1995	1,10,198	72,640	65.9
1996	1,13,627	70,308	61.9

World export of cashew kernels and India's share in the total world's export is presented in Table 27.3. It is observed that except in the year 1980 India occupied dominance in the international trade and transacted above 60 per cent share of the world export. In the year 1980, the share was 47.4 per cent in the year 1985 with the export of 40,553 MT, India captured 61.4 per cent share of total world export. Recently in the year 1996, India captured 61.0 per cent share of world export by exporting 70,308 MT of cashew kernels.

The percentage share of cashew export in agricultural export and country's total export were calculated and given in Table 27.4. It is found that in the period 1981-85, the share of cashew export was 1.75 per cent and 5.47 per cent in the country's total export and agricultural export was maximum, *i.e.*, 9.51 per cent in the period 1986-90. In the year 1995-96, the share of cashew in the country's total export was lower, *i.e.*, 1.16 per cent. However, in the year 1999-2000 the share was increased to 1.51 per cent and 8.43 per cent to the country's total export and agricultural export, respectively.

It was cleared that cashew constituted about 6-9 per cent of the total export earnings from agricultural and allied products. In the total country's export on an average cashew constituted about 1.5 per cent share.

Cashew Kernel

It can be seen from Table 27.5, that the simple growth rate of total

TABLE 27.4: Average Export Earnings from Cashew and Share in Total Agricultural Exports and Country's Total Export

(Figures in crore of rupees)

Sl. No.	Particulars	1981-85	1986-90	1991-96	1995-96	1998-99	1999-2000
1.	Total export	8,967.00	17,382.20	56,541.40	1,06,353.00	1,41,604.00	1,62,738.00
2.	Agricultural export	2,879.30	3,161.82	9,148.39	21,138.00	26,200.00	29,082.00
3.	Cashew export	157.50	300.84	829.82	1,237.00	1,637.00	2,451.00
4.	Percentage share of India's cashew export in						
	(a) Total export	1.75	1.73	1.46	1.16	1.17	1.51
	(b) Agricultural export	5.47	9.51	9.08	5.85	6.24	8.43

TABLE 27.5: Simple and Compound Growth Rates of Total Cashew Export for the Period-I, *i.e.*, 1980-81 to 1989-90, Period-II, 1990-91 to 1999-2000 and Overall (1980-81 to 1999-2000)

	Particulars	Simple Growth Rate			Compound Growth Rate (%)		
		I	II	Overall	I	II	Overall
(i)	Cashew kernel quantity (MT)	1,160.60*	4,140.94**	3,081.83**	3.18*	6.51**	6.08**
(ii)	Value at current prices (Rs. lakh)	2,509.00**	16,153.71**	9,544.72**	11.69**	15.97**	16.62**
(iii)	Value at constant prices (Rs. lakh)	201.95NS	2,072.67**	1,491.05**	1.32NS	7.28**	6.45**
(iv)	CNSL export quantity (MT)	456.20NS	–426.20*	–207.50**	–9.24*	–15.50*	–-6.43**
(v)	Value at current prices (Rs. lakh)	–17.22NS	–10.07NS	3.21NS	–3.05NS	–4.21NS	1.61NS
(vi)	Value at constant prices (Rs. lakh)	–36.31NS*	–12.36NS	–14.15**	–11.80NS	–11.38NS	–7.54**

* Significant at 5% level.
** Significant at 1% level.
NS Non-significant.

quantity of cashew kernel exported was maximum, *i.e.*, 4,141 MT in period-II which indicated that during the second period the per annum quantity exported increased, significantly by 4,141 MT. However for the period-I, simple growth rate of the quantity exported was 1,161 MT and it was significant. At the overall level, the quantity of cashew kernels exported, increased significantly by 6.51 per cent during the period-II. However, percentage increase was 3.18 per cent during period-I and at the overall level the quantity exported significantly by 6.08 per cent. Current prices increased significantly by Rs. 16,154 lakh. However, for the period-I the value received at current prices, increased significantly by Rs. 2,509 lakh per annum. At the overall level at current prices increase was Rs. 9,545 lakh per annum. In terms of percentage the value received at current prices increased significantly by 15.97 per cent during the second period. However, percentage increase was 11.69 per cent for period-I. At the overall level, the value received at current prices increased significantly by 16.62 per cent.

At constant prices, the simple growth rate was maximum for period-II, *i.e.*, Rs. 2,072 lakh, which indicated that during, the second period the per annum value received at constant prices increased significantly by Rs. 2,072 lakh. However, for the period-I, the value at constant prices increased significantly by Rs. 201.95 lakh. At the overall level, the value increased significantly by Rs. 1,491 lakh. In terms of percentage, the value received at constant prices increased significantly by 7.28 per cent during period-II. However, percentage increase was 1.32 per cent for period-I. At the overall level increased significantly by 6.45 per cent in value terms.

Thus, it can be concluded that in the case of cashew India was virtually a monopoly supplier to the world. The area under cashew in India grew at the annual rate of 5.65 per cent during period under study. The production increased at an annual rate of 4.55 per cent. The positive growth rates in respect of production and export of the cashew were due to the increased domestic production and increased demand for cashew kernels in the international market. The increased domestic production and exports were due to the several measures taken by the Directorate of Cashewnut Development, Cashew Export Promotion Council, State Agricultural Universities on improved methods of production, assistance for export of cashew kernels by setting up facilities for upgrading quality and technical advice on scientific post-harvest operations and processing. Though India had a monopoly position in the international market, the countries like Brazil, Vietnam, Mozambique were emerging as competitors in the recent times. This would certainly erode a portion of our market share. Hence, we should make efforts to retain our monopoly position in future by updating our cultivation practices by using of high yielding varieties and maintaining quality prescriptions in international trade.

Cashewnut Shell Liquid (CNSL)

It can also be seen from Table 27.5 that, the quantity of CNSL exported, decreased by 456.20 MT and 426.20 MT during the period-I and II, respectively. But these figures were non-significant. At the overall level the quantity of CNSL exported, decreased significantly by 207.56 MT per annum. In percentage terms, the quantity decreased by 9.24 per cent and 15.50 per cent significantly for the periods-I and II respectively, while the figures was significant. At the overall level, the quantity of CNSL decreased significantly by 6.43 per cent.

At current prices, the value received, decreased by Rs. 1,721 lakh per annum during the period-I. However, during the period-II, the value decreased by Rs. 10.07 lakh per annum. With overall increased in value by Rs. 320 lakh but all these figures were non-significant. In terms of percentage, the value of period-I, was decreased by 3.65 per cent. During the period-II, the value decreased by 4.21 per cent. At the overall level the value received at current prices increased by 1.61 per cent. However, these figures were statistically non-significant.

At constant prices, the simple growth rate of value received for the period-II, was Rs. 12.36 lakh, *i.e.*, indicated that the value decreased by Rs. 12.36 lakh per annum during the period-II. So also, during the period-II, it was decreased by Rs. 36.31 lakh per annum. At the overall level the value decreased by Rs. 14.15 lakh per annum. In terms of percentage, during the period-II, the value at constant prices decreased by 11.38 per cent per annum and during the period-I, it decreased by 11.80 per cent. At the overall level, the value decreased by 7.54 per cent. But all the values for simple and compound growth rates were non-significant except at overall level it was found to be significant.

From Table 27.5, it was observed that, there was fall in the export of CNSL quantity. As there are mainly two process involved in cashew processing, *viz.*, oil bath process and drum roasting process. In the oil bath process, extraction of CNSL leads to scorching and breakage of nuts, thus reducing their market value. In the other process, the nuts were not damaged much and they command better market price.

However, the entire quantity of shell liquid gets burnt. It was this factor, which weighs in the mind of processors in deciding about the extraction CNSL. The price of CNSL should be so high as to compensate for the loss in the low quality of kernels obtained in the oil bath process, then only the processors will go in for extraction of CNSL. Also our traditional market Republic of Korea had decreased its import of CNSL, these two factors might be responsible for fall in the export of CNSL quantity.

Following strategies will be useful to enhance cashew industry in a positive manner and keep the pace with growing competition.

1. This unit requires huge amount of raw material and gives around 90 per cent of female employment in the region. The rapid growth of this unit need to be encouraged. In view of this, wasteland in the region need to brought under plantation crops by inducing large scale cultivation and processing operation as well. It is essential to promote productivity level also. Due to this benefits that could be derived are import substitution, development of waste land, improvement of ecology, prevention of soil erosion, employment generation etc.
2. Grading and sorting of raw nuts should be done at the production site. Since grading in vogue is improper. This could help the processors to procure superior quality cashewnut to recover higher percentage of superior grade cashew kernel. It will also reduce the cost incurred on dryage and facilitate to earn more return.
3. All the unit owners are facing the problem of credit. The credit is not available in adequate quantity and at proper time. Because of this problem, they cannot purchase sufficient quantity of raw material which is available in nearby area. To overcome this problem, it is necessary to make modifications in lending policy of financing institute. Another important aspect related with credit is that reduction in the interest rate. This may give good encouragement to the industry.
4. The Maharashtra Government has not imposed any purchase tax on cashew. Therefore, processors of southern neighbouring States purchase raw nuts from this district by offering higher price to the producers. In such situation, processors are forced to purchase raw material for higher prices whereas processors from this State go for purchasing raw cashews in adjoining States they have to pay four per cent purchase tax. So there is

need to impose 4 per cent purchase tax by the State government so it will help to increase Government revenue as well as safeguard the processors interest in the Konkan region.

5. Low capacity utilisation by the cashew processors is hindering the progress of this industry. To overcome this problem, government may improve electricity supply failure, revise credit policy by way of providing easy credit availability and lower interest rate.
6. Cooperative cashew processing units are required to be established in this area for getting additional benefit in income and employment.
7. In the global agrarian sector, India is largest producer, largest processor, and exporter and also largest area. By the ever increasing trend in exports, there has been an increased trend in import as well, but increase in domestic production has helped in reducing the imports but the import has helped to increase domestic consumption only. Though there was marginal reduction in imports, they continued to play a major role in export, thereby around half of the export earnings were drained towards import. Also the development of cashew processing in importing countries was bound to affect the availability of raw nuts processing by India. Hence extension services should be geared for further strengthened and to ensure the adequate supply through increase in production. The findings of the research institutions in modern technologies in production generated should reach the farmers without time lag.
8. In four southern State of Andhra Pradesh, Karnataka, Tamil Nadu and Kerala together have nearly 61,000 sq. km of wasteland (Nayar, 1999). If out of these areas, at least 10.0 per cent of additional could be brought under area cashew plantations and dependence on import could be decreased. Central Wasteland Development Board has also included cashew as one of the crops for waste land development. Considering the cashew as plantation crop, the step should be taken by the government to exempt it from land ceiling act. Sufficient number of grafts should be made available to farmers to help them to meet their replanting and area expansion programmes. The States should need to initiate cultivation of cashew on large scale in the public sectors. Some States have also initiated such steps.

9. India has a near monopoly in seventies, however with increasing competition from cashew processing and producing countries in recent years, had affected our exports. Our exports scenario showed that over dependence on U.S.A. and Netherlands. Such type of dependence on two-three countries for export market is not desirable in international trade. This had put certain limitations on capacity to bargain for better price. The prime marketing strategy should be to strengthen non-traditional markets and explore new one and also it is a time for promotion of value added products from cashew kernels.
10. India is a signatory to the World Trade Organisation agreement. As per the terms and conditions of this agreement, liberalisation in import has been announced under EXIM Policy of Government. Liberalised import under OGL is helping in acquiring more nuts by imports but the economy getting drained for import hence the unit should consider a balancing approach to meet on export and domestic need. So efforts like development and application of grade regulation, ISO-9002 would be helpful in future. To face the competition from the newly emerging cashew growing countries, it is need of time to accelerating for the overseeing and formulating the development programmes for promotion of export.
11. The instability in export earning from the selected countries is mainly due to fluctuation in production in other producing countries and increased value of Indian cashew in world market however there is a need to maintain this upsurge because of the emerging threat from other processing countries of the world.

APPENDIX-I

World Production of Cashew (1999) Production (MT)

Country	*1999*
India	4,60,000
Vietnam	70,000
Indonesia	20,000
Brazil	1,60,000
Tanzania	80,000
Mozambique	40,000
Others	70,000
Total	9,00,000

APPENDIX-II

Area and Production of Cashew for the Year 1999-2000

(A—Area in '000 ha; P—Production in '000 MT; APY—Average Productivity per hectare in kg.)

State	*Area*	*Production*	*Productivity*
Kerala	122	100	850
Karnataka	91	60	700
Goa	54	30	610
Maharashtra	121	125	1,470
Tamil Nadu	85	45	540
Andhra Pradesh	103	100	1,100
Orissa	84	40	670
West Bengal	9	8	900
Other	17	12	800
Total	686	520	7,640

APPENDIX-III

Export of Cashew Kernels from India during 1995-96 to 1999-2000

Countries	1995-96		1996-97		1997-98		1998-99		1999-2000	
	Qty (MT)	Value (Rs. Crs.)	Qty (MT)	Value (Rs. Crs.)	Qty (MT)	Value (Rs. Crs.)	Qty (MT)	Value (Rs. Crs.)	Qty (MT)	Value (Rs. Crs.)
U.S.A.	21,190	351	24,877	456	29,678	536	34,793	722	46,275	1,202
Netherlands	11,462	210	16,708	322	17,827	333	14,222	303	17,761	481
U.K.	2,404	43	4,180	80	5,078	96	5,375	117	6,718	189
Japan	5,544	101	5,142	99	5,261	97	4,925	109	4,725	129
U.A.E.	2,044	33	2,156	41	2,437	45	2,900	65	2,820	76
Australia	657	12	990	18	1,939	35	1,920	40	1,710	46
France	2,707	50	2,891	55	1,865	30	2,313	46	1,803	46
Canada	179	3	427	8	630	11	447	10	1,412	38
Saudi Arabia	518	11	705	13	1,013	18	1,110	25	1,020	28
Israel	707	14	1,115	22	1,170	24	1,116	26	904	26
Italy	247	4	374	7	883	17	904	19	946	26
Spain	553	10	411	8	783	15	571	13	691	19
Singapore	1,279	23	1,580	29	1,102	19	1,037	22	734	19
Germany	1,051	19	1,506	29	1,339	23	1,110	22	613	16
Hongkong	1,242	24	1,075	21	700	14	957	23	581	16
New Zealand	238	4	303	6	349	6	382	8	475	13
Russia	14,905	263	664	9	1,190	16	36	85	410	7
Korea Rep.	330	6	305	4	115	2	108	2	58	1
Total	70,374	1,241	68,663	1,260	76,593	1,396	77,075	1,630	92,461	2,451

Source: D.G.C.I. & S., Calcutta.

APPENDIX-IV

Export of Cashewnut Shell Liquid from India during 1995-96 to 1999-2000

Countries	*1995-96*		*1996-97*		*1997-98*		*1998-99*		*1999-2000*	
	Qty (MT)	*Value (Rs. Crs.)*	*Qty (MT)*	*Value (Rs. Crs.)*	*Qty (MT)*	*Value (Rs. Crs.)*	*Qty (MT)*	*Value (Rs. Crs.)*	*Qty (MT)*	*Value (Rs. Crs.)*
Korea Rep.	261	0.38	363	0.44	616	1.11	185	0.46	383	0.99
Japan	145	0.30	848	0.92	1,072	1.45	233	0.41	191	0.47
USA	15	0.05	0	0.00	2,095	2.72	100	0.12	0	0.00
UK	108	0.40	357	1.09	586	1.71	923	2.49	0	0.00
Korea DP Rep.	128	0.18	0	0.00	0	0.00	16	0.03	0	0.00
Malaysia	7	0.02	10	0.02	0	0.00	0	0.00	0	0.00
China P. Rep.	0	0.00	1	0.00	23	0.05	0	0.00	0	0.00
Hongkong	96	0.12	0	0.00	0	0.00	0	0.00	0	0.00
Germany	0	0.00	96	0.10	0	0.00	0	0.00	0	0.00
Others	0	0.00	0	0.00	54	0.13	465	0.33	164	0.74
Total	760	1.46	1,735	2.77	4,446	7.17	1,912	4.21	754	1.84

Source: D.G.C.I. & S., Calcutta.

28

HARNESSING FLORICULTURE FOR EXPORTS

— *Anita Jhamtani* and *Rachana Sharma*

Flowers have been an integral part of Indian culture since time immemorial. Man has traditionally used flowers for expressing his innermost feeling to God and deities, to the beloved ones and for complimenting anyone and versifying any conceivable emotion. Earlier flowers were marketed either in loose form or in the form of garlands. Now-a-days there is a marked change and more of cut flowers in the form of bouquets and flower arrangements are being sold by the florists shops. The traditional flower growers and sellers are facing tough competition from modern cut-flower industry. These kinds of products are mainly sold by florist shops. This situation is already hanging with respect to the increasingly important role being assumed by the big retailers. The role of consumers has changed, they have become professional buyers, and demand for different types of flowers like Gladiola, Gladiardia, Chrysanthemum, Rose etc., has gone up.

World—Current Status

Floriculture activity consisting of cut-flowers cultivated under controlled conditions mainly for export is growing in the world at around 6-10 per cent per annum. The major importing countries of cut-flowers are Netherlands, USA, Germany, Italy, France, UK, Switzerland, and Japan. North America, Europe and Japan have a great demand for floriculture products. The Netherlands and the USA have reached the threshold level. The developing countries such as Columbia, Israel, South Africa and Kenya have emerged as additional production centres in the past few decades. The most important flower trade in the international market is still held by the Roses, however, others like Chrysanthemum, Carnation, Gerbera, Dahlia, Poinsettia, Orchids, Lilium, etc., are also marketed. In most of these countries climatic condition does not favour round the year cultivation.

The global floriculture trade is estimated at $50 billion. Currently more than 10 countries are engaged in floriculture. The international trade in cut-flowers and potted plants is growing every year. From the total production of cut-flowers and potted plants, only a small part is exported to the international market. Because of the high transport costs per unit, the export of this type of product is usually to neighbouring countries. The main exporters are the Netherlands, Denmark and Belgium. The Netherlands has the largest area under production, and under glass with intensive production. In Ecuador, the area under production is growing explosively, increased by hundreds of hectares per year. Cut-flower production in Africa is highly competitive to any professional grower in the world. The US with about one-third of the consumption, has the largest share of the total world consumption, followed by Germany, Italy and France.

It is quite evident from Table 28.1, that Netherlands will face stiff competition from Latin American, African and Asian countries in the near future. India, which entered commercial floriculture quite late, has a bright future.

TABLE 28.1: Factors contributing to Competitiveness

Factor	*Netherlands*	*Latin America*	*Africa*	*India*
Geography	very good	moderate	moderate	good
Climate	good	very good	very good	very good
Land labour/raw materials	poor	very good	very good	very good
Capital	very good	good	moderate	moderate
Infrastructure	very good	very good	moderate	poor
Knowledge	very good	good	good	good
Domestic market	good	poor	poor	good
Network	very good	good	moderate	moderate
Government	very good	moderate	poor	good

Source: Commercial Floriculture, FAO/DAC, MOA.

India—Current Status

Inspite of a long tradition of agriculture and floriculture, India's share in the international market for flowers is negligible. The Indian floriculture industry is a more recent development as export started from 1990. The industry is growing at annual rate of 20 per cent per annum. The area under the production of flowers is 82,000 hectares. The red revolution came with the help of Dutch consultants who seized the opportunity for selling the expertise, greenhouse technology and especially the planting material. The mixed euphoria of growing flowers was nursed by big corporate houses for which investment in floriculture as compared to steel and fabrics was very low. This resulted in a spurt in floriculture units in India.

India has achieved the distinction of being the biggest supplier of flowers to Japan. Indian exports have shown an increase of 25 per cent in the last five years. Indian flowers especially roses are well received in major countries of the world who prefer rote rose type of varieties in pastel shades with short stem (about 5 cm) and longer vase life and fragrance (confetti). Of the total area under flowers more than half is under traditional crops jasmine, scented rose, small flowered chrysanthemum tuberose, crossandra and aster. The annual growth rate of domestic trade in floriculture products is estimated to be 15-20 per cent.

TABLE 28.2: Export of Floricultural Products from India since 1995

Products	*1995-96*	*1996-97*	*1997-98*	*1998-99*
Cut-flowers	100.33	180.86	189.41	252.70
Dried flowers	364.56	323.40	581.10	695.88
Live plants	81.48	91.26	33.73	13.44
Dried plants	35.83	25.62	48.38	NA
Bulb, tubers etc.	19.21	12.84	29.81	28.23
Total	601.41	633.98	882.43	990.25

Source: Commercial Floriculture, FAO/DAC, MOA.

Consumption of flowers in the southern States is much higher than in the northern States. During the last few years there has been drastic change in flower trade especially of modern cut-flowers in Mumbai, Pune, Bangalore, Chennai, Hyderabad, Delhi, Chandigarh, Lucknow and Calcutta. Mizoram, having a large Christian population, has traditional use for flowers, which find wide application in the social activities of the people. Usually, a flower auction takes place in Aizwal every Saturday. In Guwahati, the demand for cutflowers, potted plants and orchids is rising due to market penetration by Oil companies and due to existence of institutional buyers like hotels, airports and a few corporate houses.

A Promising Venture for India

Most of the flowers are grown under protected conditions in covered structures like greenhouses and poly/glass houses in European and other countries. Due to intense cold, high energy cost, production in these countries is limited during winter months (October to April), thus they have to depend largely on import to meet their domestic demand as most of the festivals fall during this period when the demand of flowers is at its peak. Hence, the prospects are good for new entrants like India because of following reasons:

- India is a treasure house of temperate plants, particularly orchids and roses.
- India is situated centrally in comparison to major flower consumption centres like Europe, Japan and Middle East countries.
- Labour is cheap and cost of inputs like water, electricity, fertiliser, pesticides, etc. is less.
- Climatic conditions are ideal during winter months when flower demand is at peak.
- Wide choice of cutflower species and cultivars due to different climatic regions.

In India, however, flowers are largely grown in the open fields along with other foodgrain crops. These are grown for local consumption and do not meet the requirements of export quality. The export requires cut-flowers of high quality and standard. These pose a challenge to the traditional flower growers as it requires different set of skills and resources for which farmers need to be informed and trained. There is need to develop the individual farmers capabilities so that they get specialised and produce high quality flowers. According to industry experts the main reason why Holland is so good at growing perfect quality is that growers specialise in one flower variety. They have activated their greenhouses and developed a system whereby growers take their products to the auction everyday. The Dutch have also specialised in some lesser known flower varieties and they produce high quality in large quantities. This enables them to capture the attention of Americans who look for unusual flowers and buy them. Considering the advantages India has, it is important that we too develop our floriculture industry and grow diverse varieties of high quality which can create interest among the buyers.

Problems of Floriculture Industry

Despite the plus points, India's performance in the global floriculture market is miserable mainly because the traditional flower growers are not trained for quality production and those who entered the industry during its earlier years were mostly those who had nothing much to do with cultivation of flowers. Those who had never grown a blade of grass come into floriculture simply to avail the support provided by the government through various organisations. These industrialists had no emotional attachment with cultivation of flowers and worked mechanically employing expensive foreign technology brought through the exorbitantly expensive consultants. During these intervening years since efforts have been made to train up the traditional floriculture for going cut-flowers under cost effective greenhouses. Despite these efforts there still remain several problems which would need to be looked in, if we wish to do better in the export markets. These are:

- dearth of warehousing space at the airports
- limited availability of cargo space in planes for the bulky products like flowers
- certification procedures related to phyto-sanitary and custom formalities are tedious
- no direct air service from India to flower marketing areas
- lack of co-ordination between government and floriculture industry
- peak season for flowers (December to February) coincides with that of garment exports from India
- unstreamlined export-import policy
- non-availability of basic inputs
- small and scattered production units (without proper linkages between producers and exporters)
- absence of quality control mechanism
- absence of cold chain
- unorganised domestic market and lack of established marketing channels.

Government Programme and Policies

A number of steps have been taken by the Central and State governments to encourage floriculture industry. The Agriculture and Processed Foods Export Development Agency (APEDA) also proposes to set up wholesale market-cum-auction-centre for flowers in major cities *viz.*, Delhi, Mumbai and Bangalore.

Some of the import schemes are:

- Central scheme "Commercial Floriculture", Ministry of Agriculture
- Scheme for Export Promotion and Market Development
- Scheme for Packaging Development
- Scheme for Assistance to Promote Quality and Quality Control (APEDA)
- Scheme for Infrastructure Development
- Scheme for Feasibility Studies Surveys Consultancy and Database Upgradation.

It is also significant to note that floriculture export industry has been given the status of export-oriented unit. Custom duty on export of plant and nursery and other technological inputs required for floriculture has also been completely removed.

Suggestions to Increase Flower Exports

Government of India should concentrate to reduce the high rate of import duty on Indian cutflower, contemplate duty exemption on the import of materials for greenhouses and tissue culture labs considering the huge capital inputs, setting up an export promotion council. Policy initiatives should be taken to provide incentives to flower growers. The National Banks for Agricultural and Rural Development needs to take up provision of loans at low rates of interest for floriculture entrepreneurs to set up for greenhouses.

The Agricultural Extension system should undertake activities for floriculture and impart training to farmers and entrepreneurs. Institutions like Indian Institute of Horticultural Research, Bangalore and Indian Agricultural Research Institute, New Delhi should act as focal institute for Research and Development in the area and link up with the extension system especially for floriculture. There is need to create appropriate infrastructure for production, post-harvest handling and transportation of floricultural products. To enhance the export and reduce the losses due to post-harvest handling there is need to set up cold storage units or terminals at international airports.

Make provision for refrigerated vans for road transport and adequate warehousing space at airport. Provide direct air services to flower marketing areas like Amsterdam, etc. For floriculture extension there is need to encourage cooperatives and NGOs to provide services, technical advice and market assistance. Encourage a "consortium approach" among the units for branding, grading, packaging transporting, quality control, supply assurance, market promotion, research and development.

REFERENCES

CBI, Exporting Cutflower and Foliage.

Degroot, N.S.P. "Tropical—seeds.com: Flowers and Ornamentals", *Agricultural and Economic Research Institute* (LEI-DLO), The Netherlands.

Jhamtani, Anita (2000). "Socio-economic Implications and Opportunities for Women in Floriculture" in H.P. Singh and N.K. Dadlani, ed., *Commercial Floriculture*, FAO/DAC, MOA.

Kapoor, Manish. "Floriculture—A Promising Forex Earner", *The Tribune*, Chandigarh, India.

North East India. National Informatics Centre, Assam.

"The Floriculture Industry", *Agri Yellow Pages Resources* (Floriculture).

PART IV

RE-ORGANISATION OF EXTENSION MANAGEMENT SYSTEMS AND ROLE OF INFORMATION TECHNOLOGY

29

EVOLUTION OF RESEARCH, DEVELOPMENT AND EXTENSION IN BRAZIL
A Case Study of Embrapa Dairy

— *Sergio Rustichelli Teixeira, Shankariah Chamala* and *Tom Cowan*

The Brazilian National Research & Development Corporation (Embrapa) is the premier institution for agricultural research. Embrapa Dairy is a unit of Embrapa for the dairy industry and has the institutional mission *to provide solutions for sustainable development of the Brazilian agribusiness by generation, adaptation and transference of knowledge and technologies for social benefit* (Embrapa, 2001). Extension corporations in Brazil have a similar mission and the integration of actions involving R&D with E has always been more a desire than a real fact. The weak links between R&D with the farming community and the incomplete feedback from Extension (Fig. 29.1) are phenomena that need to be studied in R,D&E (Ban and Hawkins, 1996) in order to develop strategies to enhance their effectiveness.

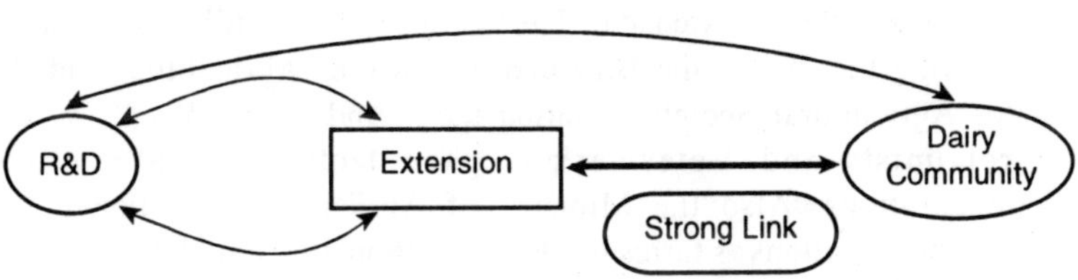

FIG. 29.1: Links of R&D, Extension and Dairy Farmers Community

In practice scientific researchers continued to have total control of a particular technology, but without a shared vision of where it could be applied, it is unlikely that technologies will be adopted successfully

(Teixeira, 2000). Equally important, the literature shows that R,D&E, does not have a major influence on farmers' decisions (Hamilton, 1995). To deal with the present reality Embrapa is investing in training researchers in extension science (Cezar, 1999). As illustrated in Fig. 29.2, Embrapa's centres are building a stronger bridge between extension and production systems.

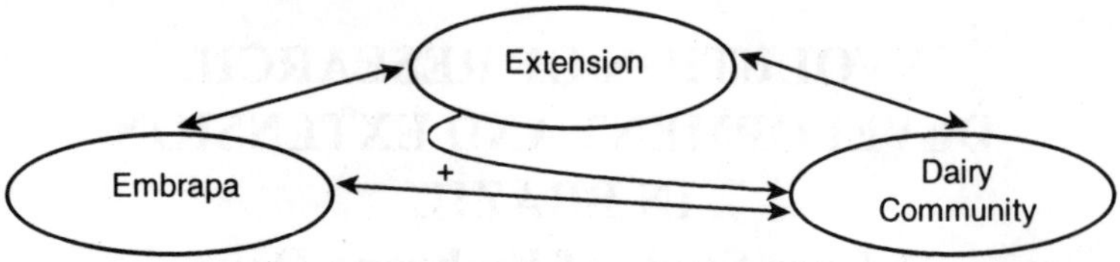

FIG. 29.2: Links to be Reinforced, Mainly Research and Farmers Community

R,D&E are at a turning point of their evolution where the problems of communication between them are well known. But it is a challenge for them to work in partnership. Although this is not the only problem, the resources for R,D&E are also decreasing and the production systems are becoming more complex. These scenarios require the involvement of more stakeholders in the planning of R,D&E actions and a new extension management system. To suggest a new direction to the approach of R,D&E with production systems, it is necessary to understand the root and evolution of the present situation to avoid repeating old mistakes.

R,D&E Evolution in Brazil

The extension activity in Brazil started in the Agricultural High School of Viçosa (ESAV), now known as the Federal University of Viçosa (UFV), by Peter Henry Rolfs. In 1930, the "Farmer Week" was organised to show agricultural practices. This was the main educational approach used by Agricultural Extension. The "Farmer Week" still exists today.

During the 1940s, the Brazilian Ministry of Agricultural and the States Agricultural Secretary started the "Land Users Week", where Agronomists and Veterinarians gave lectures with practice demonstrations. Also the Ministry of Agriculture established the Agricultural Stations as farms for demonstration of agricultural practices and to produce plants/seeds for distribution. More than 200 Agricultural Stations were established, but did not achieve practical objectives. In 1949, the Ministry of Agriculture, Education and Health created the Rural Missions, which were staffed with Agronomists, Doctors, Sociologists, Psychologists and Social Assistants. The teams travelled in the same vehicle to rural areas to give assistance to communities. It did not last long. The decade finished with a good Extension experience at Santa Rita do Passa Quatro—São Paulo, under the influence of professor John B. Greefing from the Federal University of Viçosa. Also in the 1940s the Credit and Rural Assistance Association of Minas Gerais (ACAR-MG) was created. This association has been considered as the embryo of the second biggest Extension service in the world (Olinger, 1996). ACAR put into practice a new philosophy of rural intervention based on educational principles (Oliveira, 1987).

In 1954, the Brazilian Government, in partnership with the U.S.A. Government, created the Agricultural Technical Office (ETA), in Rio de Janeiro, to execute projects that could contribute to the development of Brazilian agriculture. The Project 2 had the objective of "development of a Rural Extension programme in harmony with supervised rural credit". In 1956 a group of Extension leaders from different Brazilian States decided to create a National Association to coordinate the Rural Extension services in the country. The Brazilian Government and ETA supported the idea of the National Association. In 1956, ACAR was created after some agreement between the Brazilian republic President Juscelino Kubichek together with Nelson Rockefeller from U.S.A., to centralise the extension work in Brazil. The legal entity was a non-governmental, non-profit institution, operating as a private corporation (Olinger, 1998). The aim was to transform the Brazilian small farmer into an American farmer (Figueiredo, 1981). Medium and small farmers were going to be attended individually or collectively through cooperatives and projects. The extension staff accessed the communities in a Jeep to: stay in touch with farmers' problems, organise meetings to discuss actions, conduct practical demonstrations, and assist farmers in accessing and securing credit from banks (Olinger, 1998). The technological era was planned and the farmer was encouraged to change from organic to chemical fertilisation and to use more motorised equipment instead of animals. From 1948 to 1967 most of Extension State organisations were created (Oliveira, 1987).

During the 1960s, these models of extension grew with the creation of Brazilian Credit and Rural Assistance Association (ABCAR) as a Federal 'Head Office' organisation of public utility. A five-year plan was also created with the aim of increasing productivity. The whole country had programmes and plans to support the expanding Brazilian economy as listed in Table 29.1.

In the 1970s an increasing number of Veterinary, Agronomy or Animal Husbandry students were enrolled in Universities. Until the end of 1970s, it was easy to find a job as a rural professional. Rural extension had about 80 per cent of all graduated people working in agricultural activities (Olinger, 1998). During the 1970s, Research Institutes (IPEAC) conducted research in many States of Brazil. In 1974 Embrapa was created to drive agricultural applied research in Brazil, whereas the universities took care of the basic research (Olinger, 1996).

TABLE 29.1: Brazilian Programmes and Plans from 1964 till 1979

Programmes and Plans	*Period*
Programme of Government economic action	1964-1966
Strategic programme of development	1967
Strategic programme of development	1968-1970
I Brazilian National Plan of development	1972-1974
II Brazilian National Plan of development	1975-1979

A model of integration involving research—extension—farmers was developed to ease the flow of information to and from the farmers. In 1975 Embrater (Technical Assistance and Rural Extension Brazilian Corporation) was established with a core business of technology transfer. The new Corporation was government supported and together with Embrapa was to create technology packages.

During the 1980s almost every Brazilian State had an extension and research company. In the dairy industry productivity became an important issue and there was a strong sense that better integration with the farmers was necessary. Torchelli (1984) wrote that to develop successful agricultural research programmes, mainly to the small and medium farmers, it is necessary to develop field biological and social research to identify realistic needs.

In the 1990s, the Federation of Associations and Syndicates of Rural Extension Workers (FASER) proposed a constructivist free extension service, paid by the government, and focusing on the farmer family. Olinger (1998) concluded from the Brazilian National Seminar of Technical Attendance and Rural Extension (1997) that a public rural extension dealing exclusively with the small farmer was necessary. During the 1990s, the available dairy stock of technologies at the Brazilian National Agricultural Research System (SNPA) was enough to make the country self-sufficient and an exporter of milk. Due to several reasons, such as the policies regulating the dairy industry, the stability of the economy and the extension methodologies, this self-sufficiency with sustainability, was not reached. On the other hand, new regional projects like Silvânia, an integrated research-community project in a small town in the west of Brazil, started to show good results. The basis was the inclusion of the community focus in the technological issue (Gastal, 1997).

According to Carvalho (2000), Brazil has sufficient number of scientists to produce good agricultural research, even with low resources. However, the results are not often shared with farmers and the extension agents that assist them. On the other hand, agricultural extension in Brazil is a science that needs to be more studied with a realistic view of a constructivism approach. An external consultant, Tom Cowan, Director of the Australian Tropical Dairy Institute, visited Brazilian dairy industry, in particular the activities of Embrapa Dairy, and concluded that Embrapa Dairy has a very strong programme of research and need to develop an excellent information base. This base can be the beginning of major change in the industry. However, projects should be written to cover both the research and extension components (Cowan, 1995).

Examples of Success and Failure in Technologies Transference in Brazil

The above good advise by Cowan (1995) can be supported by analysing some cases where the inclusion or not of an extension component resulted in the success or success of practical use of technologies/practices.

Production of Improved Beans and Corn Seeds in Paraná—Brazil

From 1986 to 1989, the Brazilian Agronomic Institute of Paraná (IAPAR) developed a research project on the "viability of the production of improved beans and corn seeds at farmers communities level" at Rio Azul—Paraná. Initially, the researchers did not exchange knowledge with the relevant community. They believed that the good quality of technology was relevant to every category of farmer, and, therefore, the extension agents only explained the "technological" package. The first results showed non-adoption of the new seeds. Besides the economic and agricultural findings, the researchers concluded, when dealing with the community, that it was necessary to first discuss and analyse the community technological needs. This highlighted the importance of following sound methodological processes when developing projects of that nature (Neumaier, 1990).

Carrot Named "Brasília"

Embrapa conducted the research with farmers and seven scientific researchers involved in the "Brasília" study, a variety of carrot generated through research led by genetic researchers. The results indicated that the variety adopted by the farmers was: tolerant to diseases, that it increased the productivity of the carrot, and the profits of the producers. The process of transfer of technology of the carrot "Brasília" was good. It succeeded, mainly because the analyses of the needs was made in a participatory process involving researchers, farmers, technicians and taking into account the consumers' demands (Macedo, 1998).

Rotation Pasture in Dairy

In 1991, the Central Cooperative of Rural Producers (CCPR-MG/Itambé) started a project with elephant grass pasture rotation in partnership with 102 farmers. The Department of Animal Husbandry at the Agricultural Superior School of Luiz de Queiroz (ESALQ) gave the orientation. A group of extension agents was trained to give technical support. Later, influenced by the industry, other farmers adopted the same technology. In 1999 an evaluation showed that 44 farmers did not reach their expected level of production, especially from a financial perspective. Alves (1999) wrote that there was an excess of investments in the technology but little attention was given to farmer skills and community knowledge. In addition, he concluded that when the farmers are not well equipped the technology would fail. It is fundamental to know the limitations of all the links that form the community knowledge base. The paper suggested that the experience of CCPR-MG highlighted the importance of taking into account quantitative and qualitative issues rather than solely focussing on the technological side.

Coopavel—Rural Show

Coopavel is an agricultural cooperative in the Paraná State. In 1998 the extension agents of the dairy cooperative got in touch with Embrapa Dairy to start a partnership for the Rural Show event. The cooperative had the intention of organising a dairy farm at the event incorporating top technologies. Embrapa proposed to first took at the issues and needs of the region. A survey and interviews were organised to join a team of researchers and extension agents of the cooperative with the communities. The data collected, together with the opportunity to talk to the community, gave the broad view of the 'reality' and formed the basis for the structure of the dairy farm and the programme of the day at the event. The technologies showed: where individual heifer's shelter; correct pasture rotation; and simple barns (Teixeira, 2000).

As a result of the Rural Show 1999, from 367 farmers that provided milk to the cooperative, 25 started to adopt individual shelters for heifers, 150 adopted correct pasture rotation and many milking plants were requested and used by farmers. The researchers and extension agents' analysis and findings were fundamental in defining the information and technologies to be used in the stations and the field day at the Rural Show 1999. To minimise the risk of losing any progress made, complementary stages of the same technologies were shown at the Rural Shows of 2000 and 2001.

Trends of R, D, E & Farming

Research and Development Trend

Carberry (2001) wrote that agricultural scientists are under pressure in Australia to deliver increased benefits to industry and community clients. The same situation is true for Brazil. The financial resources for R&D are changing in different countries, shifting from a former position of government paying wages and enough resources for research, to an industry demanding and financing part of the research.

In New Zealand, in the late 1990s, the only government funding for the dairy industry was money that went into research. The government contributed $1.00 for every $1.00 that the industry put into research. The Livestock Improvement Corporation Ltd., which is part of the New Zealand Dairy Board, which belongs to Cooperatives is moving out of all aspects of research training and extension and becoming a commercial organisation (Wegener, 2000).

In Australia, the Dairy Research Development Corporation (DRDC) in the 1990s highlighted the need to develop new methods of communicating information to the dairy industry. The Regional Development Programmes or RDPs have been developed in each State of the country to strengthen the link between research and the farming community. The Subtropical Dairy Programme (SDP) in northern Australia was formed in 1995 (SDP, 1999). A structure (Fig. 29.3) with seven Sub-Regional Teams (SRT) informs DRDC of the farmers' needs and the types of research and development that are relevant to the region (SDP, 1998). SDP is directly linked to the source of funds for R&D, that is DRDC. Farmers' members see themselves as investors in R&D, since they pay levies with this objective.

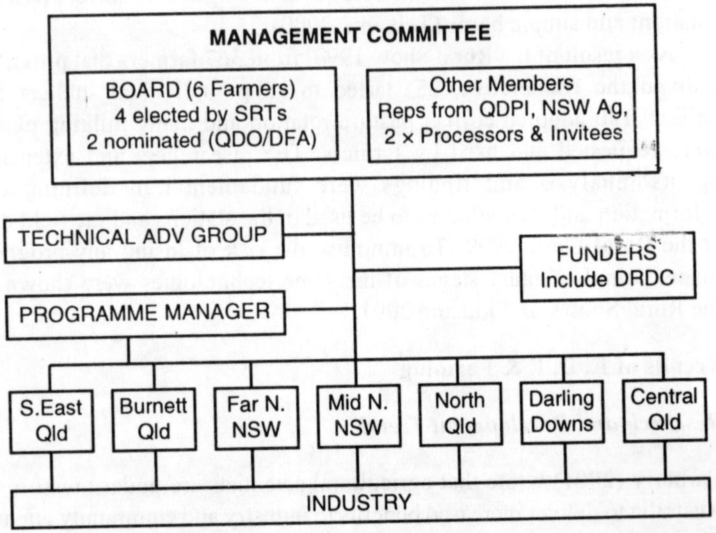

FIG. 29.3: Subtropical Dairy Programme to collect Queensland Technological Needs

In Brazil the decrease of resources for R&D is making Embrapa enhance the communication with different stakeholders of the dairy and related industries to develop cooperative work based on contracts. In most of the contracts the organisation pays for the research and Embrapa pays for the wages of the scientists. However, the large size of the country makes it impossible to be present in regions which are more than 1,500 km away from the research station. In addition, the milk production has moved to the west, north-east and south of the country. The new strategy is to move researchers to each of the three new milk frontiers. Researchers at the research station are aligning the projects and actions of Embrapa Dairy closer to the needs of the communities.

Extension Trend

In many countries of the developing world, the Ministry of Agriculture or State Government established the extension offices, employ the people and the services which become increasingly involved in carrying out all types of governmental activities at the local level (Swanson, 1989). As a result of this, in several cases the effectiveness of the government extension organisations decreased through the interference of politicians in the selection of staff members, target groups and messages to be delivered. An extension agent would have potential problems if he/she refuses to deliver party information on behalf of a local member (Ban, 2000). Chamala (1999) conducted a literature review of the trends of extensions in Australia and prepared Fig. 29.4 showing a historic evolution of extension and the trends for the future.

In Brazil an electronic survey in November 2000 conducted by *Milkpoint*, an electronic journal that has a large participation of dairy stakeholders asked: "What is limiting the enhancement of technical indicators in milk production?" The answer presented in Table 29.2 shows that Agricultural Extension is the second most limiting influential factor.

TABLE 29.2: What is Limiting the Enhancement of Technical Indicators in Milk Production?

Limiting Factors	*Number of Respondents*	*%*
Applied research	3	3.37
Agricultural extension	26	29.21
Rural credit	9	10.11
Milk price	47	52.81
Others/I don't know	4	4.49
Total	89	100.00

Source: *Milkpoint* (2000b).

Farming Trend

In Australia, in 2000, where deregulation of the market has a major impact on any farmer's decisions, the Queensland Dairy Organisation (QDO) maintains its intention of working together with other stakeholders of the dairy industry instead of being left out of decisions made which affect them directly. The present of QDO said: The days of processors and supermarket negotiating away Queensland farmers' livelihoods must stop (Chamberlain, 2001).

In Brazil the Specialised Agricultural Electronic Journal *Agrosoft* (2001), published a research paper from the Brazilian National Agriculture Confederation (C.N.A.). C.N.A. spoke to 39.904 landowners with properties bigger than 50 hectares throughout the whole country, and collected data to show that only 20 per cent are less than 45 years old. It was also found that 62 per cent of the landowners have only completed their first grade of education, 4 per cent are illiterate, and 53

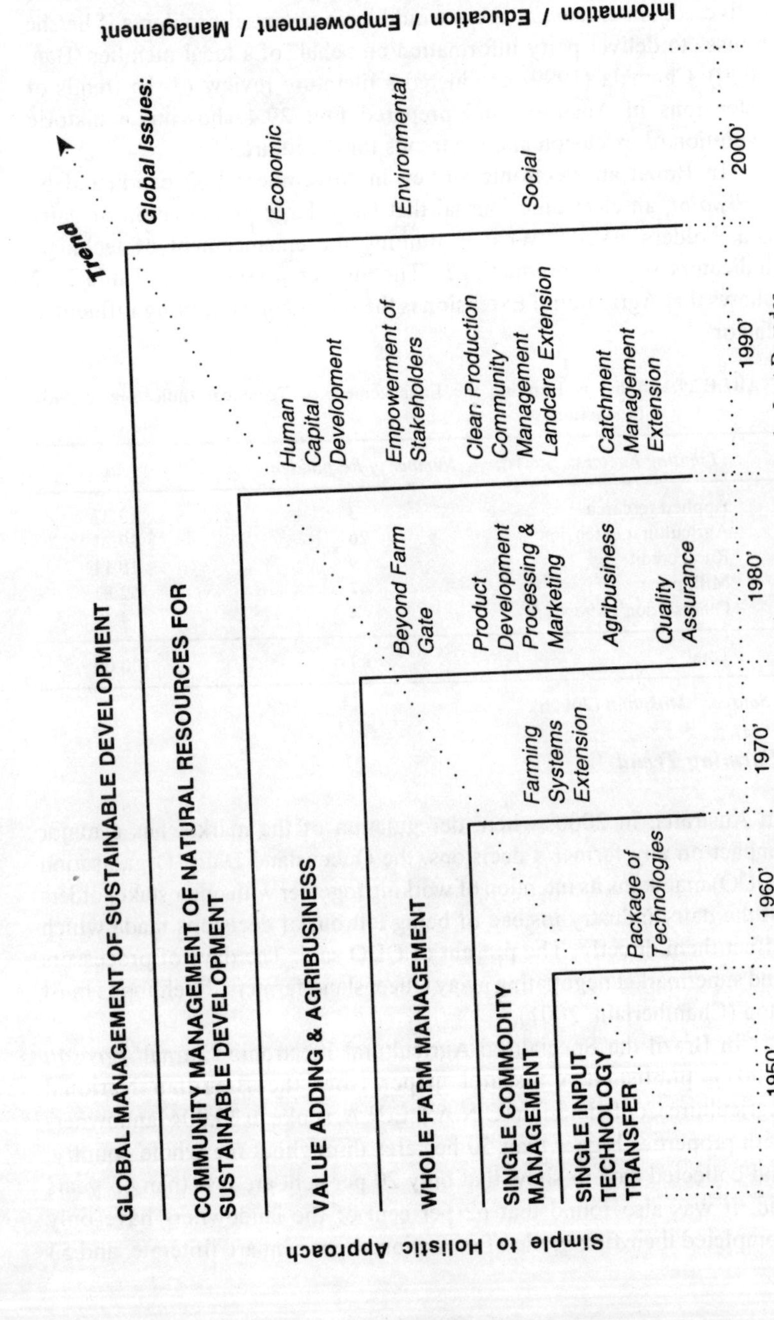

Chamala, 1999.

FIG. 29.4: Changes in Rural Extension Approaches over the Last Decades

per cent have a monthly income of R $500 or US $185.[1] The main difficulty according to 34 per cent of land owners is the rate of the rural credit, 44 per cent do not have electricity, only 6 per cent have access to irrigation systems, 90 per cent do not have a milking machine and only 17 per cent have access to a computer. According to C.N.A. the trend of the landowners is to look for a job out of farming and 64 per cent are doing it. On the other hand, some farmers are reacting to the present situation. In Juiz de Fora, Minas Gerais State, dairy farmers have organised a farmers' co-operative to sell their milk, the result in 1999 showed that they received R $7 cents (US $4 cents[2]) more than the dairy farmers that were not taking part of this cooperative scheme. Also in Brazil, in 2000, at the community of Vale do Rio Doce, the association of farmers is increasing the milk price by R $6 cents (*Milkpoint*, 2000a). In addition, supported by extension services, they are organising themselves to buy goods and machinery at low prices.

Conclusion

Brazilian agricultural R,D&E, following a world trend, are in a transition period from a government sponsored budget to a share stakeholder's budget. From this perspective it is possible to conclude that stakeholders will keep a close observation on how effectively their money is being used to meet their needs. R&D will be forced to include Extension components in their projects.

A greater involvement of clients and stakeholders of the whole dairy industry and better communication with tax payers is beginning to build an effective R&D service in Brazil. This position was reached because it was concluded that when the production system was not analysed in a quantitative and qualitative way, the risk of non-adoption by farmers was high. R,D&E was using good technology with bad results highlighting that the complexity of any production system must be considered in the identification of technological needs.

R&D is becoming stronger in Brazil than the extension system and in many cases are actually absorbing extension organisations. The different focus of R&D (technological) and Extension (free, social and to small farmers) created barriers in relationships. As a consequence R&D are starting their own extension activities.

Many land owners will stop farming in the near future, meanwhile other farmers are assuming stronger positions in the Brazilian and Australian dairy industry looking for their rights and changing their position of a weak link to active participants in the industry and in R,D&E.

NOTES

1. Considering the exchange rate of US $1.0 : R $2.7 in October 2001.
2. Considering the exchange rate of US $1.0 : R $1.75 in 1999.

REFERENCES

Agrosoft (2001). '*Como a Agricultura Chegou a 2001? (How the Brazilian Agriculture has Reached 2001?)*', info@agrosoft.com. Available at.

Ban, A.W. v.d. (2000). "Different Ways of Financing Agricultural Extension", *Agricultural Research & Extension Network*, Vol. 106b, No. Agricultural Extension, pp. 8-19.

Ban, A.W. v.d. and Hawkins, H.S. (1996). *Agricultural Extension*, 2nd ed., edn., Blackwell Science, London, ISBN 063204053X (pbk).

Carberry, P.S. (2001). "Are Science Rigour and Industry Relevance Both Achievable in Participatory Action Research?", in *10th Australian Agronomy Conference*, Hobart, 1/2/2001, p. 11p.

Carvalho, M.P.d. (2000). "Novo Canal Para Divulgacao de Pesquisa em Andamento", *Milkpoint*. Available at.

Cezar, I.M. (1999). *Sistema Participativo de Geracao e Transferencia de Tecnologia para Pecuaristas: Modelo Conceitual—o Caso Aplicado a Embrapa Gado de Corte* (Participatory System for Generation and Transference of Technology for Farmers: Conceptual Model—The Case of Embrapa Beef Cattle), S., S. (Ed.), p. 16p.

Chamala, S. (1999). *Review of Research, Development and Extension Models*, University of Queensland, Brisbane, ISBN 0949414913.

Chamberlain, A. (2001). "Farmers Fight for the Right to Negotiate", in *Queensland Dairyfarmer*, Brisbane, p. 1.

Cowan, T. (1995). *Report on Symposium and Consultancy in Brazil*, Department of Primary Industries, Mutdapilly, p. 25.

Embrapa (2001). *Embrapa Dairy Cattle*, http://www.cnpgl.embrapa.br/. Available at.

Figueiredo, R.P.d. (1981). "A Extensao Rural Face a Problematica da Pequena Producao no Brasil" (The Rural Extension Facing the Problem of the Small Production in Brazil), in *Terceiro Seminario sobre Meio Rural e Educacao (Third Seminar of Rural Regions and Education)*, (Ed., Vargas, F.G.), Fundacao Getulio Vargas, Rio de Janeiro, 9/11/1981.

Gastal, M.L. (1997). *Mudanca Tecnologic, Modernizacao da Agricultura ou Desenvolvimento Rural?* (*Technological Change, Agricultural Modernization or Rural Development?*), 1 edn., Embrapa Cerrados, Planaltina Br.

Hamilton, N.A. (1995). *Learning to Learn with Farmers*, 1 edn., Wageningen. ISBN 90 5485 476 6.

MAARA, M.d.A.A.e.R.A.-ÿ (1972). *O Fato: A Portaria Ministerial n. 143* (The Fact: The Ministry "Determination", n. 143).

Macedo, M.M.C. (1998). "Pesquisadores, Agricultores e Ciência e", *Cadernos de Ciência e Tecnologia*, Vol. 15, No. 1, pp. 29-62.

Milkpoint (2000a). "Associação Muda Perfil de Produtores de Leite (Association Change Dairy Farmers Profile)", *Milkpoint*. Available at.

———— (2000b). "O Que Mais vem Limitando a Melhoria dos índices Técnicos de Produção de Leite No Brasil? What is Limiting Enhance of Technical Indicators at the Milk Production?", www.milkpoint.com.br. Available at.

Neumaier, M.C., Yu C.M., Freire, C.J. (1990). *Avaliacao Socio-economica da Producao Comunitaria de Sementes em Rio Azul-Parana* (*Socio-economical Evaluation at the Community Production of Seeds in Rio Azul-Parana*), 1 edn., IAPAR, Londrina Br.

Olinger, G. (1996). *Ascensão e Decadência da Extensão Rural no Brasil*, Epagri, Florianópolis.

———— (1998). *Extensao Rural: Verdades e Novidades* (*Rural Extension Truth and News*), 1 edn., Epagri, Florianópolis BR. ISBN 8585014 38 5.

Oliveira, A.G.d. (1987). *Origem e Evolução da Extensão Rural no Brasil: Uma Análise Histórico-crítica* (*Origin and Evolution of Rural Extension in Brazil: A Historic-critic Analysis*), Unpublished Ms. Thesis, Extensão, UF Viçosa, Viçosa-MG, p. 138.

SDP (1998). *Sub-Tropical Dairy Programme Ltd.*, DRDC, Brisbane.

SDP (1999). *Sub-Tropical Dairy Programme Business Plan*, DRDC, Brisbane.

Swanson, B.G., J.B. Claar (1989). "The History and Development of Agricultural Extension", in FAO, ed., *Agricultural Extension: A Reference Manual*, FAO, Rome, pp. 1-21.

Teixeira, S.R.N., L.P. (2000). "Partnership COOPAVEL and Embrapa Dairy: A Case of Success in Transference of Technologies", in *4th International Conference on Technology Policy and Innovation*, Vol. 1 (Ed., PUCPR, P.U.C.d.P.), Pontifícia Universidade Católica do Paraná-PUCPR, Curitiba, PR.

Torchelli, J.C. (1984). "Interacao Pesquisador-Produtor: Um Enfoque Inovador na Pesquisa Agropecuaria" (Interaction between Researcher-Farmer: A New Focus at Agricultural Research), *Cadernos de Difusao de Technologia*, Vol. 1, pp. 27-41.

Wegener, P. (2000). *Visit to New Zealand Dairy Industry*, University of Queensland, Brisbane, p. 7.

30

PRODUCT DIVERSIFICATION IN COCONUT
Problems and Prospects

— *S. Kalavathi* and *T. Ranganathan*

Although India has made tremendous achievements in production and productivity of coconut, the development in the post-harvest processing has been rather slow when compared to other coconut growing countries like Philippines, Indonesia, Sri Lanka, Thailand and Malaysia. In India, it is confined to copra making, oil milling and manufacture of coir and coir products. But the household level processing is negligible even in south India, where a number of small-scale copra-making units and coconut oil mills are operating. As such, any advantageous position in the price of coconut oil is not often transferred to the coconut growers and is mostly enjoyed by traders (Aravindakshan, 1996).

In India, the price of coconut is ruled by the price of coconut oil, as our coconut-based economy is solely dependent on coconut oil. Further, price of coconut oil is subjected to violent and unpredictable fluctuations, depending on its demand and supply together with availability of substitute oils. The Commission for Agricultural Costs and Prices (1997) forecasted a downward pressure on the price of coconut and its products, in accordance with an expected growth rate in the production of coconut to the tune of 10 per cent.

Based on this, they have highlighted the need for emphasis on the diversification of coconut use, in forms other than milling copra to be increased at a rate well in excess of 10 per cent per annum. Singh *et al.* (1999) and Thampan (1999) also suggested farm level processing as the best possible measure to ensure better price and stabilisation of coconut economy, by way of value addition at farm gate level. Undertaking processing activities through farmers' co-operatives is another better option to ensure better and stabilised price for coconut and its products (Aravindakshan, 1996).

Even though several technologies are available with the Coconut Development Board for the manufacture of coconut-based products like desiccated coconut, coconut cream, coconut milk, aseptically packed coconut water, coconut vinegar, nata-de-coco, coconut spread, spray-dried coconut milk powder and coconut liquor, their commercialisation has begun only recently. There is also a large potential for utilisation of by-products such as shells, husk, and mid rib. In addition, the timber value of coconut wood can be well exploited. Simultaneously, there is a felt need to evolve appropriate technology for overcoming the problem of disposal of waste generated from consumption of raw coconut.

Coconut, thus assumes importance as a crop, which offers highest potential for diversification of maximum number of products and byproducts. But its potential still remains untapped and this situation needs to be critically analysed. With this background, a minor study was undertaken to assess the level of awareness of farmers on diversification of coconut products, their extent of practice or adoption and the problems as faced by the entrepreneurs.

Methodology

For assessing the level of awareness and extent of adoption of product diversification, a total number of 200 coconut growers were selected from Alappuzha and Palakkad districts of Kerala and Coimbatore and Kanyakumari districts of Tamil Nadu. Fifty farmers each were selected randomly from each of these districts. But as the field level processing was observed to be very low, the situation of product diversification could not be generalised. Hence, ten units engaged in the manufacture of diversified coconut products were analysed so as to understand their problems and prospects for further growth. The customers' preference for diversified products was also analysed by making random observation of their response in a food processing *mela*, where most of the coconut products had been displayed.

Findings and Discussion

Awareness and Adoption of Product Diversification

The data collected at field level was too inadequate to generalise about product diversification. However, the available data is presented in Table 30.1.

TABLE 30.1: Product Diversification at Field Level

(*n* = 50 *each*)

Diversified Products	*Alappuzha*		*Palakkad*		*Coimbatore*		*Kanyakumari*	
	n	%	*n*	%	*n*	%	*n*	%
No products	40	80	47	94	30	60	47	94
Copra/Oil	10	20	3	6	9	18	3	6
Tender nuts	0	0	0	0	11	22	0	0

Majority of the respondent farmers did not practice any product diversification in all the four districts except for few farmers making copra/oil in Alappuzha (20%), followed by Coimbatore (18%), Palakkad and Kanyakumari (6%). Apart from that 22 per cent of the coconut growers practiced tender nut sales in Coimbatore. Thampan (1999), based on a Farmer Participatory Survey conducted in Kerala, reported that at farm household level, not more than 5 per cent of the farmers were engaged in copra making for sale and harvesting of tender coconut for sale was not being reported from any of the study areas. Shortage of palm climbers for regular harvest and marketing problems had been stated as the major constraints.

But the small-scale processing of these traditional products presents a still better picture, when compared to the household level processing. The availability of copra in the country had increased to 8.5 lakh tonnes in 2000 from 5.5 lakh tonnes 10 years ago. In Tamil Nadu and Andhra Pradesh, tremendous growth in copra production was noticed over a decade, while it was almost stagnant around 4.5 lakh tonnes in Kerala. In Tamil Nadu, it had grown up to three lakh tonnes in 2000 form a meagre 30,000 tonnes in 1990. Similarly, in Andhra Pradesh, copra production went up to 50,000 tonnes in 2000 from 6,000 tonnes in 1990 (*The Hindu*, 13th October, 2001). The growth pattern of the coconut small-scale processing sector is further revealed by analysing the consumption pattern of coconut in India (Table 30.2), as presented by Singh and Markose (1999).

Data presented in Table 30.2 reveal that the utilisation of coconut in our country is mainly confined to mature nut, copra and oil. Further, the use of coconut oil, especially in the soap industry has decreased from 23 per cent of the total production during 50's to 5 per cent during 1993-94, due to substitution with cheaper vegetable oils. This situation emphasises the need to promote industrial, edible and toiletry uses of coconut oil so as to stabilise its price.

TABLE 30.2: Consumption Pattern of Coconut

	Kerala (%)	*Other States (%)*	*All India (%)*
Raw nuts	41.02	78.50	61.60
Mature nuts	38.20	57.30	48.70
Tender nuts	2.90	16.80	10.50
Desiccated coconut	Neg.	4.40	2.40
Copra	58.50	21.50	38.40
Milling copra	53.00	12.80	30.90
Edible copra	5.80	8.80	7.50
Oil			
Edible	54.80	—	39.40
Toiletry	16.40	—	46.50
Industrial	5.60	—	14.10
Moved to other States	22.90	—	—

Even though all farmers are well verse with copra/oil making, only 12.5 per cent of them had just awareness about products other than copra/oil. Anithakumari and Kalavathi (2001), in a study conducted in Alappuzha district observed a cent percent non-adoption and a meagre one per cent knowledge on the post-harvest technologies of coconut.

The products manufactured by the units under study included copra, coconut oil, coir and coir products, desiccated coconut, coconut cream, coconut milk, mature nut and tender nut water concentrate, coconut milk powder, vinegar, shell-based and wood-based handicrafts, furniture and coir pith compost.

Observations from the present study revealed that the potential for fast establishment in the field of coconut-based industries was more with conventional products like copra, oil, coir and coir products. This finding is in agreement with that of Poduval and Pillai (1998), Singh (1999) and Thampan (1999). Still there is vast scope for further expansion in these fields, especially in the coir industry. The available data on the present husk utilisation pattern revealed that full potential has not been tapped in coir industry, even if coir and coir products constitute the major items of our export trade. The present husk utilisation pattern is as given below:

Kerala	38%
Tamil Nadu	23%
Karnataka	22%
Andhra Pradesh	16%
Orissa	15%
Others	10%
Mean	21%

The untapped potential of coir industry can be exploited well by the entrepreneurs by ideally investing in geo-textiles and manufacture other coir products.

The studies conducted on product diversification revealed that the potential for fast establishment in the field of coconut-based industries was more with conventional products like copra, oil, coir and coir products. Manufacture of desiccated coconut and coconut shell-based handicrafts are the other industries among the non-conventional products, which could establish to some extent. The other products such as coconut cream, coconut milk, mature nut and tender nut, water concentrate, coconut milk powder, vinegar, shell charcoal, activated carbon and coir pith compost are yet to get a firm footing in the processing and by-product utilisation sector. The constraints related to the slow rate of growth in the processing of coconut have been analysed.

Constraints in Product Diversification

Lack of modern technologies within the country for large-scale manufacture of value added products is one of the major handicaps for the retarded growth of coconut industry in India. A number of other constraints are found to hinder the growth of the already established units, some of which are discussed below.

- Inadequate institutional support in strengthening the existing units.
- Lack of awareness about the product and related market problems.
- Difficult marketing in retail domestic market, especially that of the traditional coconut growing areas.
- Inadequate research and development efforts to replace or refine the inefficient technologies.
- Improper selection of site for the industry in terms of infra-structural facilities and easy availability of inputs.
- Capital deficit or inadequate credit for strengthening the existing units.
- Inadequacy or absence of quality control facilities.
- Minor shortage of labour and electricity.
- Procedural difficulties in inter-State trade and export.
- Procedural difficulties in the licensing of food products of small-scale units.

- High transportation charges in the marketing of bulk items like coir pith compost.
- Low and unsteady profit margins due to violent market fluctuations.
- Low profit margins due to credit sales and delayed payments from retail domestic markets.

Most of the coconut industries are facing one or more of the above problems or even other unidentified problems based on the nature of the product dealt with and also based on the capacity of the unit. From small-scale to large-scale units, the major problem was with marketing of products, especially non-conventional products. Hence, majority of the firms call for institutional support in terms of product popularisation rather than providing financial assistance. But some of the successful entrepreneurs in this area deserve appreciation for their own effort to solve most of the constraints.

The success mainly depended on the professional ability and coordinated efforts of the management committee of those enterprises. The special features of such units were that they always strictly adhered to quality standards in addition to their effort in retaining natural flavour. Many of them developed own techniques or modified the existing techniques so as to save labour, time and resources. Methods for effective utilisation of waste/by-products were developed as their own and adopted by these units. Wherever this was not possible, proper disposal of waste was being done without causing environmental pollution. The major strategies for development of coconut industries should include the following:

- Strengthen research and development efforts to substitute the inefficient technologies and to evolve improved modern technologies.
- Provide market support for popularising non-conventional products.
- Select the location of the industry in a proper place, considering the infrastructural facilities and cheap and easy availability of inputs.
- Arrange centralised or cooperative marketing for easy movement of the products and for assuring stabilised price.
- Provide easy and timely credit facilities.
- Promote the export of potential coconut products.

- Broaden the demand base in the non-traditional area by attracting the consumers with new variety products.
- Prescribe strict control measures for maintaining quality of products.
- Avoid procedural difficulties in licensing and inter-State trade.

Customers' Response towards Diversified Products of Coconut

Customers' response was collected from a food processing *mela* (World Food Tech-1999), held at Kozhikode during October 1999 with the joint effort of various organisations for the popularisation and sales of novelty food products of various small and large-scale industries. Different food products made of coconut by a number of firms attracted most of the customers, as these seemed to be new. The response of the customers regarding various coconut products, both conventional and non-conventional was collected by direct observation and interaction with them. The general observations are given below:

- More unemployed youths were coming forward to enquire about coconut technologies when compared to other products.
- Regarding coconut-based small-scale handicrafts, sales of costly wood items seemed to be less, when compared to shell items.
- Enquiries on furniture items made of coconut were found to be more.
- The demand for non-conventional items like coconut cream, milk powder and tender nut water concentrate for household uses was low, but not discouraging. Among these, milk powder was having higher demand.
- Most of the enquiries about food items were on their shelf, life and natural flavour.
- When all the coconut products taken together, the most prospering enterprise was found to be branded coconut oil industry.

31

THE USE OF www-BASED ELECTRONIC SURVEYS IN RURAL EXTENSION

— *Airton Spies, Mal Wegener, S. Chamala* and *Bob Beeton*

Electronic surveys (e-surveys) are becoming more popular as the Internet continues its exponential growth and is widely spread among all classes of society (Teo *et al.*, 1997). The days of using paper as the sole means of surveying people's opinion are gone. An electronic mail survey, which does not require stamps, envelopes or paper, instead using e-mail and/or a Website where the respondents can access the questionnaire, has the potential to yield high responses rates quickly at low cost. For example, in a survey among university students, Dillman (1999) identified the same response rate in an electronic survey as those obtained by a four-contact paper mail strategy. Many computer applications designed to assist in the construction, application, and processing of questionnaires have been developed and are now available in the market. Recent publications reviewed these software packages for e-surveys and concluded that they are becoming more popular and more user-friendly (Struhl and Kuever, 1998; PCToday, 1997; Schindler, 1998; Marketing Masters, 1999). It seems that the traditional Total Design Method (TDM) for surveys, based on mechanically applied, paper-based questionnaires, is not the only tool for assuring high quality responses in the 21st century (Dillman, 1999). However, regardless of which way the survey is submitted to the respondents, it needs to be tailored to the needs of each case. Well-designed surveys can create respondent trust and motivation to respond.

E-surveys can be used in a range of environments such as www via internet browsers, mailed diskettes, local networks of personal computers, using portable laptop computers, touch screen personal computer systems, and stand-alone desktop computers (Marketing Masters, 1999). It is important to note, however, that the people surveyed by this means may not be representative of the population as a whole, because of limited access to the Web by some parts of the population.

The basic question in designing a self-administered survey is "who is being surveyed and what is the topic?" The resources available (time and financial resources) also play a key role, as does the location of the respondents. We used an e-survey in phase 1 of a two-stage study to investigate issues related to sustainability of the pig and poultry industries in Santa Catarina State (SC), South of Brazil because it was a cost-effective, and more time-efficient way to canvass opinion than sending 350 paper questionnaires and manually collating the results. The main objective was to identify the pig and poultry industries stakeholders' views and perceptions on principles of sustainability, and to select a set of core indicators of sustainable production. Therefore, the respondents ("who") for this survey were stakeholders from the pig and poultry industries in SC, who have access to computers and the Internet. The "topic" was sustainability indicators for the pig and poultry industries, and the stakeholders' views and perception about the principles and issues of concern that are central to the sustainability of these industries. It also focused on how the sustainability indicators should be measured and how they could be used to assist the industries achieve a more sustainable future.

Principles applied for the Construction of the web Survey

Most of principles applying to paper-based questionnaire design also apply to electronic surveys. There is a need for good presentation and organisation to encourage respondents and navigation needs to be as easy as possible. The principles for constructing web questionnaire suggested by Dillman (1999) were applied in this case. These principles were:

- Introduce the web questionnaire with a welcome screen, that is motivational, emphasises the ease of responding, and instructs respondents about how to proceed to the next page.
- Provide a PIN number for limiting access to people in the sample.
- Choose an item for the first question that is likely to be interesting to most respondents, easily answered, and fully visible in the welcome screen.

- Present each question in a conventional format similar to that normally used in paper, self-administered questionnaires.
- Restrain the use of colour so that figure/background consistency and readability are maintained, navigational flow is unimpeded, and measurement properties of questions are maintained.
- Avoid differences in the visual appearance of questions on screen that may result from different screen configurations, operating systems, browsers, partial screen displays, and wrap-around text.
- Provide specific instructions on how to take each necessary computer action for responding to the questionnaire, and give other necessary instructions at the point where they are needed.
- Use drop-down boxes sparingly, consider the mode implications and identify each with a "click here" instruction.
- Do not require respondents to provide an answer to each question before being allowed to answer any subsequent ones.
- Provide skip directions in a way that encourages marking of answers and being able to click to the next applicable question.
- Construct web questionnaires so respondents scroll from question to question unless order effects are a major concern, or when telephone and web survey results are being combined.
- Exercise restraint in the use of question structures that have known measurement problems with paper questionnaires, such as check-all-that-apply and open-ended questions.

Case Study of the Use of e-survey—Pig and Poultry Industries in SC

The pig and poultry industries in SC are based on a strongly structured, vertically integrated system. The development of these industries began in the 1960s, but moved towards the vertically integrated systems they currently have about 1972 (Altman, 1997). In this system, producers contract their production to a meat firm, which in turn, provides most of the inputs, and determines the production programme. Farmers are suppliers of labour and capital. The industries have grown rapidly and are now among the most competitive in the world (Spies *et al.*, 2001). Despite the technical and economic progress, the industries long-term sustainability may be questioned, as the negative externalities flowing from the current production system are becoming more apparent. Pressure from the market and from the communities affected by unsound practices is building up and, therefore, the pig and poultry industries are seeking sustainable ways of producing their product.

The need for indicators of sustainable production was identified and an e-survey to identify stakeholders' views was designed. The e-survey was chosen because it fulfils the requirements to get high quality and high rate responses at the lowest cost, in the shortest time. One of the drivers for adopting this method was the fact that respondents were in SC, Brazil, while the researchers were in Australia. Therefore distance was a key factor in selection of the e-survey as the method of collecting data. Fig. 31.1 outlines the phases and activities involved in the design and implementation of the survey.

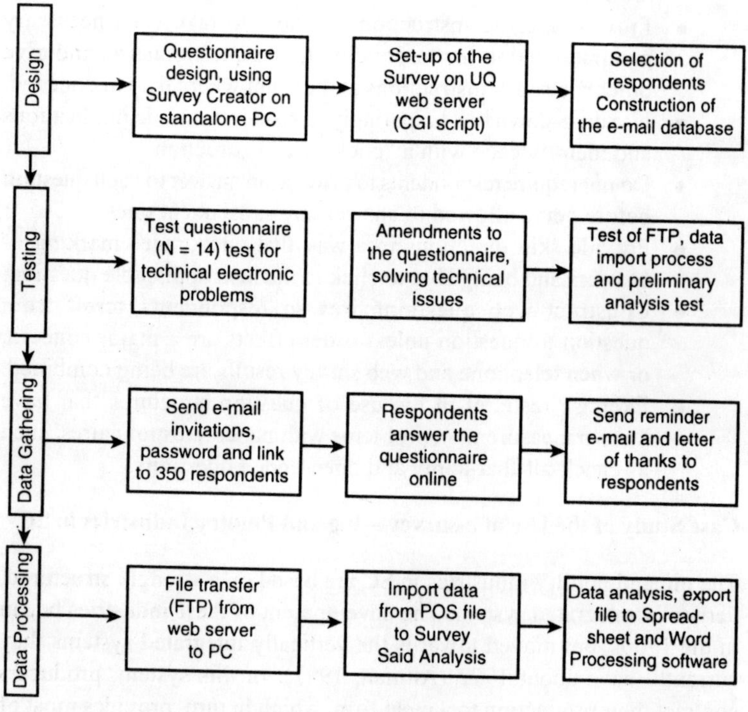

FIG. 31.1: Framework of the Steps to Design and Implementation of the e-survey

Subjects for Study

Primary data were to be collected from key stakeholders:

- Research scientists from Brazilian National Agricultural Research Corporation (EMBRAPA), Santa Catarina State Agricultural Research and Rural Extension Corporation (EPAGRI) and Universities.

- Rural extension agents from Epagri and local councils.
- Private consultants.
- Meat company extension and veterinary service staff.
- Meat company managers (Sadia, Perdigão, Ceval, Aurora and Chapecó).
- Cooperative directors.
- Policy-makers (Councils, mayors, government organisation managers).
- Non-governmental organisations.
- Pig and poultry producers.
- Suppliers to the pig and poultry production chain.

A sample of these groups was selected, aiming for 350 respondents in the phase 1 survey and 252 usable responses were received. An additional 62 face-to-face interviews with pig and poultry producers were subsequently conducted, as the number of these stakeholders reached by the Internet was low due to their lack of access to the web. To complete the study and to define the set of key sustainability indicators, four focus group meetings were organised to fine-tune the results of Stage 1. In addition, secondary data on pig and poultry farm technical and economic performance were collected from EPAGRI's farm accounting network database in Stage 2. Other sources of secondary data for the calculation of environmental impacts using Life Cycle Analysis (LCA) and externality costs were gathered from Instituto Cepa's (ICEPA) database as well as from EMBRAPA's National Centre for Pig and Poultry Research in Concórdia (CNPSA).

Design and Implementation of the web Survey

The commercially available *Survey Said*[1] software (Marketing Masters, 1999) was used to create, collect, transmit and process the data. The selection of stakeholders was based on their interest in the pig and poultry industry and the availability of an e-mail address. The home pages of private and public institutions from SC were consulted to identify e-mail addresses. Some organisations and people were contacted directly by telephone to obtain their e-mail and permission to include them in the list of participants. Once the survey was set up on the web-server at the University of Queensland, invitations were sent by e-mail messages to all participants using a mail merge facility. The e-mail message was sent with an invitation to access and respond to the questionnaire on the home page. A hyperlink was provided and full instructions on how to proceed were included in the e-mail message. An access code (password) was provided to ensure that only authorised respondents would have access to the survey questionnaire. No financial incentives or rewards were provided for respondents. A reminder message was sent to selected participants who did not reply after two weeks and a letter of thanks to those who did. The survey was open to respondents for 30 days and yielded 252 responses, providing a response rate of 72 per cent. The final result was 228 usable questionnaires, as some respondents may have experienced Internet connection problems while they were answering and an incomplete questionnaire was sent. Immediately after completing the questionnaire and clicking on the "send" button, respondents received a confirmation of successfully sending it and a message of thanks from the research team. This is another positive feature of e-surveys, because the survey software can generate such a reply automatically. Finally a conventional e-mail message of thanks and a brief summary of the results, such as the overall response rate and a summary of respondents' demographic profile, was sent to all participants who replied and answered the questionnaire.

Data Processing and Analysis

Soon after the invitation messages were sent, responses began to arrive back in the web server's database. These data were transferred on a daily basis by file transfer protocol (FTP) to the computer that hosted the Survey Said software. Immediate data analysis was possible, as well as monitoring the number of responses.

Once imported, data from the electronic survey were analysed using techniques such as frequency analysis, ranking, cross-tabulation as well as verbatim and multi-field analysis. For some questions, the Likert scale was used, to provide a composite measurement of standardised response categories. Categories such as *strongly agree, agree, neutral, disagree, and strongly disagree* as suggested by Babbie, (1998, p. 4) were used. Data was also exported to a spreadsheet compatible file, which allowed additional statistical analysis, using Pivot tables.

Illustrative Results from the e-survey

To illustrate how the e-survey was used in the study of sustainability indicators for the pig and poultry industries in SC, some of the results are presented next. These are results obtained directly from the e-survey software. Additional analyses were performed with other software and methods to explore more carefully the answers provided by the respondents.

From an operational perspective, the results obtained in this e-survey were surprising good. Some initial testing was carried out to ensure that no technical problems would upset respondents' attempts to answer the questionnaire and that the data were properly sent to the web server and, from there, imported with integrity to the personal computer (PC). Soon after the e-mail inviting the respondents to access the questionnaire was sent, the completed questionnaires began to arrive. The first day yielded approximately 25 questionnaires. The demographic profile of respondents is presented in Table 31.1. It is important to notice that 62 farmers were added by direct face-to-face interviews, using the same questionnaire as used in the e-survey, because of their limited access to the Internet.

TABLE 31.1: Demographic Profile (n = 288)

Characteristic	*Number*	%
(1)	*(2)*	*(3)*
Gender		
Male	250	86.81
Female	28	9.72
Missing cases	10	3.47
Age (years)		
< 25	7	2.43
25—39	111	38.54
40—49	110	38.19
50—64	52	18.06
> 65	2	0.69
Missing cases	6	2.08
Institutional link		
Private sector—agribusiness	114	39.58
Public sector—Federal, State or local level	162	56.25
NGO	6	2.08
Missing cases	6	2.08
Occupation (multiple choice accepted)		
Farmer	71	24.65
Farmers' union or cooperative leader	11	3.82
Research scientist	82	28.47
Rural extension agent or private consultant	116	40.27
Veterinarian services supplier	5	1.74
Teacher or university lecturer	34	11.81
Environmentalist	12	4.17
Student	13	4.51
Federal, State, local authority, community leader	10	3.47

(Contd.)

TABLE 31.1 (*Contd.*)

(1)	*(2)*	*(3)*
Business owner, manager or farm supplier	35	12.15
Other and missing cases	30	10.42
Highest Educational Level		
Primary	47	16.32
Secondary	21	7.29
Polytechnic/diploma and others	9	3.13
Tertiary—bachelor degree	46	15.97
Post-graduation diploma—specialisation	52	18.06
Master degree	88	30.56
PhD	19	6.60
Missing cases	6	2.08

Data from the demographic profile show a majority of males and highly educated respondents. Respondents holding a Master or PhD degree counted for more than 48 per cent of the total. To some extent this is an indication that these professionals tend to have more access to the Internet facilities and are also keen to respond to a questionnaire. Stakeholders' views and perceptions about the five principles of sustainability defined by CSIRO (1998) were tested. These five guiding principles adapted for the pig and poultry industries were:

1. Physical productivity of pig and poultry systems is sustained or enhanced over the long-term (Fig. 31.2).

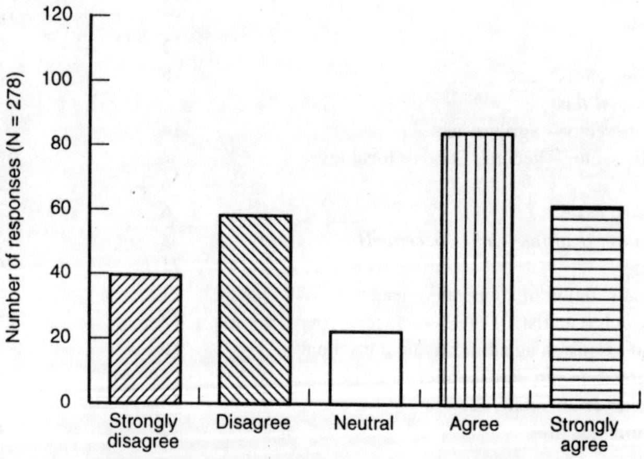

FIG. 31.2: Stakeholders' Perception of the Principles of Sustainability for the Pig and Poultry Industries in Santa Catarina—Principle 1: *Need to Increase Physical Productivity*

2. Adverse impacts on natural resource base by the pig and poultry production and its associated eco-systems including feed production are ameliorated, minimised or avoided (Fig. 31.3).

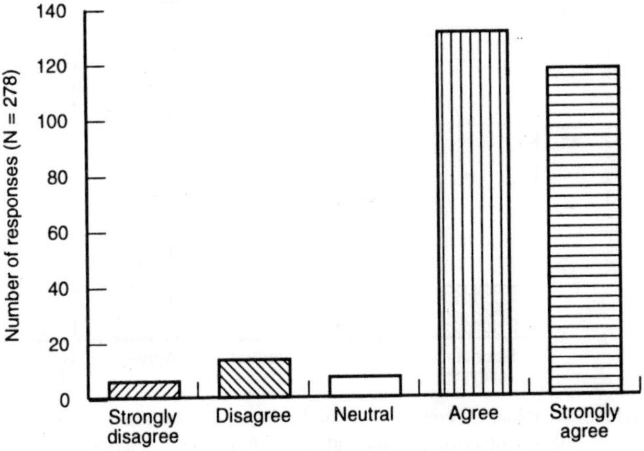

FIG. 31.3: Stakeholders' Perception of the Principles of Sustainability for the Pig and Poultry Industries in Santa Catarina—Principle 2: *Current Production Systems Impact Adversely on the Natural Resource Base*

3. Residues resulting from the use of chemicals in pig and poultry production and its associated activities are minimised (Fig. 31.4).

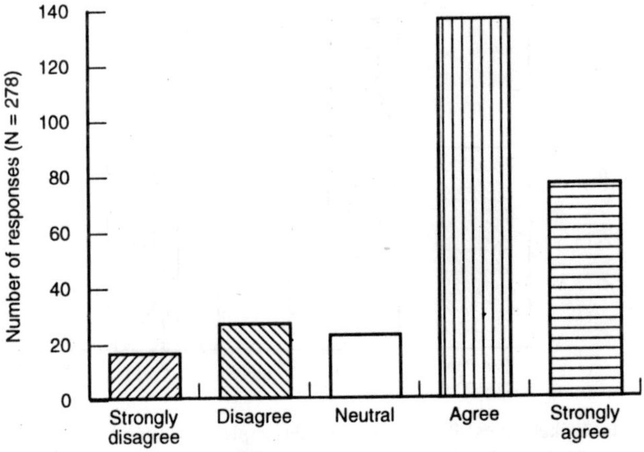

FIG. 31.4: Stakeholders' Perception of the Principles of Sustainability for the Pig and Poultry Industries in Santa Catarina—Principle 3: *Residues resulting from the Use of Chemicals in the Pig and Poultry Industries are a Major Environmental Problem affecting Sustainability of these Industries*

4. The net social benefit derived from the pig and poultry industries is maximised (Fig. 31.5).

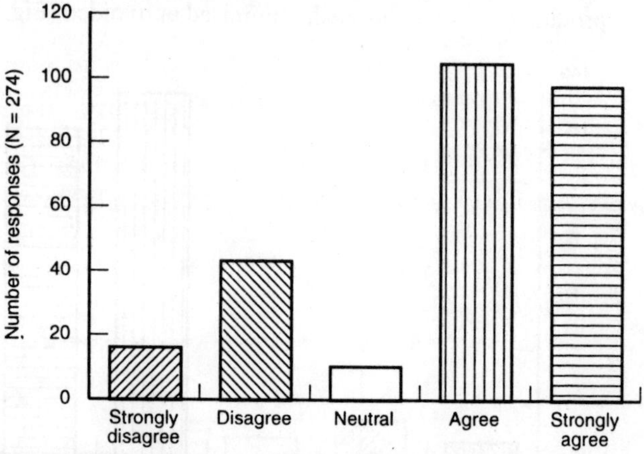

FIG. 31.5: Stakeholders' Perception of the Principles of Sustainability for the Pig and Poultry Industries in Santa Catarina—Principle 4: *Need to have a Net Social Contribution*

5. Pig and poultry production systems are sufficiently flexible to manage risks associated with the vagaries of climate and markets (Fig. 31.6).

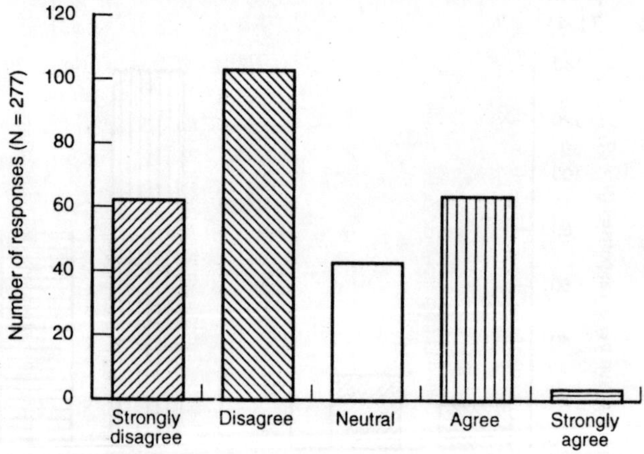

FIG. 31.6: Stakeholders' Perception of the Principles of Sustainability for the Pig and Poultry Industries in Santa Catarina—Principle 5: *Current Production Systems are Sufficiently Flexible to Cope with Risks associated with the Vagaries of Climate and Markets*

The results for the principle 1 yielded a mixed response. To some extent, it reveals that stakeholders do not agree with the principle that there is a need for continuous pursuit of increased physical productivity. However, principles 2, 3 and 4 yielded more responses towards "agree" and "strongly agree", which reveals stakeholders perceptions that these industries need to adopt environmentally sound practices and to benefit the community as a whole. The results for the last principle reveal that the current pig and poultry production systems are not sufficiently flexible to cope with risk associated with the market and climate. We also asked stakeholders' opinion about the concentration of pig and poultry production in fewer, but large-scale operations and these results are shown in Fig. 31.7.

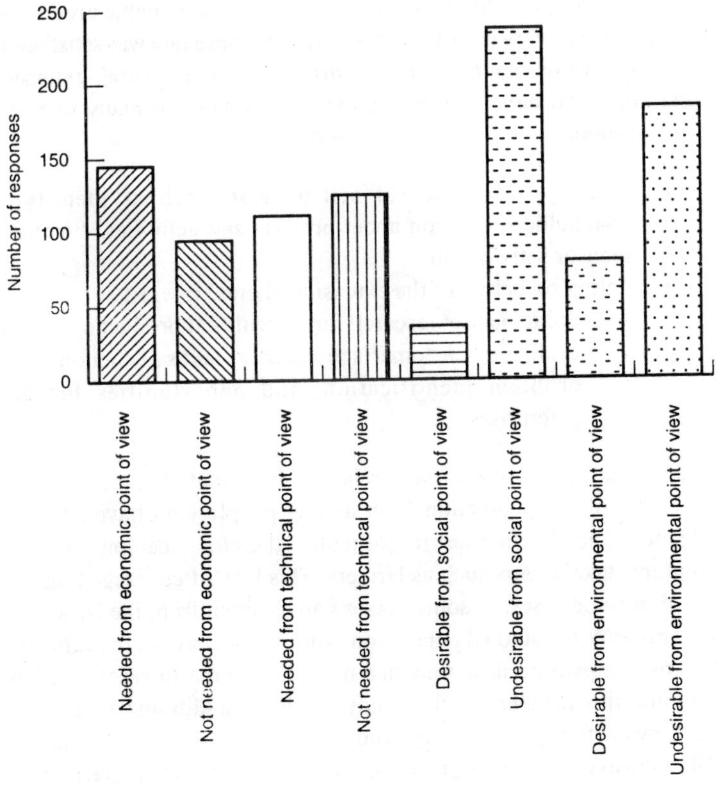

FIG. 31.7: Stakeholders' Views about the *Process of Concentration* of Pig and Poultry Production in a Smaller Number of Producers and Large Scale Operations

Discussion and Conclusion

Results of this study suggest that polling through e-mail or from a website can replace traditional paper-based methods of gathering information, as these methods can give quicker access to stakeholder opinions. The growing acceptance of computers, the Internet, and electronic mail opens the possibility for researchers to take advantage of alternative technologies to conduct surveys. For researchers, the most noticeable benefits are cost-effectiveness, broader distribution, timeliness, potentially higher response rate, and faster data processing and analysis. There are benefits for respondents too, which include more attractive question formats, user-friendly questionnaire design, and hassle-free answering and sending procedures (Klassen and Jacobs, 2001).

In this study, we identified the key issues for sustainable production for the pig and poultry industries in SC. The response rate was satisfactory and results were obtained in a short time frame. Rural extension programmes can benefit from the use of e-surveys in many different ways and some example are presented below:

- Investigating issues of self-interest, such as identifying stakeholders' opinion about projects and activities, client and sponsor satisfaction.
- Acting on behalf of the extension clients' interest
  - — doing market research for agricultural production
  - — evaluating customer's preferences and satisfaction
  - — problem identification, and opportunities for new enterprises.

The weaknesses of e-surveys relate to the potential bias in selecting the sample of the population from among people who have access to computers and the Internet in particular, therefore leaving out some important stakeholders such as farmers. This lack of coverage might be critical in some cases, as some persons are omitted from the target pool that represents the studied population. They would have zero probability of being sampled because they had no access to the Internet. We have overcome this problem in this study by doing additional face-to-face interviews with pig and poultry producers. This provided their inputs as well. However, in some situations, it might have an impact on the results, as their answers were collected in a different way than those who responded to the self-administered online questionnaire. The disk-by-mail or e-mail file system can be used when respondents have access to computers, but not to the Internet. Despite these limitations, there is great potential for using e-surveys in rural extension activities.

NOTE

1. See http://www.surveysaid.com

REFERENCES

Altman, R. (1997). *A Agricultura Familiar e os Contratos—Reflexões sobre os Contratos de Integração a concentração da Produção e a seleção de Productores*, Gráfica e Editora Pallotti, Florianópolis.

Babbie, E. (1998). *The Practice of Social Research*, 8th edn., Wadsworth Publishing Company, Belmont.

CSIRO—Commonwealth Scientific and Industrial Research Organisation 1998, *Sustainable Agriculture: Assessing Australia's Recent Performance*, CSIRO, Collingwood.

Dillman, D.A. (1999). *Mail and Internet Surveys: The Tailored Design Method*, Second edn., John Willey & Sons., Inc., New York.

Klassen, R.A. and Jacobs, J. (2001). "Experimental Comparison of web, Electronic and Mail Survey Technologies in Operations Management", *Journal of Operations Management*, Vol. 19, pp. 713-28.

Marketing Masters (1999). "Survey Said Reference", Marketing Masters, DePere, WI.

PCToday (1997). *Surveying via the Net*, PCToday Supersite. Available at: http://www.pctoday.com/editorial/goingonline/970403a.html. Access: 09.10.01.

Schindler, E. (1998). *Online Survey Software—Ask Anything*, Zdnet. Available at: http:/www.zdnet.com/sp/issues/980720/337576.htm. Access: 09-10-2001.

Spies, A.; Spies, C.I.S.; Chamala, S.; Wegener, M. and Beeton, R. (2001). "The Need for Environmental Extension Education in Facilitating Sustainability of Pig and Poultry Industries in Santa Catarina State, Brazil", in *APEN 2001 International Conference*, APEN, Toowoomba, Australia, pp. 175-82.

Struhl, S. and Kuever, C. (1998). *High-tech Surveys have Arrived—A Review of Software for Interviewing on the worldwide web and with a PC*, Quirsk's. Available at: http://www.quirks.com/articles/article.asp?arg_AticleID=3. Access: 09-10-2001.

Teo, S.H.; Lim, V.K.G. and Lai, R.Y. (1997). "Users and Uses of the Internet: The Case of Singapore", *International Journal of Information Management*, Vol. 17, No. 5, pp. 325-36.

32

ROLE OF INFORMATION TECHNOLOGY FOR PROMOTING AGRIBUSINESS

— *P.O. Ingle, A.D. Makesar* and *S.S. Marawar*

For the socio-economic development of the farmers, it is necessary to equip them with the latest agricultural and allied knowledge. Lots of technologies and related information is generated by the research institute. However, it is observed that most of the rural population still have no accessibility to the crucial information in order to cultivate the crops, and subsequently marketing of the agricultural produce. This has led to non-remunerative agricultural produce. The producer's shares in foodgrain is only 65 per cent and vegetables it is even less than 50 per cent. It is mainly because of non-accessibility to information on market trends. Profitability of marketing of agricultural produce is immensely important for the remunerative agriculture.

In context to the Vidarbha agriculture which is characterised by 90 per cent rainfed farming and mostly cotton, sorghum and pulses based cropping pattern, it is necessary to provide agribusiness information which helps the farmers for effective marketing of their produce.

The information is also be essential for guiding and decision-making about diversification of the rainfed farming.

Marketing Infrastructure in Vidarbha

The situation about the marketing infrastructure in Vidarbha has tremendous scope for establishment for regulated market in future. At present regulated markets in the forms of Agriculture Produce Marketing Committee (APMC) exist for only foodgrains and functioning at all districts and Taluka places. The agricultural and allied commodities such as fruits, vegetables, flowers, livestock do not find any systematised and regulated system of marketing throughout the Vidarbha region. Of the eleven districts in Vidarbha only the regional market at Nagpur and Amravati are the only one competitive in nature and, therefore, vast majority of the farmers have to depend on uncertainty of prices, pulse and pressure of the Taluka market. Most of the markets are dominated by the middle men with practical no voice to the producer in choosing the place land the price for selling his commodity.

The market information about arrival and prices market intelligence for prediction of prices and demand and supply are not available to the farmers and hence the role of information technology will have immense importance in creating effective linkages between the producer, consumer and the market.

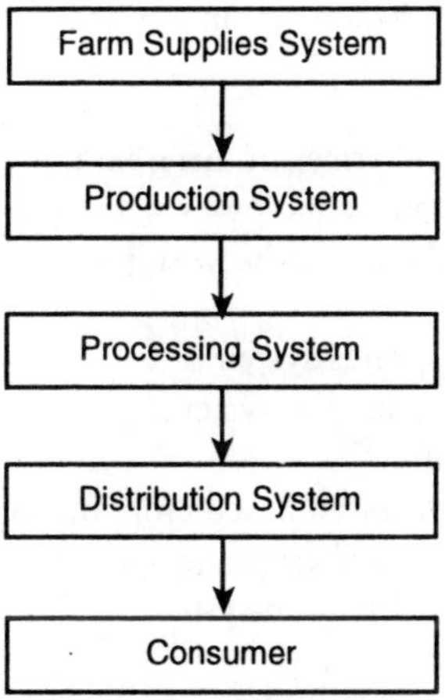

FIG. 32.1: Agribusiness—Nature and Importance

Agribusiness System in Vidarbha

Agribusiness is defined as the sum total of all operations involved in manufacture and distribution of farm supplies production activities on the farm and storage, processing and distribution of farm commodities and items made from them (Davis and Ray, 1957).

In view of this integration of the information on all the four above mentioned components of agribusiness needs to be attended. At present these four systems of agribusiness do not have any fruitful linkages and in fact the efforts have not been made purposefully for generation, processing and distribution of market information in the past and it is quite obvious constraints was non-availability of the potential and apt information technology system in the region but now lot of hopes can be gathered to exploit this system in promoting the agribusiness for the Vidarbha region in particular.

Present Information Status

The present information status is observed to be very poor with regard to printed literature, information on Radio and T.V. and other modern methods like Internet. There is lot of duplication, at the same time lack of communication amongst State Agricultural Universities, Development Departments, Co-operative Sectors, NGOs and private agencies.

As per the agribusiness and rainfed farming is concerned the information on the following aspects needs to be provided. Cropping strategies for farmers based on integrated information on various production factors.

1. Availability and prices of different inputs.
2. Crop cultivation technologies with agro-ecology specifically.
3. Information about post-harvest, processing, value addition and storage.
4. Agriculture supplies system.
5. Agriculture production system.
6. Process system
   (i) Based on cotton based cropping system.
   (ii) Oranges based cropping system.
7. Agricultural distribution system.

Present Information System is characterised by the following

Practically there is no use of Internet in the rural areas of Vidarbha for the information on agribusiness Cyber cafe are restricted to major cities and towns of the region. However, it is not very difficult to establish cyber cafes in every village in the Gram Panchayat Office via telephone link through connecting of VSNL server.

Presently the use of other important systems is very poor.

(i) *Extension contact intensity is very low*—The average contact intensity is 0.38 hours/target population and average technical manpower cultivator ratio is 1 : 1623. The extension expenditure intensity is only Rs. 2.47 per hect. (Salaman & Sadamate, 2000).

(ii) *Inadequacy of printed information*—It is observed that handful of exclusive agricultural publications are produced in Vidarbha by SAU, Development Department and supply agencies. However, there is no linkage between different sources information about prices and arrival of various agricultural commodities are published in the newspaper, regional newspapers. However the contact intensity of such publications is very low, *i.e.*, 1 : 500 copies/person.

(iii) *Air and Doordarshan*—The mass media like AIR and Doordarshan are very effective in dissemination of information, however their contact intensity needs to be improved. At present only 2 to 4 per cent time is devoted for agricultural information daily as against in China separate radio and T.V. channel broadcast agricultural information for farmers 24 hours.

(iv) *Multimedia*—Computers all over the world have taken very big strides over past few years and are providing to be the most effective, fast and cost effective media for the dissemination of information.

Internet in a period of 5 years, has reached 30 million users world-wide (*Manorama*, 2000). Information is becoming a fourth important in addition to capital, labour, and resources. Hence a craze for computer aided and related electronic gadgets is seen developing fast. In India this has not reached in rural areas so far but it may be hoped in next millennium for that the tremendous rate of growth of computer aided media plans of the government to promote it.

Production of CDs on different crops and related subjects have not taken place to a large extent in India. Launching of full fledged and effective of agricultural and allied websites by various agency and institutions for farmers is yet to be seen on large scale.

Future Strategies for Use of Information Technology in Vidarbha

There are 11 districts in Vidarbha comprising 110 Talukas. Each Taluka approximately 12,996 villages. Understanding the present scenario of information system, it is suggested that an effective competitive information network needs to be developed as presented in Fig. 32.2.

The website on market information should be launched for State, regional, district and taluka market information. Internet facilities at Gram Panchayat level through telephone connectivity needs to be provided. As suggested the agribusiness information based on four

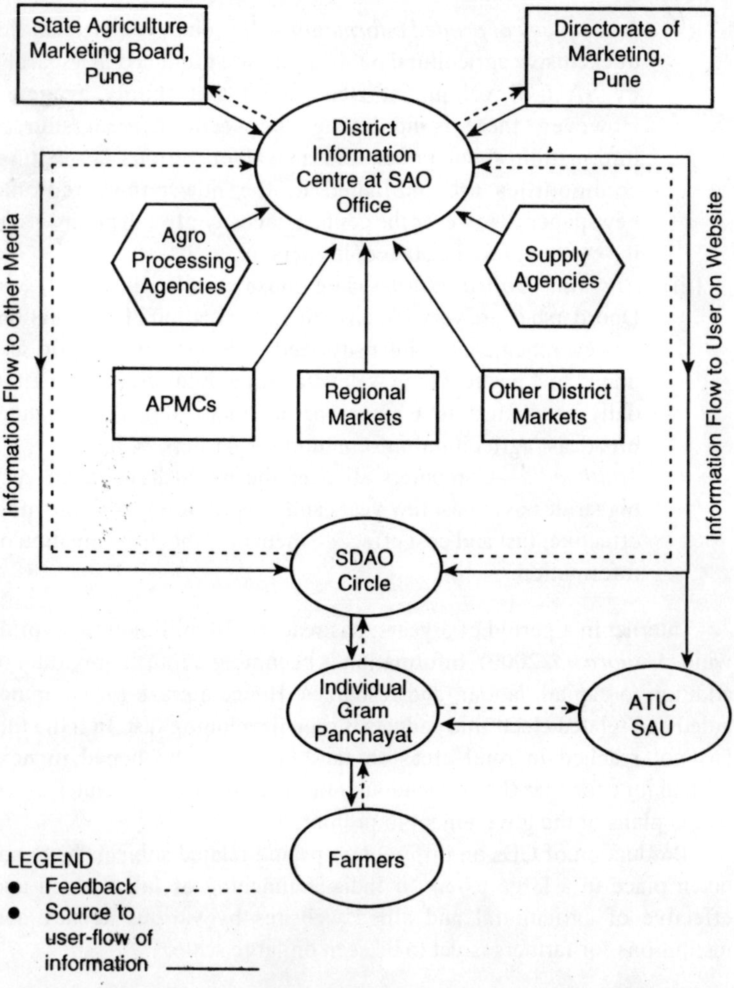

FIG. 32.2: Diagrammatic Presentation of the proposed Information System for Promoting Agribusiness in Vidarbha

components will have to be monitored at district level and then provided through website to the farmers. The monitoring will have to be made at each taluka level for providing feedback to the district centre. It is expected that the agribusiness information in every village on website and it can be retrieved at Gram Panchayat level by the Gram Panchayat Secretary daily and supplied to the potential users. Efforts can also be made for promotion of audio and video programmes, CDs and cassettes.

The information can also he retrieved at ATIC Centre of SAU and can be provided to the farmers.

Information Printed Media

Use and production of printed media can also be enhanced and made effectively developing strong linkages amongst the four components of agribusiness system already discussed. In conclusion it may be said that there is tremendous scope for use of information technology in bringing about the change in agribusiness scenario of Vidarbha region of Maharashtra State but definitely there is a long way to go.

REFERENCES

Chaudhari, M.M. (2000). *CEC Newsletter*, New Delhi, Vol. 5(1).

Government of India (1999). *District Socio-Economic Survey Report, 1998-99*, Government of Maharashtra, Directorate of Economics and Statistics, M.S.

Government of India (2000). Publication of Government of India, New Delhi, 287.

Ingle, P.O. (2001). *Mass Media Strategy for Effective Farm Communication*, Paper presented at MSEE Seminar held at Rahuri, MPKV, Rahuri, Maharashtra.

John H. Davis and Ray A. Goldberg (1957). *A Concept of Agribusiness*, Harward University, Bostan.

Manorama (2000). *Year Book Malayalam Manorama*, Kottayam, 85-89.

Rasheed, S.V. and V.V. Sadamate (2000). *Privatising Agricultural Extension in India*, NCAEPR, New Delhi, 23-25.

Sharma, R.D. (1998). "Agricultural Communication in the World of Tomorrow", *Indian Farming*, August, 13-17.

33

MOUSE CLICK FOR MARKETING

— P. Venkataram Muni Reddy, K. Madhavi Latha and *T. Krishna Prasad*

Farmers have adopted to the rapid transformation in crop production technology, post-harvest management, processing etc., to increase the productivity. Yet to obtain high market returns through agribusiness, in the existing marketing scenario, he must update his knowledge through information technology, which acts as a super highway in getting quickest and reliable up-to-date market information from any corner of the earth. The world over transactions are shifting from shops to cyberspace, and in future with the heavy competition in agribusiness farmers cannot achieve targets without the help of information technology.

Information technology is playing an important and vital role in bringing the whole world together. Internet connects one another all around the world, which helps to communicate almost instantaneously with one another. Internet is gaining tremendous popularity all over the world. Many websites are now available which provides encyclopedia of information on anything. Of particular interest are the increasing number of websites which pertain to agriculture and notably to markets. In agribusiness internet provides an infrastructure which helps to communicate and share information between buyers and sellers.

Agribusiness requires an infrastructure by which farmers can expose their output like the best quality fruits, vegetables etc., produced by them, where consumers have the choice to purchase, where banks or financial institutions transfer the funds between consumers and farmers. This infrastructure consisting of farmers, consumers and bankers is possible via intranet, extranet and over the internet (Fig. 33.1). Marketing through internet provides an opportunity for the farmers and consumers to communicate and share the information.

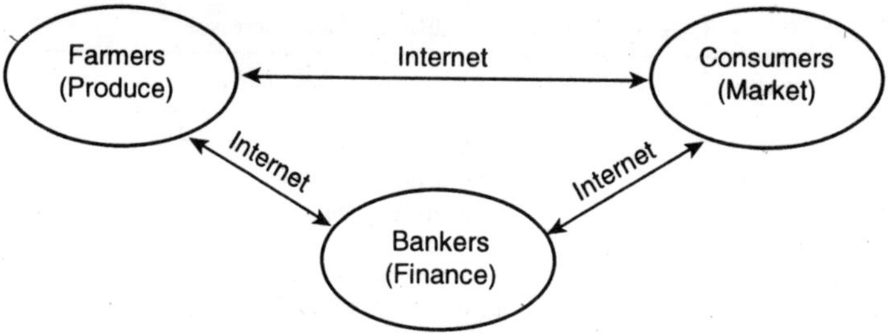

FIG. 33.1: Infrastructure for Marketing

Though Indian farmers produce good quality fruits and vegetables, regarding exports, our share is hardly one per cent of the world. This is just because of non-availability of information about latest technology and market demands. From this it is very clear that to promote agribusiness in India, we have no alternative but to adopt to it. To market the produce, we have to identify the buyers and the market where the demand is more, then establish the contacts, fix the price etc. At each and every stage an effective and efficient information technology must be available. And here the internet plays an important role in providing such latest information. In the future, agriculture sector with good websites will bring a magnificent change in a very significant way for those who adopt this technology.

Internet allows farmers to save time on order and delivery and getting customer feedback. In the existing competition, there is a need to rapidly attract new customers as well as retain existing customers. For this, marketing through internet is the need of the hour, which provides new and exciting ways to interact with global market. The portals must offer market data and information from daily closing prices to weekly, monthly and annual market reports, covering different regions and commodities. Reports must cover both domestic and international markets.

Several agri-portals must be launched which works as a platform for various agricultural activities including marketing aspects. The information revolution, which is creating an upstream in all fields of life has touched only a fringe of this needy segment. This is because no proper software is developed which gives latest information which will serve the farmer in the process of farm business management.

Now-a-days, most of the agricultural research institutes around the world have their own websites, which provides information on everything related to agriculture (Table 33.1).

TABLE 33.1: Some Useful Web Page Addresses

Site	*URL Address*
ACIAR	*www.icic.com/ac/aer/articles/083.html*
Asian Development Bank	*www.asiandevbank.org*
Asgrow Seed Co.	*www.asgrow.com*
Cargill Hybrid Seeds	*www.agriculture.com/contents/cargill/index.html*
Crucilferous Vegetables	*www.dole5aday.com/crucil.html*
Resources	*www.ent.iastate.evu/list/complete.html*
FAO Plant Genetic Resources	*www.icppgr.fao.org*
Food and Nutrient Infor Centre	*www.nal.usda.gov/fnic*
Garden Web	*www.gardenweb.com*
Genetics Glossary	*Helios.bto.ed.ac.uk/bto/glossary*
Horticultural Net	*www.hornet.com/*
ICRISAT	*www.cgiar.org/icrisat/*
Inst for Agric. and Trade Policy	*www.mtm.org/water*

The information available on portals should be made available to the farmers in the local languages along with English. Farmers should be provided with information in first instance to manage the business more effectively and competitively.

The basic concept and process of marketing remains essentially the same. Farmer can advise his product and buyer can identify the produce he needs, terms can be finalised and payments can be made through bankers. All these transactions can be performed through internet within a few minutes. Deal can be made between the buyers, sellers and the bankers from anywhere in the world. Based on the demand for the product, the acreage under that particular crop can be increased. Furthermore, with the availability of internet in homes, the farmers can now work from home and other remote locations without any loss of productivity or time. Thus the new information system and technologies will play a major role in determining the way in which a product can be marketed.

These web pages also provide information on the likely price distribution of key commodities over the coming years. Such information helps farmers and traders make decisions on when and in what ways to market their grain. Whether to sell at harvest or store on-farm in anticipation of higher prices later in the season. When combined with enterprise budget data, the information can also be used in deciding which crops to produce in the coming season.

In India, most of the portals in agricultural sector are only work in progress. To log on to a website and order a farm produce is still far away. According to a survey at the *agritechnica 1999*, every third farmer in Germany with a farm size of from 50 to 100 hectares is already online and every second on larger farms. But in India, where most of the farmers own small holdings, this seems to be difficult. This may be made possible by adopting the corporate farming system, which is the need of the hour with advent of new agricultural policy. By taking up corporate farming, a group of farmers can put up a computer and any educated young man from that group can undergo training of how to browse the internet. He can provide the farmers current commodity, analysis reports on world markets and trade for different commodities. Food market overviews provide valuable information about some of the most important export market.

One of the best example is “Agricon Infotech Services (AIS)”, Nasik has launched a portal website named http:/www.agricamplus.com. AIS works for improving the Indian agriculture. Through this portal they provide the users all the required information in local languages along with English. It involves farmers, agriculturists and boosts the agribusiness companies to associate with them. Such portals should be started and should be directed towards all the end users in agribusiness and work for sharing latest know-how and the information, for technology development in agribusiness. And because these services are basically just electronic exchanges, they can be delivered at *very low cost* and just *on mouse click* within *no time*.

34

INFORMATION HIGHWAY FOR RURAL DEVELOPMENT

— *N. Anandraja, J.V. Prabu* and *K. Chandrakandan*

Tracing the reminiscence of transfer of information from its place of origin to other places took a long time. Consequently the messages intended become obsolete and lost its utility and necessity to the actual users. During the historical era, even birds were used for communication. As the metamorphosis took place in the communication development, print media, cinema, documentary etc., have been growing to a larger extent to reach the mass of people.

The invention of radio waves brought significant changes towards better communication too. Subsequently, satellite arrival provided scope for wireless communication.

Information Highway—What is it?

In the eve of new millennium, the emerging information highway can bring a sea change for the growth and promotion of all fields. The day-to-day fiascoes can be put to a cliche if the information technology will be properly tapped and utilised in agricultural field.

To trace peck the precedent communication scenario, the informations teaching the intended target group were not as much rapid as of todays. The efforts of Sam Pitredo has brought out the spread of telecommunication even in the remote villages, which facilitates the transaction of information easier. After the advent of television the visualisation process has developed and the information was visually seen and understood.

The realisation of computers brought out a quantum change in the telecommunication process. Geo-Synchronous Satellites, had been instrumental in making a drift from radio wave communication to wireless communication. In the oncoming futuristic communication growth, it is hoped that information technology will bring a highly developed agriculture by its worthwhile contributions to the society by norrowing down the enormous gap between the scientists and farming community.

Bridging the Gap: The Prime Aim of IT

As the farming community of India is normally endowed with fragmented land holdings, containing marginal and small sectors to the maximum, the affordability, accessibility and possibility to travel a long distance and to hunt the information is seldom possible. Moreover the required message at appropriate time and season again is a bottleneck, as because the access of radio alone cannot serve the purpose.

As the traditional agriculture is transforming into hitech agriculture, the need for updated information is also essential for agricultural production and productivity, eventually giving a lucrative yield and income to the farming community. The distance to reach the actual source of reliable information is at very slow pace.

At this predicament, the computer network can be resorted to for clearing the querries/doubts. This network can be the better answer and to absolve the current problems that arise here and there. The younger generation farmers shall rely on this information machine solely by keeping in touch with 'Online Clarification' and the telecom hikes up to upgrade and improve their farming with scientific technologies. A well framed database with a tentacle of 'expertise in agricultural fields' should be available as a database which could be the best credible source of information, placed in research institutions and SAUs situated elsewhere in the country.

Cost Reduction through IT

Generally farmers are graced only by the income generated by the farm, involving cropping and allied enterprises. So, their economic status is not always stable. Consistent income to get anything in a periodic fashion is also not that much possible. Due to this situation, a regular subscription for any journal/magazine may be highly expensive for the farmers.

On the other side, for each and every practice, farmers may not come to TNAU or its sub-stations for any clarifications, consultation, references, advises and guidances. Because of this handicap, the farmers are desperated to seek updated information. Eventually they will be demotivated to practice recent advancements in agriculture. This would lead the farmers to adhere to only local sources of information, who are normally lagging in the technical expertise, and may not be coping up with the advancements in technologies.

If the computer aided network could be relied upon for subsidising the problem in lieu of other sources, the expenses can be drastically reduced on travel costs and accessibility of information can also be made possible, even by staying in their locality itself. *www.mssrf.org.*

New Hopes for School Dropouts

The employment opportunities in the government and private are shrinking day by day due to heavy competition and burgeoning educated population. Apart, the avenue for continuing the studies are very difficult in villages, because of inadequacy of finance, over burdened household chores, lack of motivation, exploitation, disparity and related socio-economic problems. So, these potential groups are deprived of education from primary to higher levels. Similar is the case for women too.

Colossal development in agriculture could be obtained only by the strenuous effort of youth. To achieve this, they must be geared up to have 'hands on training' on computer operations in a very simple way.

Apart from their skill acquisition, this computer knowledge would also supplement their earning if they could work as computer operators to process and analyse the data, culling the needed information to the needy farmers, doubt clarification etc.

Gearing up the Extension Machineries on IT—The Need of the Day

Today's extension machineries are not keeping them abreast of new scientific advancements. The reasons they specify are over loaded official work, engagement in a lot of miscellaneous activities which seemed to dilute the efficiency of their task.

Periodically, those sources who are expected to render good service to the farming community are lacking in their day-to-day activities. In fact, they should be the sole authority and authentic sector for approaching all sorts of farm problems.

For that matter, computer wisdom is again an indispensable qualification for them to pump up the information to the needy in time, at seasons and in exigencies. This would also lessen their bothering on transport facilities. Staying in the office and networking their advises and ideologies, would bring a lot of changes in the rural scenario.

In situ storage and retrieval of information could reduce a lot of stress and burden of the extension worker in his field visits. This would pave a way for them to inculcate a sense of zeal to answer the querries of the farmers at any time.

Advertisement and Market Promotion—A Facilitating Role

Most of the farmers reside in the localised area without sensing what is going around in the environment. Though they are endowed with quality agricultural produces at their disposal, their handicap is to seek for a proper means to sell the same for good price. If the internet could be properly accessed, the advertisement showing their availability of produces for a good market shall be carried out without any constraints. Their would also fetch right price and right time of sales for the needy who really demand for the quality produces.

More so, the marketing seems to be a mammoth impediment for the farmers, who are solely on the grace, the co-operatives and also under the clutches of middlemen. If he has had an internet, he can trace out a huge sources of procurers for their commodity without waste of time and for an attractive price without equipping any sort of loss by burning the hands. The uncanned fire emerge from their stomach could be very easily quitted away by means of usage of the net with utmost possibility. *www.ottanchatram.com*

Crop Planning—Reducing the Despair

Toiling of the farmers is not uncommon in the farming operations for want of remunerative income since a long time due to the fluctuation in prices, surpluses felt in specific crops in a specific time, storage difficulties to heap and sale. Might be due to lack of prediction to forecast the demand of the particular crop. Lack of proper planning again fails to countervail the loss, it incur because of inadequate marketing of that crop. To mitigate these problems, the NET would clearly finalise the prospects for a crop and its potent demand in a very perspicacious way. This would give a concrete idea for the individual to decide his crop in the succeeding seasons based on the predicted demand he would bear in mind with the statistical data and explode details from various regions across the country.

Role of IT in TNAU—A Revitalisation Process Needed

TNAU is the integral unit for technology generation and technology dissemination to its ultimate users. Efforts are adequately made by the institute to render its services to a gender extent to the farming community.

To enrich its activities in new dimensions and to facilitate its rapidity, the information technology, bearing the nomenclature of 'Cyber Extension' can do a lot for the rural masses.

If the cyber extension is established in the institute, the technology transfer would be hastened. The reach of the message to the unreach could be possible within a very short span of time. *www.tnau.edu*

Focus on Multi-Media Packages

In this, the focus should be to make audio and video cassettes, bearing the lessons on package of practices for various crops. Then cassettes will be sold either in the campus centres or sent to Village Information Centre through the Department of Agriculture. From the information centres, the cassettes can be subscribed by the user farmers on the needy technologies. Otherwise with the help of computer operators, the cassettes will be played and the messages can be learnt in the Rural Community Centres. This kind of approach will help the farming community to get a detailed information on various technologies.

Trust on Digitised Information Network—A New Dimension

This kind of compacted messages utilising the Digital camera, would allow more time to picture an act/event. It is of superior quality than video graphed messages because, its capacity to condense more information within a less pace. Apart, silica based optical cables can be used to store umpteen messages because of its low bandwidth. The technology of this kind is already been available in USA, Japan etc., but want of cheaper optical cables, due to high cost of silica, this method of storage is felt expensive. If the silica could be extracted and sieved from the seashores properly with advanced technologies, the information could be well disseminated without any constraints.

Web Learning Processes and Speedy Learning

The website and web learning are today's talk in various fields for getting access the messages from worldwide and sundry sources.

However, in the agricultural field, this web learning has not been thought of so far. Yet, a lot of ground work needs to be done to facilitate this learning. To achieve this vision, a web centre has to be created in the institute. Doubts emerging from the villages will be sent to the centre. Subsequently an 'On Line Tutorial' established as a link to the centre will take all the doubts and will be sending to the concerned faculty/department/directorate/staff for awareness and for further action. If the doubts are made known to the concerned, the answers will be made ready and floated in the satellite. From the satellite the messages can reach the village Information Centre where it can be received in the computer web page and retrieved for barring and clarification. Hitherto, information tasks force has not realised the importance at village level to solve agricultural fiasco. If they could streamline their efforts on the rural areas, mammoth development in agricultural spheres would occur for a better progress. This network not only links all parts of India, but also globally.

Besides these, 'Heuristic learning' process will also be facilitated. In this process, the farmers can experiment certain things, enhance trial and error methods of learning. Because of this the indigenous wisdom of farmers and the scientific wisdom could be blended properly for practical problem solving. Heuristic learning also would develop inbuilt confidence among farmers to advent new methodologies and ideologies to be applied in their areas.

Launching of Information Technology Training Cell (ITTC)

Due to this effort, the basic computer operation skills shall be taught to the farming community and village youth as well. The unemployed youth aspiring for computer literacy to operate a website or to browse the Interact for messages will be selected. A short-term training may be given to them, which would equip the computer learning skills in general and agricultural information in particular.

So much so, the agricultural problems could be solved, if this centre is launched. More so, to employment opportunities will be adequately generated in the rural areas. Ultimately, the transfer of messages could be made very prompt and easier than any other methods.

Imparting Simulation and Animation Learning—A Forewarning Approach

Since the climate and rainfall is erratic as far as Indian Geography is concerned, no prediction can be made about what will happen and what to carry out by the farmers. Precision farming, if done, can bring lot of sea changes in the agricultural fields. Forecasting and forewarning phenomena will make the farmer cautious and to take up corrective measures. Accordingly accurate farm plans can be made and the crop suitability for the appropriate seasons will be assessed.

At this predicament, the predictions on climatology by the scientists may not be clearly taught to the farmers, because of its complicated processes. To simplify this for better understandability, animated and simulated lessons telling the consequential effects out of a practice and other intricacies will be taught very conspicuously. These lessons would give adequate awareness and learning for proper planning in cropping systems.

Strategies

To recapitulate, the scope and effective use of the information technology is innumerable. The lacunae are improper training, and inadequate awareness prevailing among the farming community. To promote the internet assisted farming there is a need to devise the following strategies:

1. The farmers are to be made aware of the utility of the Internet, Website and other related information regarding information super highway.
2. A good team of expertise should be always made available to communicate the message there and then to the needed sectors in demand as well as 'on service' in 'on line Tutorials'.
3. A proper strategy should be construed to check away the 'inhibition' and 'phobia' against computer operations by extending a large scale training for the school dropouts and unemployed youth and women.
4. 'A short-term course' is to be organised periodically for the extension machinery to become well versed with operations like website formulation, CD-ROM and other advanced techniques to pursue their extension work more successfully by interplaying with data base.

35

CYBER EXTENSION AND AGRIBUSINESS

— *K. Chandrakandan* and *N. Anandraja*

The economic development of developing countries like India largely depends on development in the farm sector. It has received importance after 'Green Revolution' which has brought about commercialisation of agriculture in selected areas and given business-touch. Agribusiness is a very new concept which has gained importance in our country and so also in many nations in the recent past.

With the advent of new economic policies in the country, it is necessary to understand the various aspects of the trade and commerce to understand agribusiness. Though this was confined to the small farm sector, over the years, the technological revolution had occurred in the on and off farm sectors which had spread and allocated the different functions to various farms specialising in processing, value-addition, storage, transport, cold storage, export and the like.

It is, therefore, imperative that the agribusiness is diverified and yet each sub-sector is so specialised and commercial profession.

Considering the commercialisation, value-addition, export-orientation and globalisation of Indian Agriculture, the extension approaches and methodologies have to be also changed, keeping pace with changed needs and environment.

The extension organisation and approaches what we have today are all old-fashioned and outdated.

WTO Agreement on Agriculture

The WTO obligations are being implemented in the country in a phased manner, resulting in market policy changes by the government. It would result in new opportunities along with new challenges for the economy.

A paramount concern for the agricultural sector is whether the WTO Agreement would interfere with the country's ability to follow its agricultural policies and programmes.

It is important for all those concerned with agricultural development to know about the Agreement, its implications, the status, various provisions and their impact on the country's economy and about the proposals under the negotiations in areas of market access, domestic support, export competitions and food security. The farming community also has to be made aware of the implications, besides taking advantage of various provisions of the Agreement as applicable for developing countries, especially in the context of small farmers.

The Union Agricultural Minister spelt out in the Lok Sabha in April 2001 that India's strategy on WTO would be food security and protection of domestic farmers. 'The Food Security Box' should be there for all the developing nations including India which has demanded substantial and meaningful reductions in tariff including elimination of peak tariff and tariff escalation, substantial reduction in domestic support and elimination of domestic subsidies.

The National Agricultural Policy announced recently addresses the challenges arising out of economic liberalisation and globalisation. It seeks to actualise the vast untapped growth potential of Indian Agriculture, strengthen rural infrastructure to support faster agricultural development, promote value-addition and secure a fair standard of living for farmers and farm workers.

Extension is now on the crossroads. It is more applicable to field extension, the direction and destinations of which is undefined and unanswered yet.

Omissions and Commissions of Earlier Attempts

Most of the rural development efforts taken prior to our Independence lacked government support and finance. Further, the efforts were sporadic and fragmented. The efforts initiated by the government after Independence lacked. People's support and they failed to address the basic needs of rural communities. They lacked mission and vision. To put it in a nutshell, the most of the earlier attempts on extension and rural development had failed owing to non-involvement of public. In other words, the public were not taken into cognizance and so they were mostly person-oriented and starved of funds. The rural masses remained passive spectators but not willing partners.

Taking lessons from the past efforts, and so also requirements of globalisation of Indian Agriculture and implications of WTO regime, the following model called, "Farmer-led Participatory Extension Management" is proposed and the salient features are explained hereunder:

Differences

	T&V System	*Cyber Extension with Participatory Mode*
1.	Lack of coordination among development departments, NGOs, Financial and marketing institutions	Better coordination would be ensured through partnership mode by persuasion
2.	Focus was more on crop production technologies of selected crops only	Lays emphasis on integrated farming systems and diversified enterprises
3.	Contact farmer approach	Opinion leader approach
4.	Mostly interpersonal communication	Ideal blend of modern IT and traditional channels
5.	Top-down and target-driven	Participatory and demand-driven
6.	Official planning and implementation	People's involvement in planning, implementation and evaluation
7.	Agricultural diploma or certificate holder at grass-root level	Suggests farm graduates at the cutting-edge

Special Features of the New System Proposed

- Equal emphasis on Cyber Extension and Conventional Extension.
- Equal emphasis on group approach and PRA technique.
- Coordinated efforts of development agencies through partnership mode.
- Developing sound data base.
- Commodity club and Self-Help Group promotion.
- Flexible annual action plan with mid-term correction.
- More frequent visits, *i.e.*, weekly instead of fortnightly.
- Minimises/eliminates waste of time on unproductive meeting, training etc.
- Input dealers and agencies involved through Inputs exhibition-cum-Sales *Mela*.
- Provides periodical information on marketing prices of farm produces etc.

It is, therefore, believed that 'Cyber Extension' may be the only answer which implies using the power of on-line networks, computer-aided communication, website browsing and digital interactive multimedia to facilitate transfer of technology.

Changes in Extension Management

To match the agribusiness and changed rural scenario, an appropriate model "Farmer-Participatory Extension System" through ideal blend of Cyber Extension (Modern IT) and Traditional Extension Approaches for Effective Transfer of Technology called "Cyber Extension with Participatory Mode" is suggested for future India.

Change is natural and many times inevitable and this law applies to any society/activity. There is change in every walk of life and it is often welcome, so much so, change in necessary in the field of extension too for more effective and efficient transfer of technology. The T&V system has been in vogue in many States. Broad-based Extension System has been introduced in a few States recently.

Our Field Extension System is a century old. Though it has been fairly effective, it also needs change, as change is law of Nature. We have witnessed many changes in our life style; but we are yet to make a drastic change in our extension approach.

In the recent past, an Information revolution is taking place globally. So, we cannot remain unchanged and static. We have to also move fast and pick up the advantages of changing trend. Otherwise, we will be dropped out from the main stream. The rapid growth of information in agriculture has led in past decades to the progressive specialisation and it is more so after the globalisation of Indian agriculture. Information/ knowledge transfer in agriculture faces many problems. Our New Agricultural Policy emphasises commercialisation, value-addition and export of agricultural commodities, besides globalisation.

The information/knowledge gained in research must be transferred in a meaningful and comprehensive way, so that it reaches the end-user, the farmer without much loss of time and content, avoiding seepage of information and distortion.

There is need to have ideal blend of modern IT with traditional extension approaches for effective and efficient transfer system. Further as per popular saying, "One cannot employ yesterday's method today and be in the business tomorrow".

The pilot project proposed might be implemented in any one district/ block every State of the country to develop more effective extension model than the one at present. This project may have all the vital components of 20-Point Action Plan proposed in the NATP-CGP II/59-2001-2003 "Action Research in designing Effective Communication Strategy in the Existing Extension System of Broad-based Agricultural Extension" under implementation in Tamil Nadu by the Department of Agricultural Extension and Rural Sociology, CARDS, TNAU, Coimbatore-641003, Tamil Nadu with Prof. Dr. K. Chandrakandan, as PI.

Those willing States may propose similar projects and seek for financial support from the World Bank-aided NATP—CGP mode or any other agencies.

REFERENCE

NCDC Bulletin, August 2001, Government of India, Ministry of A&C Publication, New Delhi.

36

AGRIBUSINESS MANAGEMENT
Rapid Dissemination of Agro-Information through Internet

— *H.K. Patil, R.M. Joshi* and *S.S. Bhosale*

Advancement in scientific research has given rise to most of the sophisticated new technologies in information and communication fields which are now drastically changing the concept of large size diversified world into global village. The networking of people through innovative communication and information technology is creating limitless access to all available forms in social learning and intelligence.

One of the biggest achievement of twenty-first century is the introduction of an advanced communication network, *i.e.*, Internet. The technology connecting millions of computers in a network. The Internet can be broadly defined as Worldwide Network of Computer communicating via an agreed upon protocol (rules for exchange of information). It provides access to the most diversified sources of information hosted by individuals and various organisations worldwide on a vast network of servers. Presently, the Internet gives on to world web or just web which interconnects thousands of servers created by various organisations. The web page loaded on the server provides various types of information such as plain text, picture animation, multi-media either free of cost, or for modest fee. Besides, the web page offer services like e-mail, conferencing, electronic publications and other commercial facilities. It is a new mode to trade text or information from one point on the globe to another quickly, cheaply and easily. It is not only one way communication process but also receive information in opposite direction and performs as an interactive medium.

Internet in Agriculture

The agriculture scenario all over the world particularly in the post-GATT era and after WTO Agreement is undergoing a rapid change. In order to take full advantage of the changing global agricultural scenario intermeshing of policies related to the pricing, marketing and trading of agricultural commodities are also reviewed. Simultaneously there is also need to review and revitalise the mechanism for transfer of technology under changing environment.

There is significant shift from agriculture supply driven to demand driven paradigm in new emerging and changing economic policy. It is viewed that future agricultural growth would be information driven. New information must reach to the ultimate user at the fastest speed to harness its potential benefits. Information like seed, water, nutrients and plant protection biological is one of the key inputs for successful farming. Knowledge intensive and precision farming techniques will be the guiding lines for sustainable agriculture in the future.

The rapid strides made in information technology (IT) in recent years have encouraged many development communicators to extend its benefits to agricultural and rural development. It is a universal truth that access to information holds the key for successful development. Some dedicated and enterprising scientists and technologists are making their best efforts to take the benefits of the information technology revolution to the rural poor.

Shining Success

"Access to Internet will soon be universal, and it can provide unrestricted low-cost access to information, as well as highly interactive distance learning. The Internet will not only facilitate interactions among researchers, but also quickly improve their ability to communicate effectively with the potential users of their research knowledge", says Prof. M.S. Swaminathan, Chairman of the M.S. Swaminathan Research Foundation (MSSRF), Chennai. The "information village" concept developed and implemented by MSSRF in a Pondicherry village is a path-breaking experiment to demonstrate the potential of the information revolution. On the same pattern 'Warna Wired Village' concept is being successfully implemented in Kolhapur district of Maharashtra State. This is the ideal example of use of Internet on co-operative basis, where the all member farmers are visiting on website. Telephone charges are paid by the concerned co-operative society.

Growing Acceptance

The Internet is prompting a sea change in international development thinking and major donor and multilateral lending organisations are radically reshaping their policies in the new information age. The shift follows a growing acceptance of the importance of communications in development.

Specific research on the impact of the Internet on development is scarce. However, studies conducted over the past 20 years by the International Telecommunication Union (ITU), The World Bank and the Organisation for Economic Cooperation and Development (OECD) have demonstrated that the ratio of benefits to costs of telecommunications usage rose from 5 : 1 to more than 100 : 1, due to improved efficiency in managing rural enterprises, time savings in travel costs.

Growth Prospects

The rate of growth of penetration of Internet has been spectacular. It took radio 50 years to have 50 million owners, TV 16 years and personal computers 17 years. It has taken Internet only 4 years to reach that figure after the invention of www. It is estimated that currently, there are about 3 to 4 lakh Internet users in India. Recently, in India the Internet service has drawn the attention mainly in metropolitan cities and large urban areas. Therefore, Internet connection statistics is presented in Table 36.1.

TABLE 36.1: Number of Internet Connections in Metropolitan Cities

Sl. No.	*Name of City*	*No. of Internet Connections*	*Percentage*
1.	Mumbai	86,382	35.98
2.	New Delhi	68,042	28.35
3.	Chennai	28,829	12.01
4.	Kolkata	16,773	6.98
5.	Hyderabad	11,493	4.79
6.	Bangalore	9,161	3.82
7.	Pune	6,202	2.58
8.	Chandigarh	6,060	2.52
9.	Ahmedabad	1,225	0.51
10.	Other cities	5,869	2.46
	Total	2,40,036	100.00

Source: *Vyapari Mitra*, December 2000, pp. 1289-91.

From Table 36.1, it is revealed that highest (more than 50%) number of Internet connections are found in Mumbai and New Delhi.

Reasons of Rapid Diffusion

The country stands one among the highest internet subscribers among the developing countries. There are two main reasons that support and enhance internet penetration into India. The English is the universal communicative language of the internet. A sizable number of Indian population are well versed with the English language and who has access to internet prefer communication in English rather than any other language. Also now in India, various organisations launched Hindi Web sites for better communication as national language. The other reason is rapid spread of computer network, growing telecommunication infrastructure and flourishing computer software industries. Besides, the liberal approach towards the internet subscribers and government's enhanced support for the development for IT and communication infrastructure in the country is also another reason for growth of internet in India.

A rapid diffusion of internet depends on proper telecommunication and computer networks. India's telephone density is hardly 1.3 per cent (Telephone per 100 population) against world average 11 and computer density is around 1.5 per thousand against world average 2. Certainly to make the internet popular information medium in rural area, there is need to strengthen the growth of both telephone density and PER CENT infrastructures. This will be beneficial for rapid diffusion of internet in India and ensure that country will not miss the opportunity of global communication revolution.

Glimpses of Internet Weblinks for Agriculture Use

There are only a handful of organisations which have their own websites on the Internet. Apart from the ICAR's website, there are 21 ICAR institutions having their own websites placed on internet for use by the farmers and the Extension agencies. At the national level, only State Departments of Agriculture of various States like Madhya Pradesh, Maharashtra, Andhra Pradesh, Punjab and Haryana are available on the internet, while there are five SAU's are on the internet like Haryana Agricultural University, Punjab Agricultural University, Sher-e-Kashmir University of Agricultural Science and Technology. University of Agricultural Sciences, Dharwad and the Tamil Nadu Agricultural University. Now names in the list are increasing at faster rate.

The details of web link addresses of selected agricultural institution is presented in Table 36.2.

TABLE 36.2: Direct Web Links to Some Selected Agricultural Institutions

Sl. No.	*Name of the Institution*	*Website Link or Uniform Resource Locator (URL)*
1.	Department of Agriculture, GOI	www.nic.in/agricoop
2.	ICAR	www.nic.in/icar
3.	Department of GOM	http://agri.mah.nic.in
4.	TNAU, Coimbatore	www.tnau.edu
5.	KVK, Babhaleshwar (MS)	http://www.kvk.pravara.con
6.	ICRISAT, Patancheru (A.P.)	http://www.egiar.org/icrisat
7.	USA, Dharwad	http://www.uasd.net

Some suggestions for increasing use of internet in near future:

- The transfer of technology through internet is mostly dependent on the services provided by Department of Telecommunication. In rural areas, telephone networking facilities are not being developed at desired speed. Therefore, there is need to accelerate the speed for establishing telephone net working in rural areas.
- Generally, the information provided through internet is mainly in English version. Therefore, considering farmer's level of literacy and his ability to be grasp the various topics, there is need to develop an appropriate language translation software for conversion of English version into selected local languages version.
- For internet and LAN working in all villages well equipped computer facilities is a prerequisite. In rural area, most of the farmers are illiterate and they are not aware of importance of dissemination of agricultural technology through internet. Therefore, awareness generating training programme on use of computers for enhancing wider application in agriculture should be arranged on priority basis.

Bird's Eye View of Present Status of TOT

The major responsibility of transfer of technology lies with the Development Departments in Maharashtra State. There are separate Development Departments for Animal Husbandry, Fisheries, Forestry, Social Forestry, Irrigation and recently the Departments of Horticulture,

Water Conservation and Watershed Development are merged with the parent Department of Agriculture. Since the responsibility of agricultural education and research was assigned to the SAU's. The extension component has remained with the concerned Development Departments and these Development Departments are isolated from teaching and research. The Development Departments are implementing several development projects sponsored by the State and Central government for welfare of farming community.

Development Departments should focus more in the lagging regions and encourage private sector or NGO's participation. Development Departments should have to develop a mechanism to monitor the quality of information regularly.

Input supply agencies should invariably be involved in the work of dissemination of information about different components of new technology.

Emerging Issues

- Development of credible and comprehensive database on the whole production system for planning and development should receive greater attention. Because meeting the information needs has always been an integral part of the farmers' job. Further, information provides the farmer with an important instruments for decision-making and taking action.
- The acquisition of computer based technology on farms is probably driven by many factors in common with other sectors. The type of farm enterprises and the nature of farm business in operation as well as computer support required by the farm business will determine the type of softwares to be used on farms. Changes within the structure of agriculture will probably have an impact on the selection and types of acquisition of softwares or other integrated systems made by the farmers.
- The functional needs of the farm in modification terms needs to be met by the package. Therefore, the issue of modification may arise concerning the flexibility of the package. In addition, the cost profiles of the software and hardware are also required to be taken into consideration, while formulating policy framework.
- Software packages must be thoroughly evaluated before they can be used as a foundation for new information system.

- In general, it is observed that much available software did not include comprehensive accounting for risk in agricultural activities. Therefore, inclusion of risk into farm and enterprise management systems in future may represent a new direction for software development as a further management tool into the next decade.

Forecasting

During the initial phases of diffusion of new technology through internet, rich farmers are likely to allocate more resources to the acquisition of information. Therefore, they will possess higher levels of cumulative information at any given period. Farmers that have more convenient access to information or better endowment of human capital will also acquire higher level of knowledge. Therefore, in years to come, farmers with better access to information with more computer trained human capital would adopt technologies earlier than other farmers. Therefore, farming community will be divided into two groups, *i.e.*, computer literate and computer illiterate farmers or computer based agricultural technology users and traditional non-computer based agricultural technology users. Other factors such as education age and business owners are also important factors.

Implementing Strategy

For fast diffusion of use of internet in agriculture, it requires the co-operative efforts of various agencies. This is joint task, which can be categorised into four clusters:

Awareness enhancing task, Supply side task, Demand side task and Provision of basic internet related services.

Awareness Enhancement Task

In India's agriculture sector, most of the staff working in agriculture department at village level and farmers in general are unaware about tremendous benefits that internet can provide in all walks of life. The 'leap frogging' potential must be clearly understood. The policy-makers and others have to make aware of how to reorganise their development strategies and organisational set-up in the light of what is after all, a very complex vertex like technological revolution.

Supply Side Task

Supply of the benefits of internet will require heavy investment. The infrastructure should be in order to satisfy high bandwidth requirements. Fibre optic and wireless technologies will have to be integrated and the infrastructure must also be extended to the remote regions of India to realise the full benefits.

Demand Side Task

In addition to the supply side of information infrastructure, agriculture sector will also need to invest the demand side. Investment will be required in education, computer literacy and informing the population in general. The investment in the demand side is minuscule as compared to that in the supply side.

Provision of Basic Internet related Services

Similar to telephone conversation, the internet is interactive in real time. A user can connect to an on-line service and get the result of the command almost immediately depending on the speed of the system, such as interactive session requires a continuous, open connection. Usually, there is no continuous (constant) exchange of data to utilise the full potential of the open connection. So, if the traffic (information exchange) is low, this system can be very inefficient. Basic internet services consists of electronic mail (e-mail) and file transfer using e-mail. Exchange of e-mail, can be compared to exchange of regular mail. The computer stores all e-mail and exchanges during connections, using a modem over a telephone line, such as post office collects all mail and ship it out at once, not individually. This is comparatively fast and reliable.

Epilogue

Internet is harbinger of global market beyond borders, beyond barriers, beyond boundaries and beyond minds. It provides large number of value added services to the users. Internet is knowledge intensive characterised by high speed, contraction of time and distance. The information revolution is shaping the agriculture by knowledge and knowledge is just like river. There should be continuous flow from source to the destiny. Benefits of internet will indirectly contribute to

national development by enhancing literacy, income and translate into better standard of living of farming committee. New community is being built through use of internet.

REFERENCES

Badge, R.K. and Kalava, A. (1998). *Agriculture Trend in IT*, XII-National Conference of Agricultural Research Statisticians Proceeding, pp. 171-75.

Gaikwad, B. and Hendre, P. (2001). "Agriculture Communication through IT—A Case Study", *Yojana* (45): 27-30.

Rakshit, A. and Bhadoria, P.B.S. (2001). *IT & Agriculture*, (45): 17-20.

Reddy, J.M., Jayashankar, P.C. and Gupta, B.S.S. (2000). "Towards Global Village", *Yojana*, 44(1): 18.

Salvekar, D. (2000). "E-commerce", *Yojana*, (51): 50.

Seshagiri, N. (2000). "The Information Policy in India", *Yojana*, (44): 5-10.

37

ROLE OF NEW TECHNOLOGIES IN AGRICULTURE DEVELOPMENT

— *V.R. Shete, D.S. Navadkar, S.D. Patole* and *P.S. Bonde*

The recently drafted National Agricultural Policy has accorded priority for promoting agribusiness activities at different levels. The role of Krishi Vigyan Kendras (KVKs), Non-Governmental Organisations (NGOs), Farmers Organisations, Co-operatives, Corporate sector and para-technicians in agricultural extension was encouraged for organising demand driven production systems. In the Indian context, there are four major organisational streams devoted to extension work for agricultural and allied production. These are: (1) the ICAR extension system, (2) extension system of the Ministry of Agricultural/State Departments of Agriculture, (3) extension system of the Ministry of Rural Development/State Development Departments, and (4) Development work by the NGOs, business houses, etc. The ICAR extension system comprises of mainly Research Institutes and Agricultural Universities. The agricultural scientists and extension personnel of these institutions are required to play first line extension role through organising demonstrations, trainings, etc. The concept of vocational training in agriculture through KVKs grows substantially owing to the greater demand for advancement of agricultural technologies and the growing progressiveness of the farmers. This is imperative for the rapid transfer of technology.

The Krishi Vigyan Kendra is one of the first line innovative schemes of the ICAR devoted to vocational training of the practicing farmers, farm woman and young farmers. While as a centre of excellence in the field of transfer of technology the KVK is going to play a decisive role in rural development efforts in India.

The purpose of the present study mainly intends to performance evaluation of KVK 'Pal' in terms of training programmes examining the impact of KVK on selected agribusiness in terms of income, expenditure and consumption of farmers.

Methodology

The Krishi Vigyan Kendra located at Pal in Raver tehsil of Jalgaon district was purposively selected for the present study because of its popularity and rapid progress since inception. The sampling technique adopted for the study was two stage random sampling; with villages under KVK as a primary unit and the beneficiaries of different activities of KVK was the secondary unit. Out of 30 responsive villages eight villages were selected randomly. The beneficiaries who had received training on different aspects of KVK were divided in four categories, *viz.*, landless, small, medium and large size groups. From each size group, 30 beneficiaries were randomly selected, making a total sample of 120 beneficiaries.

Data Collection and Method of Analysis

Most of the information of these 120 beneficiaries on various aspects were collected from office of KVK at two points of time, *i.e.*, before and after inception of KVK (1998-99 and 1999-2000, respectively). Moreover, elated information from the year of establishment, *i.e.*, 1984 to 1999 were also gathered. The data were analysed by the help of statistical tools. Furthermore, the compound growth rates in training programmes, demonstrations and trainees of KVK were estimated by using the following type of functional model.

$$Y = ab^t$$

Per cent compound growth rate = (anti log $b - 1$) × 100

Results

Year-wise Training Programmes of KVK, Pal

Training is an essential part of KVK in order to develop competence of clients. Table 37.1 indicates the number of training programmes conducted at KVK and other stations alongwith number of participants. It is observed that there were 2,794 training camps conducted over a period of 15 years, out of these 1,614 camps (57.77%) were conducted at other places while 1,180 (42.23%) were conducted at KVK, Pal. The maximum programmes were held during 1990-91 at KVK campus with maximum male participants (184). During the span of 15 years, 47,077 participants were trained in different disciplines, out of which 34,379 participants (75.15%) were male practicing farmers, youths and artisans while 11,698 participants (24.85%) were farm women and female youths.

TABLE 37.1: Year-wise Training Programmes of KVK, Pal for the Year 1984 to 1999

Years	*Number of Camps*			*Number of Participants*		
	On Camp	*Off Camp*	*Total*	*Male*	*Female*	*Total*
84-85	14 (1.18)	1 (6.10)	15 (0.50)	337 (0.95)	27 (0.23)	364 (0.77)
85-86	83 (7.03)	79 (4.89)	162 (5.79)	1,755 (4.96)	985 (8.42)	2,740 (5.82)
86-87	107 (9.06)	73 (4.52)	180 (6.05)	2,659 (15.99)	514 (4.39)	3,173 (6.74)
87-88	113 (9.57)	124 (7.68)	237 (7.98)	2,490 (7.03)	832 (7.36)	3,322 (7.02)
88-89	125 (1.01)	136 (8.42)	261 (9.34)	3,028 (8.55)	806 (6.89)	3,834 (8.14)
89-90	112 (9.49)	205 (14.48)	317 (10.65)	3,672 (10.37)	760 (6.48)	4,432 (9.41)
90-91	134 (11.30)	266 (14.48)	400 (13.44)	5,274 (14.90)	1,841 (15.73)	7,115 (15.11)
91-92	126 (10.67)	270 (16.72)	396 (13.31)	4,989 (14.10)	1,613 (13.78)	6,602 (12.85)
92-93	120 (10.16)	232 (14.37)	352 (11.83)	5,039 (14.24)	1,016 (8.68)	6,055 (12.85)
93-94	34 (2.88)	29 (1.79)	63 (2.11)	1,480 (4.18)	823 (7.03)	2,303 (4.89)
94-95	34 (2.88)	34 (2.10)	68 (2.28)	759 (2.14)	376 (3.21)	1,135 (2.41)
95-96	43 (3.64)	66 (4.08)	109 (3.90)	1,079 (3.04)	578 (4.94)	1,657 (3.51)
96-97	58 (4.91)	50 (3.09)	108 (3.63)	249 (3.53)	604 (5.16)	853 (3.93)
97-98	43 (3.64)	24 (1.48)	67 (2.28)	792 (2.23)	441 (3.76)	1,233 (2.62)
98-99	33 (2.79)	25 (1.54)	58 (1.95)	772 (2.18)	432 (3.69)	1,204 (2.56)
Total	1,180 (100.00)	1,614 (100.00)	2,794 (100.00)	35,379 (100.00)	11,698 (100.00)	47,077 (100.00)
Per cent	42.23	57.77	100.00	75.15	24.85	100.00

Note: Figures in parentheses indicate percentages to total.

Demonstration Programmes of KVK, Pal

The KVK, Pal conducted demonstration intended to show genetic production potentiality of new technologies and to influence both the farmers and extension agencies. That the total number of method demonstrations of KVK, Pal during the year 1984 to 1999 were 155. The maximum and minimum method demonstrations were conducted during the year 1988-89 (29.03%) and in 1984-85 (1.29%), respectively. During the 15 years 5,562 participants were involved in demonstration programmes in different disciplines, out of which maximum 1,754 participants (31.34%) were in the year 1998-99 and minimum 45 participants (0.80%) were in the year 1985-86. The overall average of 15 years indicated that the average number of participants were 371, method demonstration 10 and the average discipline involved were 3 in numbers (Table 37.2).

TABLE 37.2: Year-wise Method Demonstration of KVK, Pal

Sl. No.	*Year*	*Number of Discipline*	*Number of Demonstrations*	*Number of Participants*
(1)	(2)	(3)	(4)	(5)
1.	84-85	2	3 (2.60)	65 (1.16)
2.	85-86	1	2 (1.29)	45 (0.8)
3.	86-87	3	8 (5.16)	152 2.73)
4.	87-88	4	11 (7.09)	261 (4.69)
5.	88-89	5	45 (29.03)	1,754 (31.54)
6.	89-90	4	11 (7.09)	882 (4.69)
7.	90-91	2	6 (3.87)	188 (3.38)
8.	91-92	2	19 (12.25)	333 (5.98)
9.	92-93	4	9 (5.89)	209 (3.75)
10.	93-94	2	6 (3.87)	218 (3.92)
11.	94-95	2	5 (3.22)	220 (3.95)
12.	95-96	5	10 (6.45)	163 (2.93)

(Contd.)

TABLE 37.2 (*Contd.*)

(1)	(2)	(3)	(4)	(5)
13.	96-97	3	5 (3.22)	120 (2.16)
14.	97-98	5	7 (4.51)	125 (2.25)
15.	98-99	2	13 (8.38)	537 (9.65)
	Total	46	155 (100.00)	5,562 (100.00)
	Average	3	10	371

Note: Figures in parentheses indicate percentages to total.

Year-wise Extension Activities

As could be seen from Table 37.3 that, in all 344 programmes were conducted through 90 extension activities in which 97,975 participants were involved. At the end of 15 years, average per year 23 programmes in 6 activities 6,532 participants were involved in KVK extension activities.

TABLE 37.3: Year-wise Extension Activities of KVK, Pal (1984-1999)

Sl. No.	*Year*	*Number of Activities*	*Number of Programmes Activity*	*Number of Participants*
(1)	(2)	(3)	(4)	(5)
1.	84-85	4	12 (3.48)	390 (0.39)
2.	85-86	7	33 (9.59)	1,312 (1.33)
3.	86-87	6	19 (4.56)	2,562 (2.61)
4.	87-88	7	23 (6.68)	1,935 (1.97)
5.	88-89	7	29 (8.43)	2,206 (2.25)
6.	89-90	5	20 (5.81)	2,126 (2.17)
7.	90-91	5	25 (7.27)	1,799 (1.83)
8.	91-92	7	28 (8.14)	2,150 (2.19)

(*Contd.*)

TABLE 37.3 (*Contd.*)

(1)	(2)	(3)	(4)	(5)
9.	92-93	6	39 (11.33)	53,305 (54.40)
10.	93-94	7	32 (9.30)	11,032 (11.26)
11.	94-95	6	15 (4.36)	2,308 (2.35)
12.	95-96	6	17 (4.94)	2,413 (2.46)
13.	96-97	3	11 (3.19)	894 (0.94)
14.	97-98	5	21 (6.10)	6,419 (6.55)
15.	98-99	4	25 (7.26)	7,134 (1.28)
	Total	90	344 (100.00)	97,975 (100.00)
	Average	6	23	6,532

Note: Figures in parentheses indicate percentages to total.

Activity-wise Break-up of Extension Programmes

The major extension activities of the KVK, Pal were field days, film show, exhibition, organisation of Shivar Pheri and Krishi mela. However, field days and film show occupied the predominant shares (32.84% and 28.48%, respectively). The highest number were observed in exhibition. During the time period of 15 years, average 29 extension activities were conducted and 8,164 participants were benefited (Table 37.4).

TABLE 37.4: Activity-wise Break-up of Extension Programmes (1984-99)

Sl. No.	*Activity*	*Number of Programmes*	*Participants*
(1)	(2)	(3)	(4)
1.	Field days	113 (32.84)	8,936 (9.12)
2.	Film show	98 (28.48)	6,110 (6.23)
3.	Exhibition	19 (5.52)	61,106 (62.35)
4.	Organisation of farm	24 (6.97)	1,008 (1.02)

(*Contd.*)

TABLE 37.4 (*Contd.*)

(1)	(2)	(3)	(4)
5.	Science club	6 (1.74)	1,213 (1.23)
6.	Farm tour	6 (1.74)	138 (0.14)
7.	Exposure visit	8 (2.32)	225 (0.22)
8.	Shivar Pheri	27 (7.84)	4,304 (4.34)
9.	Kisan Mela	20 (5.81)	9,261 (9.45)
10.	Celebration of Independence day	11 (3.19)	926 (0.94)
11.	Diagnostic visit camps	7 (2.03)	2,661 (2.71)
12.	Others	5 (1.45)	2,087 (2.71)
	Total	344 (100.00)	97,975 (100.00)
	Average	29	8,164

Note: Figures in parentheses indicate percentages to total.

Lab-to-Land Programme (LLP)

The Lab-to-Land Programme (LLP) was launched as a part of celebration to adopt the small and marginal farmers and landless labours to educate about the latest agricultural and allied technologies and to adopt these technologies and give the way to improve their economic and social life. The discipline-wise LLPs by the KVK, Pal are depicted in Table 37.5. The KVK, Pal conducted total of 152 LLPs during the time period of 15 years. Nearly 58 per cent of the LLPs were conducted in crop production helpline. It was followed by the LLPs in horticultural and animal science. More than 50 per cent of farmers were benefited by LLPs in crop production. The magnitudes of LLPs in horticulture and animal science disciplines worked out to 18.43 per cent and 17.14 per cent, respectively.

Front Line Demonstrations

It is revealed from Table 37.6 that, in the case of groundnut, 20 farmers had participated in demonstration with the adoption of technology on 8 ha. area for variety T4-26 and local check of variety SB-11, the result of

TABLE 37.5: Discipline-wise Number of Lab-to-Land Programmes during the Year 1986-87 to 1996-97

Sl. No.	*Discipline*	*Number of Programmes (enterprise/demonstration)*	*Number of Beneficiaries*
1.	Crop production	88 (57.89)	1,285 (50.31)
2.	Horticulture	23 (15.13)	481 (18.43)
3.	Animal Science	17 (11.18)	438 (17.14)
4.	Home Science	13 (8.55)	134 (5.24)
5.	Agricultural Engineering	11 (7.23)	216 (8.45)
	Total	152 (100.00)	2,554 (100.00)

Note: Figures in parentheses indicate percentages to total.

the demonstration plots showed increase in yield (38.31%). In the second case blackgram, 50 farmers with 10 ha. of area had grown variety TAU-1 on demonstration plots and increase in yield was 29 per cent over local check. Whereas, 15 farmers in 5 ha. area had grown variety Vijay, which resulted in increase of plot yield by 24.62 per cent. The increase in yield was 45.72 per cent in the case of green gram whereas 12 farmers in 5 ha. had grown variety PM-9339.

TABLE 37.6: Result of Front Line Demonstrations of KVK, Pal for the Year 1999-2000

Crop	*Variety*	*Farming Situations*	*Number of Farmers*	*Area (ha.)*	*Yield*				*Increase in Yield (Per cent)*
					Demonstration			*Local Check*	
					Highest	*Lowest*	*Average*		
Ground nut	TG-26	SB-XI	20	8	17	12.75	14.80	10.70	38.31
Black-gram	TAU-1	T-9	50	10	2.50	1	1.55	1.2	40.00
Bengal gram	Vijay	Chapha	15	6	20	9.75	14.02	11.25	24.62
Green-gram	PM-9339	Kopar-gaon	12	5	10.50	5.00	6.83	3.75	45.72

Impact of KVK, Pal on Income, Expenditure and Consumption Pattern of Beneficiaries

The KVK is a grass-root level institution designed to impart need based and skill oriented short and long-term vocational training to the farmers. The main objective of KVK is to provide training supports for bringing about production break-through in agricultural and agribusiness activities. The impact of KVK, Pal on income, expenditure and consumption of beneficiaries over the time period of 15 years is studied and depicted in Table 37.7. It was observed that, at the overall level the income, expenditure and consumption of the selected farmers were increased from Rs. 27,770.96, Rs. 15,908.37 and Rs. 1,873.07 to Rs. 35,126.95, Rs. 18,655.16 and Rs. 2,483.50 respectively after the training. The remarkable changes in these indicators were noticed in the case of landless and small farmers. The income and expenditure of landless and small farmers were increased by 42.22 per cent and 60.33 per cent respectively over the base year.

TABLE 37.7: Average Annual Income, Expenditure and Consumption of Selected Farmers for the Year 1998-99 and 1999-2000

(in Rs.)

Sl.No. Particulars	*Landless*	*Small*	*Medium*	*Large*	*Overall*
1. *Income*					
B	12,747.16	14,414.00	35,353.81	49,168.87	27,770.96
A	17,276.00	19,484.76	43,759.83	59,987.08	35,126.95
% change	42.22	35.51	23.77	22.00	28.80
2. *Expenditure*					
B	6,211.90	5,356.70	22,423.00	29,641.90	15,918.77
A	10,634.10	6,131.20	25,630.49	32,824.80	18,655.16
% change	61.53	14.26	14.30	107.3	17.26
3. *Consumption*					
B	900.00	1,315.30	2,316.40	2,960.60	1,873.07
A	1,210.70	1,833.40	3,038.60	3,851.30	2,483.50
% change	34.52	39.39	31.17	10.69	32.58

Note: B—Before, for the base year 1998-99; A—After, after training, *i.e.*, year 1999-2000.

Growth Rates of Trainings, Demonstrations and Trainees of KVK, Pal

The growth rates in trainings, demonstrations and trainees of KVK, Pal for the period from 1984-85 to 1998-99 are presented in Table 37.8. The rate of year-wise increase in trainings, demonstrations and trainees of

KVK was not uniform during the entire period. The growth rates of these variables were positive and significant in Period-I (1984 to 1993-94). Similar type of trends were noticed pertaining to demonstrations during Period-II (1994-95 to 1998-99) and the entire time period as well (1984-85 to 1998-99). In conclusion, the analysis revealed that there existed wide variations in the performance of KVK, Pal in terms of change in trainings, demonstrations and trainees over entire period of time. The Period-I was characterised by increase in trainings, demonstrations and trainees of KVK as compared to Period-II and was mainly due to availability of hundred per cent grants as well as scientific staff.

TABLE 37.8: Annual Compound Growth Rates of Number of Trainings, Demonstrations and Trainees of KVK, Pal (1984-85 to 1998-99) Annual Compound Growth Rates (ACGR)

Sl. No.	*Period*	*Number of Trainings*	*Demonstrations*	*Trainees*
1.	Period-I (1984-85 to 1993-94) N = 9	33.18*	19.27*	5.97*
2.	Period-I (1994-95 to 1998-99) N = 6	–1.19	12.69	7.86
3.	Entire period (1984-85 to 1998-99) N = 15	–3.09*	2.72*	–2.52*

* Significant at 5% level of significance.

Summing Up

Based upon the findings of the study it can be summarised that the progress of KVK is remarkable. The KVK, Pal succeeded in accomplishment of its objectives to maximum extent by conducting different programmes, *viz.,* trainings, demonstrations and various extension activities. The KVK has made positive impact on income, expenditure and consumption of the beneficiaries. The results of frontline demonstration indicated increase in the yield of oilseeds and pulses over local checks. As such the policy should be devised for improvement of agribusiness.

REFERENCES

Anonymous (1998). *Transfer of Technology Annual Report, 1997-98*, Indian Institute of Horticulture Research, Hessaraghatta Lake, Post Bangalore, p. 154.

Anonymous (1998): *Krishi Vigyan Kendra Selsura*, 29th Annual Report, Punjabrao Krishi Vidyapeeth, Akola.

——— (1999). *Transfer of Technology*, Annual Report, 1998-99, Indian Council of Agriculture Research, New Delhi, p. 120.

Naik, K., Venkata Raja, Natraju, M.S. and Crowder Loyvan (1996). "Training Farmers through Krishi Vigyan Kendra for Sustainable Agricultural Development", *Agriculture Extension Review*, Vol. 8, No. 6, p. 4.

Sangramsingh, S.P., Kanugo, A.P. and Mahapatra, B.P. (1993). "Impact of KVK on Socio-economic and Infrastructure Utilisation of Farmers", *Orissa Journal of Agriculture Research*, 31-38.

Sinha, M.N., Ramamurthy, R.S. and Vinayagam, S.S. (1994). "Prosperity of Farming Community through Krishi Vigyan Kendras", *Indian Journal of Dairy and Biosciences*, 61-64.

38

FUNCTIONING OF 'HOPCOMS'

Awareness of Member Producers and Opinion of Consumers

— *G.T. Ravishankar* and *N.R. Gangadharappa*

Professional Management to augment production of fruits and vegetables, efficient marketing to encourage producers to increase production and ensure retail distribution to make it accessible to consumers at affordable price, and lastly promotional effort to increase consumption for healthy nutritional status, are three inter-related, interdependent programmes. They are the leaping steps towards uphill task of achieving food security, healthy nutritional status and ultimately to poverty alleviation in developing and under-developed countries.

In India, no doubts, that the agricultural development has been accomplished in nation food security but without a significant impact on socio-economic development.

India produces around 12 per cent of world's vegetables and 8 per cent of world's fruits. Presently, all horticultural crops put together covers about 12 million hectares, which is approximately 7 per cent of the cropped area of the country.

The estimated annual production of these crops is about 103 million tonnes, contribution of over 18 per cent the gross agricultural output of the country.

The importance of fruits and vegetables in human diet needs no emphasis. Our planned efforts envisage a balanced diet for every family, necessitating a considerable increase in production and productivity of fruits and vegetables. The question now arises is the proper balance between the incentive price to farmers and reasonable price to consumers. This focuses on the dual issue of reducing the price spread between the primary producer and the ultimate consumer without impairing the quality of goods and services.

Present vegetable and fruit marketing is not favourable and profitable to farmers, as there is a wide gap between consumers price and producers price. Intermediaries are exploiting farmers as well as consumers. The apparent solution to increase growers share and to evolve a suitable marketing system to protect the benefits both growers as well as the consumers lies in the development of co-operative marketing organisations.

Efforts have been made to set up organisations like Growers Co-operative Marketing to undertake direct marketing of fruits and vegetables from farmers to consumers to avoid intermediaries. They mainly aimed at reducing marketing cost, ensuring fair returns to growers, quality supply of fresh commodities at reasonable price to consumers.

The co-operative societies/farmers associations are one of the important institutions which are meant for the betterment of the farming community. In Karnataka, the Department of Horticulture with a burning desire to end the age old exploitation by middlemen took a vow not to rest until an alternative arrangement is made for the marketing of fruits and vegetables. The result is the birth of HOPCOMS during the year 1959, the enlarged form of which is the Horticultural Producers Co-operative Marketing and Processing Society Ltd. The Society had jurisdiction over Bangalore Urban and Rural, Kolar, Mysore, Tumkur, Mandya, Mangalore, Dakshina Kannada, Chikkamagalur and Shimoga.

From its, inception HOPCOMS is providing different facilities to producers and consumers. So the present research study was conducted to find out how the producer farmers are aware of different facilities extended to them by the HOPCOMS and also what the consumers feel towards facilities provided to them and to seek perceived suggestions for the improvement of this co-operative institution.

Methodology

From the nine districts in which HOPCOMS is operating 3 districts were selected for the study purposefully since in these districts there is more area under fruits and vegetables and also through their proximity to urban centres they are the mainhinter lands to supply fruits and vegetables. From the selected districts, one taluk has been selected considering the criterion that taluks having highest membership in HOPCOMS. 30 member-producers from each taluk, *i.e.*, Devanahalli from Bangalore Rural, Siddlaghatta from Kolar and Mysore taluk from

Mysore district were selected. So that the total sample size of member-producers comes to 90, and also 3 outlets from the above mentioned taluks considering the criterion that outlets having highest turnover in previous year. 30 consumers visiting the outlets were selected and interviewed to elicit the consumers opinion. So that consumer sample comes to 90. In all, the total sample size was 180 consisting of 90 each producers and consumers.

Results

Table 38.1 shows the awareness of member-producers regarding the functioning of HOPCOMS.

TABLE 38.1: Awareness of Member-producers towards the functioning of HOPCOMS Ltd.

(*N* = 90)

Sl. No.	*Awareness Statements*	*Aware (%)*	*Not Aware (%)*
1.	Awareness about new procurement centres	99.00	1.00
2.	Supply of fruits and vegetables to bulk buyers	93.00	7.00
3.	Processing of fruits and vegetables	96.00	4.00
4.	Providing manures, fertilisers, plant protection chemicals to producers	96.00	4.00
5.	Price fixation after grading the produce	98.00	2.00
6.	Modalities of price fixation	79.00	21.00
7.	Canteen facility at the premises of HOPCOMS	97.00	3.00
8.	Banking facility at the vicinity of HOPCOMS	100.00	0.00
9.	Lodging facility to producer farmers	96.00	4.00
10.	Procurement facility to the nearest point	91.00	9.00
11.	Technical support at procurement centres	45.00	55.00
12.	Cold storage facility at headquarters	58.00	42.00
13.	Supply of fruits and vegetables to consumers through societies outlet	58.00	42.00
14.	Recently launched mobile sales facility	73.00	27.00
15.	Year of starting of HOPCOMS	28.00	72.00
16.	Founder person of HOPCOMS	88.00	12.00
17.	Objectives of HOPCOMS	97.00	3.00
18.	Number of districts in which HOPCOMS is operating	27.00	73.00
19.	Organisational set-up of HOPCOMS	59.00	41.00
20.	Types of membership classes	84.00	16.00
21.	Progress of HOPCOMS over last 10 years	94.00	6.00
22.	Whether, HOPCOMS is running profitably?	96.00	4.00

It is clearly evident from the results that about 80-100 per cent of member-producers had high awareness regarding new procurement centres started activity of supplying fruits and vegetables to bulk buyers processing of fruits and vegetables, providing manures, fertilisers and plant protection chemicals, knowledge of grading before price fixation, modalities of price fixation, facilities like canteen, banking, accommodation and regarding activities like Technical support at procurement centres, cold storage facility, organisational set-up of HOPCOMS about 45-60 per cent of member-producers had awareness. They had higher awareness regarding recently launched mobile sales facility, founder person of HOPCOMS, objectives and progress of HOPCOMS and for whether HOPCOMS running profitable, the awareness ranges from 73-96 per cent.

Number of districts in which HOPCOMS is operating and year of starting of HOPCOMS member-producers had low awareness, *i.e.*, 27 and 28 per cent respectively.

It was found that majority of the consumers satisfied with the facilities like public transport, convenient parking place in the premises, correct weighing, provision of carry bags, immediate service, timings of outlets, removal of unwanted leafage, roughage etc. Removal of spoiled, infected and diseased parts, safety for consumption, freshness of fruits and vegetables, retail price, attention of sales personnel to consumers and quality service rendered and also an significant per cent of consumers expected improvement on the following items like Hygienic condition, disposal of spoilage and wastage, refrigeration facility, grading and sorting, packaging of clean produce in right quantity, availability of required item, display of price tag (Table 38.2).

Discussion

The member-producers had high awareness with respect to certain activities/functions of HOPCOMS. It may be attributed to the reason that they involved personally with the day-to-day activities of HOPCOMS. However only 45 per cent of member-producer were aware of technical information provided at procurement centres, 28 per cent are aware of year of starting of HOPCOMS and 27 per cent were aware of the number of districts in which HOPCOMS is operating. The probable reason for less awareness on the above said items might be due to the fact that items demanded specific knowledge to be recalled by the respondents, which may be difficult for them to remember.

Majority of the consumers expressed their satisfaction regarding majority of the items considered for the study. While a significant per cent of consumers opined for improvement which has to be taken care by the society for the welfare of both producers and consumers.

TABLE 38.2: Consumers Feeling towards the Facilities provided by the HOPCOMS

(N = 90)

Sl. No.	*On Five Marketing Issues*	*Percentage*		
		Satis-factory	*Needs Improve-ment*	*Not Satis-factory*
I.	*Premises of the sales point*			
	(a) Hygienic condition	36.00	57.00	7.00
	(b) Disposal of spoilage and wastage	41.00	58.00	1.00
	(c) Convenient parking place in the premises	73.00	27.00	0.00
	(d) Public transport facility	99.00	1.00	0.00
II.	*Types of service*			
	(a) Correct weighing	98.00	2.00	0.00
	(b) Provision of carry bags	68.00	32.00	0.00
	(c) Refrigeration facility	4.00	94.00	2.00
	(d) Immediate service	86.00	14.00	0.00
	(e) Timings of outlets	94.00	6.00	0.00
III.	*Products sold*			
	(a) Grading and sorting	21.00	79.00	0.00
	(b) Removal of unwanted leafage roughage etc.	96.00	4.00	0.00
	(c) Removal of spoiled, infected and diseased parts	96.00	4.00	0.00
	(d) Packaging of clean produce in right quantity	13.00	87.00	0.00
	(e) Safety for consumption	92.00	8.00	0.00
	(f) Freshness of fruits and vegetables	69.00	31.00	0.00
	(g) Availability of required item	36.00	52.00	12.00
V.	*Price of the products*			
	(a) Display of price tag	39.00	55.00	6.00
	(b) Retail price	83.00	13.00	4.00
V.	*Sales personnel*			
	(a) Attention of sales personnel to consumers	96.00	4.00	0.00
	(b) Quality service rendered	94.00	4.00	2.00

Conclusion

The results obtained indicated that some of the facilities provided by the HOPCOMS like technical information at procurement centres need to be popularised to make more number of member-producers to aware of the facility which serves to remove the major obstacle in carrying highly expensive crops like fruits and vegetables. And regarding other items also mass media can be effectively used to show the existence of the society for non-members to bring them under the fold of co-operative movement. And consumers opinion for improvement needs careful attention by authorities concerned to meet the standards expected by the consumers to satisfy and to attract more number of consumers which in turn helps society, producers and consumers of the State.

39

MANAGEMENT OF PUBLIC SECTOR SEED INDUSTRY

— *R.R. Suryawanshi, V.P. Solanke* and *P.M. Kapase*

It is well recognised that the achievements of progressively rapid growth rate in agricultural production is the most important prerequisites for economic development in India. The expected gains in agricultural production can only be achieved by application of science and new modern technology. The new strategy comprises of balanced use of fertilisers, plant protection measures, farm mechanisation, advanced irrigation method and most particularly the use of high yielding varieties and quality seeds of crops. Strong backward linkages are necessary for successful agribusiness activity. As such the production and marketing of seeds have assumed vital importance in the modern days especially with changed economic scenario under WTO regime. In this chapter efforts are made to cover management issues of public sector seed industry in Maharashtra.

Objectives

Public sector industry like *Mahabeej* in mostly engaged in production and distribution of high yielding varieties of cereals, pulses, oilseeds, self pollinated crops and also produces seed of hybrids and vegetables. The main aim of *Mahabeej* is to meet and satisfy targeted customers needs and wants. Main objective of paper was to study production and marketing management of selected *Mahabeej* seeds. The issues like problem faced by seed growers in seed production, marketing mix, marketing cost and farmers and dealers opinion regarding the use of *Mahabeej* seeds has also been covered.

Methodology

Akola and Washim districts were selected purposively and Cotton, Jowar and Soyabean are the major crops of these districts were covered in the present study. Production and processing plants of Akola and Washim were selected. On the basis of highest area under the selected crops and higher sale of *Mahabeej* seeds, three tehsils from Washim and two tehsils from Akola district were selected. Two villages from each selected tehsils, nine farmers from each selected villages, *i.e.*, three farmers from each size groups, *viz.*, small, medium and large were selected randomly. Five *Mahabeej* dealers from each selected tehsil were selected to know their behaviour towards *Mahabeej* seeds. Thus the total sample comprised of 90 farmers and 25 dealers. The data pertaining to year 2000-2001 were collected and analysed by appropriate statistical tools.

Results

Features of Maharashtra State Seed Corporation Limited (MAHABEEJ)

The M.S.S.C.L. was established on 28th April, 1976 whose headquarters is at Shastrinagar, Akola. It has six heads of departments, *viz.*, production, processing, marketing, quality control, administrative and finance with 9 regional and 26 districts offices. Its share capital structure constitutes 35 per cent from Government of Maharashtra State, 35 per cent from Universities and farmers and remaining 30 per cent from NSC with financial turnover of more than 150 crores. In the production and distribution sector 1,400 dealers and 40,000 seed producing farmers were involved. The seed production programme of 52 crops and 225 varieties on one lakh hectares area has undertaken with production of 60,000 MT of certified and foundation seeds, contributing 55 per cent share in seed market. M.S.S.C.L. started research and development activities having 1993 and concentrating on bio-technology establishment such as tissue culture, transgenic hybrids and varieties of crops.

Socio-economic Factors influencing Farmers Behaviour

Socio-economic factors of selected farmers such as age, education, family size, annual income, land-use pattern, cropping pattern, area irrigated by different sources were studied so as to verify the direct or indirect influence of farmers' behaviour towards the use of *Mahabeej* seeds or adoption of hybrid, HYVs and new varieties of *Mahabeej* seeds in total. Factors such as education, land holding, annual income of selected farmers showed positive and significant association with adoption and use of *Mahabeej* seeds.

Opinions of dealers regarding the use of *Mahabeej* seeds were also studied. About 64 per cent of dealers reported that *Mahabeej* sale was more due to quality seeds, while 68 per cent reported that M.S.S.C.L. gives better consultancy. Overall experience of dealers about M.S.S.C.L. seeds was satisfactory and they indicated that M.S.S.C.L. seeds are available at reasonable prices, good demand, best quality, better services and moderate thrust.

Seed Production Management

Mahabeej do not produce seed on its own field but it manages seed production programme on farmers fields by fixing the targets. Further booking, seed distribution, reserved programme, inspection, seed production, collection, lab test, processing and lastly packaging activities are carried out.

At the first stage *Mahabeej* fixes its seed production strategy as to how much to produce and where to produce on the basis of market research such as responses of farmers and dealers, trend and requirements of an area. Based of this total seed production targets for region and districts were finalised. The interested farmers have to pay Rs. 100 acre as booking fee in the district office at the block level.

Agriculture universities supply the foundation seeds to *Mahabeej* which are further distributed to the farmers as per area allotted by taking cash values of seeds. Only breeder seed production programme was implemented on experienced and guaranteed farmers fields by *Mahabeej*. Field officers of *Mahabeej* visit the seed plot 3-4 times for proper inspection and then seed sample is sent to laboratory for different tests such as purity percentage, germination percentage, etc.

Marketing Management

Marketing management aspect identifies the customers needs, develop products and services to meet their needs, establishes pricing policies and promotional programmes and implement a system of distribution to the consumers. Marketing management involved decisions on four important areas which often referred as marketing mix. These are decisions on product, price, promotion and physical distribution.

M.S.S.C.L. had started its own research and development activities from 1993 and uptill now total 9 varieties are released in the market. Besides this, improvement in product packing has been made so as to increase the extent of sale. *Mahabeej* had constructed special storage structures in order to prolong the life stage and to maintain quality of seeds.

Mahabeej fixed the procurement price on the basis of survey of APMC during the marketing period after handling as average price in area plus 20 per cent more. Moreover, if possible bonus is provided to seed growers by *Mahabeej*. They adopted skimming the market pricing method for price fixation. For sale price of seed, market and cost pricing method is used.

Promotional activities like advertisement in newspapers, radio, T.V., demonstration, poster campaigning and incentives to dealers, etc. are undertaken by the *Mahabeej*. About one per cent amount is spent on promotional activities.

Four marketing channels were identified in marketing of seeds by *Mahabeej*. Seed is supplied through authorised dealer, retailer, co-operative society, Krishi Seva Kendra, Department of Agriculture to farmers.

Marketing Cost of NHH-44 (Cotton), CSH-9 (Jowar) and JS-335 (Soybean)

The major crops, *viz.*, Cotton, Jowar Soybean are grown in Akola and Washim districts, which account 66.86 and 57.42 per cent area respectively of the total cropped area. The varieties of these crops which are preferred by the farmers are covered for estimation of cost of marketing. Cotton hybrid, *viz.*, NHH-44 is more preferred by large number of farmers, whereas CSH-9 variety of Sorghum and JS-335 variety of Soybean are preferred by 81.22 and 75.89 per cent farmers, respectively.

It is observed from Table 39.1 that, the major item of cost of marketing of seeds was procurement price which contributed 66.66, 60.94 and 66.58 per cent of the total selling price for Cotton, Jowar and Soybean, respectively. While remaining items of total marketing cost were variable cost and landing cost which cover processing packings, transport, storage, charges of tax, octroi advertisement, interest and service charges of M.S.S.C.L., etc.

TABLE 39.1: Per Quintal Marketing Cost of Different Varieties of Cotton, Jowar and Soybean

Sl. No.	*Item of Marketing Costs*	*Seed Value in Rupees per Quintal*		
		Hy. Cotton NHH-44	*Hy-Jowar CSH-9*	*Soybean JS-335*
1.	Procurement price (Seed purchased by M.S.S.C.L. from seed grower farmers)	22,667 (66.66)	1,950 (60.94)	1,265 (66.58)
2.	Landing costs			
	(i) Processing and packing charges	3,213 (9.45)	290 (9.06)	118 (6.21)
	(ii) Low grade, lab test and other charges	6,120 (18.00)	576 (18.00)	342 (18.00)
3.	Variable costs			
	(i) Transportation and storage	680 (2.00)	64 (2.00)	38 (2.00)
	(ii) Sales tax/octroi	1,360 (4.00)	128 (4.00)	76 (4.00)
	(iii) Advertisement	340 (1.00)	32 (1.00)	19 (1.00)
	(iv) Interest on capital, MSSCL, services and other	3,740 (11.00)	352 (11.00)	209 (11.00)
4.	Dealer's margin	2,000 (5.89)	384 (12.00)	175 (9.21)
5.	Selling price by dealers (consumer price)	34,000 (100.00)	3,200 (100.00)	1,900 (100.00)
	Producer's share in consumer's rupee (%)	66.66	60.94	66.58

Note: Figures in parentheses are the percentages to consumer's price.

Demand of Specific Seed by the Farmers

The information regarding the factors responsible for farmers decision-making process and for a demand of specific seed is presented in Table 39.2. It is seen that brand or earlier experience of crops was the major factor for farmers decision-making. Another factor like purchasing of seeds which was reported by 80 per cent farmers, followed by dealer guidance (16%) and new variety (4%). While the factors for very specific seed demand was due to earlier experience (56%) and co-farmers experience (44%). The information on technical details about seed and its package of practices was provided through companies extension activities (60%) followed by literature (28%) and farmers opinion (12%). The opinions and problems faced the dealers were studied and it was observed that sale of M.S.S.C.L. was more due to quality and advertising aspects, however, there were complaints in records to germination.

There was demand for the seeds but the support for genuine complaints from *Mahabeej* was expected by the dealers. The problems like less extension, less margin, packing etc., were reported and improvement in regard to new research, packaging, quality etc., was expected by the dealers.

TABLE 39.2: Factors Important for Farmers Decision-making and Very Specific Seed Demand

Sl. No.	*Particulars of Attributes*	*Per cent to total N = 25*
1.	Factors important for farmers decision-making while purchase of seeds	
	(i) Brand or earlier experience of crops	80.00
	(ii) Dealers guidance	16.00
	(iii) New variety	4.00
2.	Factors for very specific seed demand	
	(i) Earlier experience	56.00
	(ii) Co-farmers experience	44.00
3.	Technical details about seed and its package of practices	
	(i) Companies extension activities	60.00
	(ii) Literature	28.00
	(iii) Farmers opinion	12.00
4.	Rely on guidance (given to you)	
	(i) Yes	76.00
	(ii) No	24.00

Preferences of farmers in purchasing of *Mahabeej* seeds were studied and it was observed that farmers preferred *Mahabeej* seeds of different crop varieties to large extent compared to other seed companies. The preferences of farmers for *Mahabeej* seeds was due to reasonable prices, good quality, prompt service and less risk involvement. Besides this, problems faced by the farmers in regard to availability of *Mahabeej* seeds was also studied and it was observed that non-availability of seeds in time, lack of guarantee and credit facility from dealer, unfair means in sale of seeds etc. were reported. In opinion of farmers, the *Mahabeej* have to pay attention in regard to good packaging, minimise the price, provided seeds of new varieties alongwith technical knowledge from specialised person, proper and prompt service etc.

Conclusions

From the study it is concluded that, the public sector organisation like

Mahabeej is successfully providing backward linkage to the agribusiness activities in this part of country. The farmers intend to purchase seeds of different varieties of *Mahabeej* because of high yield potential, quality treated seed, suitable packing and distribution pattern. However, the dealers as well as farmers have to face problems like lack of technical know-how, prompt service, attention to genuine complaints etc. and non-availability of seed in time, unfair means, lack of guarantee and credit facility from dealers, respectively. Therefore, the *Mahabeej* have to improve the seed production and marketing management efficiency for better and strong backward linkage.

40

PESTICIDAL APPLICATION ON COTTON CROP IN PUNJAB
Training Needs of Dealers

— *S.S. Gill* and *M.K. Sarda*

Cotton (*Gossypium* Sp.) the 'White Gold' is a premier textile fibre crop and also the second most important non-conventional oilseed of the world. Punjab is one of the major cotton producing States in India and it produces about one-fifth of the total cotton in the country. More than ninety per cent of the area is in the cotton belt which includes the districts of Bathinda, Faridkot, Ferozepur, Mansa and Sangrur.

The intensified production technologies have escalated the pest problems. While the application of pesticide is necessary to improve production and productivity, the social aspects of such application cannot be ignored. Pesticides need to be used judiciously in view of its high social costs such as environmental pollution. The use of pesticides is beset with conflicting interests between the farmers on the one hand and society on the other. Even though social considerations would warrant the use of target specific pesticides, which degrade expeditiously in the environment, the farmers who are more concerned with profitability, would tend to use persistent, broad spectrum pesticides in a manner that is not conducive to the interest of the society.

Pesticides are used in order to save the potential yield of the crops from damage due to various insect-pests. In Punjab 44.5 per cent consumption of pesticides is only on cotton crop. It is very important that the farmers should use these costly and poisonous input properly with respect to dosage, time and method of application with the twin objectives of maximising cost benefit ratio and minimising environmental pollution. The knowledge of these important aspects must be imparted to the farmers along with the supply of these agro-chemicals. The farmer is more receptive to the technology when he is to use it and that is why he intends to know about the plant protection techniques at the time of purchase of pesticides. Punjab has a network of more than 5,000 pesticide dealers. They have to maintain a close functional linkages with the farmers, extension workers, whole sale distributors and sales representatives of the pesticide manufacturers. The place of pesticide dealers in the rural social system in Punjab is enclosed as Fig. 40.1. Thus the pesticide dealers, have an important role of rendering advice regarding pesticide application technique. In the absence of proper education and training about safe use of pesticides, they are likely to give inadequate and wrong advice. So there is a need that these dealers must have some essential technical knowledge to serve the farming community in a better way. But before organising any regular training for these dealers, their training needs must be identified.

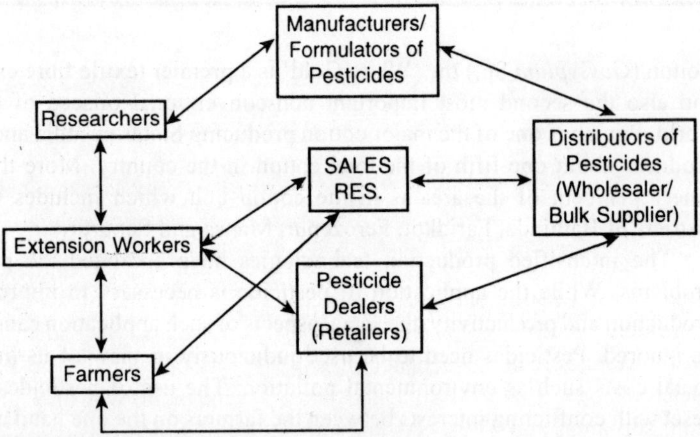

FIG. 40.1: Operational Model showing the Place of Pesticide Dealers in *Rural Social System* in Punjab

Thus realising the importance of training of pesticide dealers, the present study entitled "Pesticidal Application on Cotton Crop in Punjab: Training Needs of Dealers" was taken up with the following specific objectives.

Objectives

- To identify the subject matter areas/sub-areas, relating to pesticidal application on cotton crop, the knowledge of which is essential for dealers in Punjab;

- To find out the existing knowledge of the dealers in the identified subject matter areas/sub-areas;
- To determine the training needs of dealers on the basis of knowledge gap;
- To bring out the relationship, if any of major socio-personal characteristics with their training needs;
- To suggest a model curriculum and duration of training course for the dealers.

Research Methodology

The present study was conducted in the cotton growing areas of Punjab State in India. The cotton belt comprises the districts: Bathinda, Faridkot, Ferozepur, Mansa and Sangrur. All the five districts were selected for the purpose of this study. Out of top three cotton growing blocks in each district, two blocks were selected randomly. Thus ten blocks were selected for this study.

From the list of registered pesticide dealers, thirty dealers were selected randomly from each of the selected blocks and were invited for a group meeting on a specific date, time and venue. Of these, 206 dealers attended the group meetings and acted as respondents for this study.

From the available literature on pesticides, their sale and application, four areas; 12 sub-areas and 71 aspects, the knowledge of which was considered *prima facie* essential for pesticide dealers, were identified. For further screening, a panel of experts was consulted. The panel deleted/modified/regrouped/added some aspects. In this way three areas, four sub-areas and 27 aspects, the knowledge of which was considered essential for the dealers, were identified. A knowledge test consisting of 113 questions covering all the 27 aspects was prepared. The preliminary questionnaire consisting of knowledge test was pre-administered on a non-sample group. For the final test 65 questions covering 23 aspects were retained on the basis of difficulty index and item discrimination index. The research questionnaire was finalised after incorporating the suggestions, on the basis of preliminary administration of test and suggestions of advisory committee members.

The data for the present study was collected from the pesticide dealers by organising block level group meetings. The responses were quantified by assigning arbitrary score of one for correct and zero for incorrect response. The total score for each respondent was worked out by summing the scores for all the questions. The levels of knowledge score were categorised as low, medium and high:

Levels	*Score Range*
Low	20–40
Medium	41–60
High	61–80

In order to maintain uniformity and clarity, the knowledge score was observed out of maximum of 100.

$$\text{Knowledge score for each respondent} = \frac{\text{Total score obtained}}{\text{Maximum obtainable score}} \times 100$$

Mean knowledge scores were calculated as under:

$$\text{Mean knowledge score} = \frac{\Sigma \text{ of knowledge score of all the respondents}}{\text{Number of respondents}}$$

$$\text{Training need score} = 100 - \text{Knowledge score}$$

Suitable categories were developed for the various socio-personal characteristics of the respondents, *i.e.*, age, family background, other occupation, education, experience in sale of pesticides and exposure to agricultural components of mass media.

The data was analysed by using statistical techniques such as mean, standard deviation, frequencies, percentage and Pearson's coefficient of correlation.

Salient Findings

Profile of the Respondents

Majority of the respondents were young in age, with urban family background and only business as their occupation. Majority of the respondents had educational level of matriculation or above and business experience of four years or more, with medium exposure to agricultural components of mass media is enclosed in Table 40.1.

From Table 40.1, it is evident that agricultural graduates have not come forward to take up pesticide sale business. On the other hand, fresh agricultural graduates are facing the unemployment problem. Agricultural graduates, therefore, should be motivated to adopt this profession. The facilities like easy loan availability and preference in registration for pesticide licence should be extended to them. There is also need to encourage retired agricultural officers and scientists for taking up the sale business of farm inputs. In this way, their expertise can be gainfully utilised for the benefit of farming community.

TABLE 40.1: Profile of the Respondents

Sl.. No	*Socio-personal Characteristics*	*Category*		*Frequency (f)*	*Percentage (% age)*
1.	Age (years)	(a)	< 35	126	61.17
		(b)	35–50	70	33.98
		(c)	> 50	10	4.85
2.	Family background	(a)	Urban	151	73.30
		(b)	Rural	55	26.70
3.	Other occupation	(a)	Other than pesticide dealership	151	73.30
		(b)	Farming	39	18.93
		(c)	Service	16	7.77
4.	Education	(a)	Under-matric	8	3.88
		(b)	Matric	83	40.29
		(c)	Above matric (under-graduate)	36	17.48
		(d)	Graduate	68	33.01
		(e)	Agri-graduate	2	0.97
		(f)	Post-graduate	9	4.37
5.	Pesticide sale Experience (years)	(a)	< 4	44	21.36
		(b)	4–10	99	48.06
		(c)	> 10	63	30.58
6.	Mass Media exposure (scores)	(a)	Below (< 15) average	96	46.60
		(b)	Average (15–21)	105	50.97
		(c)	Above average (> 21)	5	2.43

Identification of Essential Subject Matter

An exhaustive list of 71 aspects was collected and rearranged under four areas and twelve sub-areas. After scrutiny, the panel of experts deleted/modified/regrouped/added the various aspects, sub-areas and areas. Finally three areas, four sub-areas and 27 aspects were identified, the knowledge of which is considered essential for pesticide dealers as given in Table 40.2.

Existing Level of Knowledge

The overall level of knowledge of the majority of pesticide dealers was medium as shown in Table 40.3.

TABLE 40.2: List of Subject Matter Areas/Sub-areas/Aspects suggested by a Panel of Experts

Areas	*Sub-areas*	*Aspects*
I. Technical Know-how	(1) Knowledge of pesticide	(i) Classification/groups of chemicals
		(ii) Formulation/active ingredient
		(iii) Mode of action
		(iv) Effect of over/under dose
		(v) Phytotoxicity
		(vi) Preparation of spray material/duct
		(vii) Waiting period
		(viii) Safe use of pesticides
		(ix) Compatibility
	(2) Damage by Pest	(x) Nature of damage
		(xi) Symptoms of damage
	(3) Plant Protection Equipments	(xii) Availability and selection
		(xiii) Use and maintenance
	(4) Application Technique	(xiv) Calibration
		(xv) Name of chemicals
		(xvi) Dosage
		(xvii) Time
		(xviii) Method
		(xix) IPM & ETL
II. Marketing		(xx) Source of pesticide availability
		(xxi) Selection of pesticides
		(xxii) Inspection of packing & labels
		(xxiii) Dispensation of pesticides
		(xxiv) Importance of original pack till last application
		(xxv) Collection of results of earlier application
III. Legal Aspects		(xxvi) Knowledge of pesticide dealers' liabilities
		(xxvii) Name of banned insecticides

TABLE 40.3: Distribution of Respondents according to their Overall Knowledge Level

Level of Knowledge	*Range of Knowledge Scores*	*f*	*%age*
Low	20–40	8	3.88
Medium	41–60	104	50.49
High	61–80	94	45.63

Area-wise Knowledge Level

Mean knowledge score was medium in all the three areas and it was highest in case of '*Technical Know-How*' followed by '*Legal Aspects*' and '*Marketing*' as shown in Table 40.4.

TABLE 40.4: Distribution of Respondents according to their Knowledge Level in the Different Subject Matter Areas

Subject Matter	*Number of Respondents under Various Knowledge Levels*						*Mean Know-ledge Score*
	Low		*Medium*		*High*		
	f	*%age*	*f*	*%age*	*f*	*%age*	
Technical know-how	69	33.49	38	18.45	99	48.06	60.86
Marketing	83	40.29	16	7.77	107	51.94	45.05
Legal aspects	60	29.13	0	0.00	146	70.87	55.18

Sub-area-wise Knowledge Level

High level of knowledge was found in case of '*Plant Protection Equipments*' and '*Pesticides*'. There was medium level of knowledge about '*Damage by Pest*' as shown in Table 40.5. No sub-area in Legal Aspects and Marketing could be identified.

TABLE 40.5: Distribution of Respondents according to their Knowledge Level in the Different Sub-areas of 'Technical Know-how'

Subject Areas	*Number of Respondents under Various Knowledge Levels*						*Mean Know-ledge Score*
	Low		*Medium*		*High*		
	f	*%age*	*f*	*%age*	*f*	*%age*	
Pesticides	73	35.44	29	14.08	104	50.48	62.53
Damage by pests	63	30.58	77	37.38	66	32.04	59.90
Plant protection equipments	32	15.53	17	8.25	157	76.22	74.15
Application techniques	109	52.91	30	14.56	67	32.52	46.91

Aspect-wise Knowledge Level

Out of twenty-seven aspects, high level of knowledge was found in case of fourteen aspects and these were: 'Inspection of packing and labels', 'Collection of results', 'Calibration of plant protection equipments', 'Use and maintenance', 'Effect of over/under dose', 'Liabilities under the Act', 'Classification/groups', 'Compatibility', 'Nature of damage', 'Phototoxicity', 'Method of application', 'Waiting period', 'Availability and selection of plant protection equipments', 'Safe use of pesticides'.

Medium level of knowledge was found in the aspects of 'formulation', 'active ingredients', 'symptoms of damage', 'IPM & ETL' and time of pesticidal application.

In the remaining nine aspects, the knowledge of the majority of the respondents was low and these aspects were: 'source of pesticide availability', 'selection of pesticides', 'dispensation of pesticides', 'mode of action', 'preparation of spray material', 'name of chemicals/weeds/insects/disease', 'dosages', 'importance of keeping original packing and banned insecticides'.

Training Needs

The training needs intensity in areas of '*Marketing*' was the highest followed by '*Legal Aspects*' and technical know-how. Sub-area-wise the training need was found highest in the case of '*Application Techniques*' followed by '*Damage by Pest*', '*Knowledge of Pesticides*' and '*Plant Protection Equipments*' under '*Technical Know-How*' area. High training need intensity was found in the aspects of, 'Source of pesticide availability', 'Selection of pesticides', 'Dispensation of pesticides', 'Mode of action', 'Preparation of spray material', 'Name of chemicals/weeds/insects/diseases', 'Dosages', 'Importance of keeping original pack' and 'Banned insecticides'. Training need intensity for 'Time of pesticide application', 'IPM & ETL', 'Symptoms of damage' and 'Formulation active ingredients' were found to be medium.

Relationship of Selected Socio-personal Characteristics with the Training Needs

It was found that the selected socio-personal characteristics like age, education, experience in pesticide sale and exposure to agricultural components of mass media were positively and significantly related with the knowledge score of the respondents. Thus it is concluded that age, education, experience in pesticide sale and exposure to agricultural components of mass media has inverse relationship with training need of the respondents.

Opinions of the Respondents about Training

Majority of the respondents expressed that minimum education for pesticide sale business should be matriculation. The duration of training course should not be more than one month. Punjab Agricultural University, Ludhiana should organise such training in the month of March/April/May and time devoted to practical should be fifty to seventy five per cent. Only one-fifth of the respondents felt that training should be imparted by officials of department of agriculture, Punjab.

Suggested Model of Curriculum

On the basis of expressed opinions of the respondents and training needs of the pesticide dealers it is suggested that the duration of pesticide training course should be at least of one month for the existing dealers. On an average in a month there are 140 working hours. Training need intensity was found to be high in nine aspects, medium in four and low in 14 aspects as shown in Table 40.6. It is suggested to allot four, five and six hours for each aspect of low, medium and high training need intensity respectively. This accounts for a total of 130 working hours. The remaining ten hours may be utilised for field visits and visits to manufacturing units of pesticide and plant protection equipments. The time devoted to practical should be at least fifty per cent. The suitable month for training are March to May. The training for technical know-how should be imparted by PAU scientists but for marketing and legal aspects the assistance from the enforcement staff of department of agriculture Punjab may be sought. Course weightage for essential subject matter areas/sub-areas are shown in Fig. 40.2.

TABLE 40.6: Training Need Intensity of Various Aspects

Training Need Intensity	*Aspects*
High	Source of pesticide availability, Selection of Pesticides, Dispensation of Pesticides, Mode of action, Preparation of spray material, name of pesticides, Dosages, Importance of original packing, Banned insecticides.
Medium	Time of spray, IPM & ETL, Symptoms of damage, Formulation/active ingredients.
Low	Safe use, Availability and selection of equipments, Waiting period, Method of application, Phototoxicity, Nature of damage, Compatibility, Classification/Groups, Liabilities under the Act, Effect of over/under dose, Use and maintenance of equipments, Calibration, Collection of results, Inspection of packing and labels.

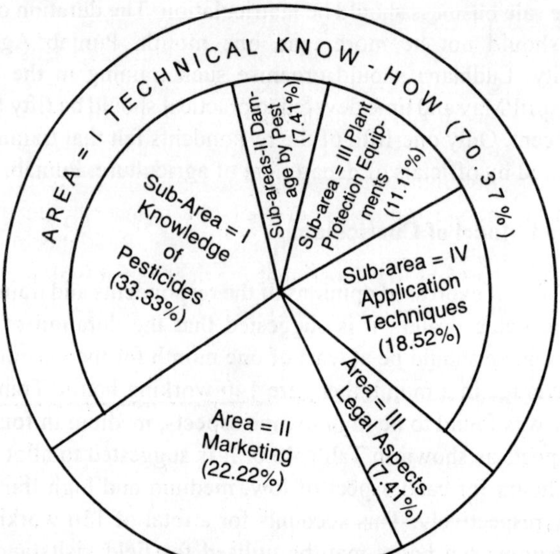

FIG. 40.2: Course Weightage for Essential Subject Matter Areas/Sub-areas

Conclusions

The overall knowledge level of the respondents was medium in the essential subject matters. It was found that there is a need to give training priority in order of '*Marketing*', '*Legal Aspects*' and '*Technical Know-How*' areas. Sub-area-wise the training priority should be given to the '*Application Technique*' followed by '*Damage by Pests*', '*Knowledge of Pesticide*' and '*Plant Protection Equipments*'.

High training need intensity was found in the aspects of 'Source of pesticide availability', 'Selection of pesticides', 'Dispensation of pesticides', 'Mode of action', 'Preparation of spray material', 'Name of chemicals/weeds/insects/diseases', 'Dosages', 'Importance of keeping original packaging' and 'Banned insecticides'. The knowledge of these nine aspects must be possessed by the dealers. In the aspects of 'Time of pesticidal application', 'IPM & ETL', 'Symptoms of damage' and 'Formulation/active ingredients' the training need intensity was medium, thus second training priority should be given to these aspects. So the knowledge of these four aspects should also be possessed by the dealers. It is nice for the dealers to possess the knowledge of remaining 14 essential aspects.

However, for future, an amendment should be made in the *Insecticide Act, 1968*, so as to make minimum training essential for registration as pesticide dealers. Minimum educational qualification for the registration should be 10+2 with science followed by with Certificate Course in Agriculture/Pesticide. Preference should be given to those possessing higher qualification like diploma/degree in agriculture. For the dealers already in business a training course of one month should be made compulsory for the renewal of Insecticide licence. Refresher courses of 2-3 days should also be started so that the dealers can voluntarily join such course. The provisions of *Insecticide Act, 1968* should also be amended that if any dealer possess lower qualification, then he should employ a person with the suggested minimum qualification.

41

RESEARCH REQUIREMENTS OF TRAINING FOR AGRIBUSINESS

— *K.R. Nadre* and *S.D. Wangikar*

Agribusiness Management refers to the application of basic managerial principles to a specific sector of our agricultural economy. The success of the business lies in the managerial performance and gauged on the basis of high profits.

With the dawn of World Trade Organisation and the propensity of the globe towards the liberalisation of economy, the realm of agribusiness has become fabulously vital. In the wake of this world scenario, dealing with agribusiness in a dextrous way and in a comprehensive manner becomes much more crucial.

In the Indian context, it could be seen that the development of agribusiness support activities is now not left to the private sector, but these activities are increasingly brought under the direct management and control of the government. There are also indications that in future, the government is likely to get more and more involved in prospering the agribusiness. Considering the overall magnitude of agribusiness sector, the new directives that are being given by the policy-makers and the plethora of activities undertaken by government, it is clear that development of training programmes for modernising agribusiness management and administration is bound to be the most difficult task. Hence, to explore the research requirements of such trainings to build them on sound footing has become indispensable.

Effectiveness of such training courses will depend upon the quality and suitability of the teaching material used in different courses. It will also be inevitable to employ a variety of educational methods in these courses, such as case-study method, small group discussions, syndicates, brain-storming, role playing and other experimental exercises. The choice of a particular method or technique should depend upon how best it fulfils the specific needs of the given situation in the course. It is felt that a somewhat greater emphasis should be placed on the case method of a study and group discussions because a case is a real life situation, written around a decision-making or problem solving situation. It describes the situation and provides facts, data and opinions to have a rationality.

The intellectual process of identification of relevant issues, analysis of the problems, development of alternatives and of making recommendations choosing the best alternative for implementation is essentially identical with the managerial decision-making process in real life.

It is often not realised that, to be effective a training programme requires a strong research base and the research has to be of applied nature.

Its objectives should be (1) Identification of training needs, (2) Development of suitable and appropriate teaching materials for the courses, (3) Identification of the difficulties encountered in the utilisation of knowledge and skills learnt through the courses in the actual field situation, and (4) Developing methods and ways to minimise these difficulties and resistance to change in organisations.

In the present context prior to development of training programmes, it is highly essential to understand:

(1) What are the major agribusiness activities which require management training? What are the priority activities and areas within activities?
(2) What is the nature of training programmes developed so far?
(3) What is the magnitude of training infrastructure, which impart training in the field of agribusiness administration?
(4) Where does agribusiness administration stand today in employing modern techniques of management?
(5) What are the management lacuna in the system?
(6) What are the skills that need to be imparted to managers and administrators to overcome these?

In the wider context, the questions that need be probed into are:

(1) What ways and means have to be devised to ensure the selection of only those who have the requisite personality, traits and aptitude for agribusiness development work?

(2) What are existing channels of communication in rural areas and the role and functions of different categories of agents of communication to further agribusiness.
(3) What is the attitude of the people towards agribusiness.
(4) How do different administrative deficiencies affect the agribusiness implementation process? At what level of organisation and at what stage of implementation are these most likely to emanate and in what form?
(5) What is the role and function of village extension worker? Is he competent to handle the responsibility of transmission of new knowledge about agribusiness to the farmers.
(6) What should be the nature of collaboration between the ICAR, SAUs and agricultural administration.
(7) What should be the nature of education imparted in Agricultural universities that would help the developmental administration in agribusiness sector.

Major functional problems which the plan of research on modern agribusiness administration and management should encompass are:

(1) Problems pertaining to role of specialists and generalists.
(2) Problems pertaining to the planning process, the levels of planning, felt needs of farm population, people's participation in planning.
(3) Problems pertaining to the execution of agribusiness programme, procedural aspects in execution.
(4) Problems pertaining to role and functions of non-officials at the planning and implementation level.
(5) Problems pertaining to communication between the rural population and administration, outlook, aspiration and motivation of rural people, people's reactions to programmes, means and channels of communication, administration and management of agribusiness programme.
(6) Problems pertaining to follow-up and evaluation of programme, evaluation techniques, methods and criteria employed.

Research in these functional problems is required to be carried out so that meaningful materials or training programmes in agribusiness management could be developed. Without such efforts training programmes in agribusiness would be barren and ineffective.

42

ROLE OF EXTENSION IN PROMOTING AGRIBUSINESS

— *Baldeo Singh* and *Vinod Kumar Chaturvedi*

Agribusiness is a thorough integration of agricultural activities of production, processing, marketing and export/import under a single institutional framework. In the present scenario of globalisation, agribusiness has a characteristic role for Indian economy to integrate with world economy. It calls for achieving sustained export growth, increased international competitiveness and greater links with international business. The recent call given by India's new agricultural policy to push agricultural exports to Rs. 10,000 crores (15% of country's overall export) gives further impetus to agribusiness to enter the world market. The significance of agribusiness gets highlighted in a situation when agricultural and allied products have enough potential to fetch export earnings. With the possibility of globalisation of Indian agriculture, production as well as marketing need to be re-oriented to meet the objectives of agribusiness. There is also a need for an appropriate extension strategy for promoting agribusiness towards meeting the challenges of globalization process.

The process of economic liberalisation which was initiated in 1991 has witnessed significant transformation of Indian agriculture. The trade surplus generated by agriculture sector is steadily increasing. The economic and social significance of agricultural export is quite higher due to scope for foreign exchange earnings and employment potential. The India's agriculture trade in post-economic reforms period is depicted in Table 42.1.

The data presented in Table 42.1 indicate that India's agricultural exports showed an impressive growth from three billion dollars in 1992-93 to six billion dollars in 1998-99. The share of agricultural exports in GDP also showed sustained trade orientation of Indian agriculture.

TABLE 42.1: India's Agriculture Trade in Post-Economic Period

(*US $ million*)

Year	*Exports*	*Imports*	*Net Exports*	*Exports as % of GDP Agriculture*
1992-93	3,028	1,618	1,410	5.22
1994-95	4,227	2,544	1,683	5.20
1996-97	6,868	2,303	4,565	7.30
1998-99	6,033	4,012	2,021	5.92

Source: *Economic Survey*, Ministry of Finance, GOI (Relevant Issues).

The Globalisation Process—GATT and WTO

The General Agreement on Trade and Tariffs (GATT) is a multilateral treaty established in 1948. India along with other countries signed the GATT on April 15, 1994 at Morocco. Subsequently, the World Trade Organization (WTO) has come into existence. The important provisions of GATT/WTO in relation to agricultural sector are highlighted below:

1. The agreement calls for a minimum market access in agriculture according to which the member-countries will have to import agricultural products to the extent of three per cent of domestic consumption. The alternative would be to export produce, which would have a higher price in international market and import cheaper food.
2. The agricultural trade before the WTO era was extremely restrictive and protected. Export and import were subject to protection through quantitative restrictions like licensing, quotas etc., extending even to complete bans on some commodities. The Agreement on Agriculture (AoA) commits all WTO members to long term reforms which would make agriculture trade fairer, distortion free, and more market oriented.

Agreement on Agriculture (AoA)

The main three elements of AoA are pointed out below:

1. Removal of all quantitative restrictions, variable levies, import bans and other non-tariff measures etc. and replace these by tariffs.
2. Restriction of subsidies granted by government to farmers to support price level with a view to keeping out imports. However,

subsidies which are not supposed to distort trade are allowed under AoA.

3. Commitments to reduce export subsidies in 10 years. (Reduction of subsidy by 36% in case of developed countries compared to 24% for developing countries).

With the current economic liberalisation and signing of GATT agreement and WTO implementation, immense opportunities have opened in India up for agribusiness to enter the global market. The obvious reasons for this are as follows:

- High economic return from enterprises.
- Increase in demand.
- Changing food habits.
- Availability of better technical support.
- Availability of better inputs.
- Availability of better credit and marketing facilities.

The importance of agribusiness has been increasing over the years. This has been mainly due to increase in agricultural production and market orientation in agriculture trade. Significance of agribusiness can be emphasised particularly in a situation when agricultural products have sufficient potential to fetch export earnings. The trend of India's export of selected agricultural commodities is presented in Table 42.2.

TABLE 42.2: India's Export of Selected Agricultural Commodities

(US $ million)

Item	*1992-93*	*1994-95*	*1996-97*	*1998-99*
Basmati rice	276.47	275.67	351.74	446.03
Non-basmati rice	60.41	108.47	542.63	1,046.54
Sugar and mollasses	122.07	19.80	303.89	N.A
Spices	135.85	195.04	338.92	387.96
Tea	337.21	310.76	292.38	538.23
Fresh fruits	N.A	60.13	68.90	63.29
Fresh vegetables	107.88	N.A	94.27	65.12
Processed fruits and vegetables	41.35	79.19	91.98	108.87

Source: Economic Survey, Ministry of Finance, Government of India (Relevant issues).

From the data given in Table 42.2, it is evident that agricultural products offer a great scope for exports. The GATT has given bright prospects for horticultural and processed products. This gives an ample scope for promoting agro-processing units as well as small scale industries for various value added products which would also generate employment opportunities.

Prospects of Agribusiness in Context of Globalisation

The pertinent issues revolve around globalisation of agricultural trade, the success of which depends upon how the country becomes efficient and competitive in production and marketing. In fact both need to be activised to meet challenges of integration of Indian agriculture with world trade. The important issues which need to be focused to bring about desired achievements related to agribusiness are discussed below.

High Level of Technological Intervention

The importance of agriculture in the international business has been increasing over the years. This has been mainly on account of increase in production of agricultural commodities arising out of technological improvements in the field of agriculture. It is imperative that farmers should accept and adopt new agricultural technology fully in their farming systems to achieve objectives of agribusiness.

Strengthening of Infrastructure including Marketing Facilities

Infrastructure development is a major factor in achieving higher performance of agribusiness activities. In order to fully exploit the export potential of agriculture sector, the infrastructure facilities like processing, storage, packaging and transportation have to be adequately developed at farm level. This is more particularly important to prevent post-harvest losses. An effective marketing system encourages increased agricultural production by offering incentive prices. Export marketing requires a sound and organised supply mechanism to ensure quality products, remunerative price and proper delivery system. An effective marketing strategy has to be developed and pursued to establish a meaningful existence of Indian agriculture in the world trade scene.

Diversification of Agriculture

Indian agriculture is considerably changing from subsistence to commercial farming. The changing food habits, increase in purchasing power and economic globalisation have thrown challenge for improving production of commercial crops to meet domestic and export requirements. The diversification within agriculture refers to shift from one crop to another or from one enterprise (*e.g.*, crop raising) to another enterprise (*e.g.*, livestock).

The horticultural crops and agri-allied activities offer a great scope for diversification. The prospective areas which deserve attention for diversification of agriculture are fruit crops, vegetable crops, floriculture, commercial crops, mushroom, poultry farming, fish farming, bee-keeping and diary farming etc. It is imperative that potential of these areas may be exploited to achieve the objectives of agribusiness in the context of globalisation.

Development of Food Processing Units

Food processing industries play a crucial role in agribusiness activities through value addition to farm products. It is necessary that steps should be taken by government as well as related institutions to develop food processing industries in the rural areas so that farmers could get remunerative price for their produce by means of value addition of products. This would also help to increase employment opportunities in the rural areas.

Restructuring of Agrarian Institutions including Cooperatives

The institutional support for providing better marketing opportunities has drawn considerable attention of government. The Central government has set up a Small Farmers Agribusiness Consortium (SFAC) to boost up agricultural production and to provide effective marketing linkages between producers and consumers. This organisation has emerged in 1992-93 as a registered society with the objective of employment and income generation in rural areas through financial support to agribusiness.

An active coordination between government and bulk consumers or exporters need to be built up to explore export potential of agricultural products and formulate strategies to match supply status. The role of NGOs in building up agribusiness holds quite a good promise. There is a need to strengthen the NGOs with a view to help them provide effective support to agribusiness activities. The cooperative movement has made progress in different social and economic fields. The cooperatives are vital to promote people's participation for common social cause. In economic field cooperatives have wide scope in rural areas. It is a common experience that organisational awareness is lacking among members of cooperatives in rural areas. There is a need to strengthen organisational structure and improve supporting services of cooperatives in order to boost up agribusiness activities.

Participation of Private Sector

The rate of investment in agriculture has declined in recent years. Adequate investment, both public and private, in agriculture sector is required to be ensured to meet challenges of globalisation. Possibilities of increasing trade in global market call for efficiency, competitiveness and innovativeness both in production and marketing. Hence, there has to be increasing emphasis on private sector initiative. The changes in economic scenario have made agriculture sector attractive to private investors. There is an imperative need that efforts of agricultural institutions are activised to provide support to private sector participation to fulfil the objectives of agribusiness.

Extension Strategy for Promoting Agribusiness

In the context of globalisation, there is a need for an appropriate extension strategy to achieve overall objectives of agribusiness. The challenge before extension is to evolve well directed and effective extension methodologies. The major issues and approaches that may help the extension strategy to promote agribusiness in relation to globalisation are delineated as under:

1. There is a need to upgrade the quality of extension and research support for development and propagation of agricultural technologies having potential for export. The extension system has to take initiative for transfer of complete package of practices of the crops which have export worth values. It includes the new package of practices in respect of high yielding varieties, techniques of propagation, use of inputs, plant protection measures and storage etc. There is an important role for extension to carry out assessment and refinement studies of available technologies for evolving new high-tech packages for export oriented crops in order to boost agribusiness activities.
2. An extension approach has to be identified which can provide continuous, relevant and latest technological messages to commercially oriented farmers through appropriate communication methods. This would mean to disseminate new cost effective agricultural technologies to potential farmers on

a regular basis with a view to help them to make constant improvements in their efforts to pursue agribusiness activities.

3. The improved agricultural technology in the form of high yielding varieties and intensive agriculture have raised the farm production and productivity which in turn have led to post-harvest problems such as storage, handling, transportation, packaging and marketing. In India the post-harvest losses are as large as 10 per cent in cereals and 30 per cent in fruits and vegetables. As fruits and vegetables are perishable, they would completely lose their quality and ultimately export value. Therefore, there is an important role for the extension functionaries to disseminate technical and managerial know-how on post-harvest technology to the potential farmers with a view to reduce post-harvest losses and achieve the stated objectives of agribusiness.
4. To compete in the world market farmers need to be informed in management and production of export worth crops. The extension system has to look into the information needs of export oriented farmers. It has to keep itself updated on the aspects of technology, export quality standards, processing, marketing and prices. In order to meet information requirements of export oriented farmers, the extension agencies have to exploit the modern communication technology including computer based technology and Internet connectivity.
5. In order to effectively organise the agribusiness activities, the potential farmers need to be provided training support so as to develop their technical competence and skills in relevant aspects of agribusiness. There is an imminent role for extension to make assessment of training needs of agribusiness oriented farmers and organise need based training programmes for them with a view to prepare them to face the emerging challenges and competition of global market.
6. In recent times, in an effort to diversify agriculture, farmers are taking to commercial and horticulture crops. There is a need for the extension system to exploit this situation. The KVKs should reorient programmes to impart skill oriented training programmes to the farmers in the areas of production and processing of these crops. The farmers are also required to be made aware about the quality standards, price structure and marketing facilities in order to boost agribusiness activities.

7. Agribusiness activities hold ample scope for entrepreneurship development which stand to promote income generation and employment opportunities. The extension agencies have to make concerted efforts to organise specific training for farmers in the relevant aspects of entrepreneurship development with particular reference to agribusiness activities in the context of globalisation of agriculture.
8. Globalisation is highly relevant to resource rich farmers. There is a little impact of globalisation on small and marginal farmers. In the fast changing agricultural scenario, the small and marginal farmers cannot be left to continue with traditional farming system. There is a need to involve small and marginal farmers in agribusiness activities by way of developing their capabilities and motivating them to make a shift in their traditional farming system in favour of commercial oriented agriculture and allied activities. This would mean to strengthen infrastructural, credit and marketing facilities in rural areas which should be accessible to these categories of farmers. In this context there is a need for the extension system to reorient itself to address to the problems of small and marginal farmers in the aspects of production, processing and marketing in an integrated manner with a view to give impetus to agribusiness activities.

Conclusion

The globalisation process of agriculture has put tremendous demand on Indian economy to integrate with world economy. In the present scenario of changing agriculture there are promising prospects for developing agribusiness activities particularly in a situation where agriculture and allied products have enough potential to fetch export earnings. In the context of globalisation of Indian agriculture, there is an imperative need for an appropriate extension strategy for promoting agribusiness. The major issues and approaches that may help extension strategy to promote agribusiness are upgradation of quality of extension and research; dissemination of relevant technologies through appropriate communication methods including computer based technology and internet facilities; carrying out assessment and refinement studies of available technologies; assessing information needs of agribusiness oriented farmers; organisation of need based training programmes for potential farmers through KVKs and other institutions; and ensuring participation of small and marginal farmers in agribusiness activities through self-help groups.

REFERENCES

Jayaramaiah, K.M. and Bisaliah, S. (1998). "Globalisation of Agriculture: Its Implication for Agriculture Extension," *IJEE*, Vol. 34, No. 1 and 2.

Prasad, J. (1995). "Promoting Agriculture Sector in Bihar: A Role Model of Agribusiness in the Liberalisation Perspective," *The Bihar Journal of Agricultural Marketing*, Vol. III, No. 4.

Rajagopal (1995). "Developing Agribusiness in India: Status and Challenges," in J. Prasad and A. Prasad, ed., *Indian Agriculture Marketing: Emerging Trends and Perspectives*, Mittal Publications, New Delhi.

Shashi Bhushan (1993). "Promoting Agribusiness: A Role Model to Augment Farm Sector in Bihar," *The Bihar Journal of Agricultural Marketing*, April-June 1993, Vol. II, No. 2.

Sharma, M.L. and Ambastha, S.K. (1995). "Globalisation of Indian Economy with Special Reference to Indian Agriculture," *The Bihar Journal of Agricultural Marketing*, Vol. III, No. 1.

Yadav, J.S. and Yadav, S.S. (1996). *Modern Trends in Agricultural Marketing*, Pointer Publishers, Jaipur.

43

INFORMATION MARKET PLACES
A Challenge to Extension

— *Baldeo Singh* and *Poonam Sharma*

The rapid technological advances in the fields of communication and information, liberalisation of economy and signing up of WTO agreement have brought agriculture at the cross roads. The uncertain and volatile market situation resulting in new emerging cropping trends, desperation and competition to raise good crops and earn more from small holdings, is causing an unprecedented knowledge and information demand by farming communities. The information driven market place and the growing competition from the private sector are two important factors compelling for introspection in the existing extension system. It is imperative to respond to these influences so that extension can facilitate the process of transforming agriculture from subsistence level to the commercial level. With this objective in mind this chapter is an attempt to address some of the relevant issues.

The Challenges of the Information Market Place

The era of information and communication advancement is bringing new challenges for our traditional extension system. Some of these challenges are:

Information and Value

Information now has real measurable value. Private-sector information providers such as crop advisors, business consultants, and private education and training providers are cropping up everywhere. The attraction is the revenue generated by refining information to meet specific audience needs.

No Longer Sole Source

Extension has spent decades as a sole-source provider in the information and outreach market. Technology has created ample opportunities to provide and promote access that was not realistic before. Because of this, the traditional education market is also more easily accessible for competitors—agribusiness and commercial crop market advisors etc. Technology combined with open access to the rapidly expanding knowledge market place entices private information providers to step up the competition for serving the clients (farmers) who previously were dependant on Extension as a sole-source of education provider.

From Distribution to Access and Empowerment

The present day rapid changes have brought a shift the "provider mentality" to "user mentality." A "Provider mentality" focuses on the process of distribution, meaning it is driven by what we want to distribute. A "user mentality" focuses on *access* and the customer, meaning we now need to anticipate customer's needs and provide them *access* to our knowledge base.

Reach vs. Richness

The *richest* form of interaction with Extension educators has always been face-to-face. But *reaching* a mass audience with face-to-face *richness* requires more resources than Extension could ever generate. But now think about Amazon.com (an internet portal for buying books on line), which uses technology to remember what individuals have ordered before and suggests other books they might find interesting, thus enhancing the *richness* of their interaction with the countless customers they *reach*. Extension adopts the same approach to meet the present as well as anticipated future needs of its clients are the demand of the hour.

Navigators in the Lead

Navigation is not a new concept to Extension. Sorting information and helping people navigate to effective answers has been the foundation of the Extension system from the beginning. However, with information explosion, the amount of information to navigate through is phenomenal. Are we up to it? Private-sector Navigators are driving fundamental power shifts among the other players. Given the ongoing privatisation of information, Extension must respond, or the private sector obviously will dominate.

Disruptive Technologies and Innovations

Christensen (2000) says, "Many of the most powerful innovations that disrupted other industries did so by enabling a larger population of less-skilled people to do, in a more convenient, less expensive setting, things that historically could be performed only by expensive specialists in centralised, inconvenient locations." Compare the distribution of traditional Extension educational programmes to the *access* model of other providers, where you can obtain what you want, when you want it, in the convenience of your own home or business. Clearly, if we do not change we could be described as having "relatively expensive specialists in centralised, inconvenient locations."

Our Competitive Advantage

Our traditional extension system though faced with the challenges of the new hi tech scenario but has certain competitive advantages, which can be exploited in its favour.

- We can leverage our brand identity of having a strong research base into these new markets. The value of science-based objective has dropped in relative importance lately, with access and timeliness moving up a higher priorities for outreach audiences. But objectivity will re-emerge as a high priority—and as our competitive advantage—when access and timeliness are offered by everyone.
- Brand identity can be maintained as ICAR and University research specialists build and offer new educational modules focused intensely on identified customer needs. Faculty and specialists can also be rewarded for their involvement, just as the private sector is currently doing, using consulting agreements.
- Having effective sensors in every community could give us a leg-up on private sector competitors. Grass-roots contacts can help us identify issues. We need to follow up with extensive and ongoing audience analysis pinpointing the needs of "communities of interest" well beyond traditional audiences.

- While attempting to cherry-pick off the best revenue-generating customers in multiple markets, private providers may not have enough foundation in any one market to gain access to the customer intelligence we can develop—if we have the right people in the field.

The Needed Strategy

A Shift from 'Extension as Knowledge Extended' to 'Extension as Knowledge Created'

The Extension system has a long and distinguished history of non-formal education focused on enhancing the well-being of individuals, families and communities. Hence, it is not surprising that Extension has developed a strong track record in extending research-based knowledge to agricultural families and communities. Extension has played an important role stimulating research and has helped identify new research agendas. Thus, Extension extends knowledge in both directions: from lab to land and land to lab—closing the feedback loop.

As communities have become more complex and integrated within the broader society and economy, Extension has had to address new challenges that do not lend themselves neatly to the disciplinary boundaries that traditionally define ICAR/university research. To meet these demands, Extension has to take a role as educator and facilitator to a new level—building community coalitions to engage in research for community problem solving. Such Extension programmes create local knowledge to be used in policy and programme design.

This reorientation reflects, in part, an international movement in non-formal education, participatory action research, designed to bring communities and researchers into a closer, more effective partnership which will result in social research for social change (Deshler and Ewert, 1995). Participatory action research not only incorporates the collective knowledge of the community, it increases the likelihood that research results will be applied by giving the community ownership over the research process and its results (Shafer, 1995).

Eldon and Levin (1991) speak of the importance of co-learning through a dialogue between practitioners and researchers. In this case, the Extension agents provided an "insiders" framework (emphasising Extension practice), while researchers provided an "outsiders" framework (providing theoretical context). As a result of the collaboration, Extension agents considered the larger theoretical ramifications of their stance, and researchers became more sensitive to and appreciative of practical details. The dialogue between the practical and theoretical resulted in new insights needs to create new forums for interaction and bridging ties among previously isolated groups.

Collaborations between researchers and Extension agents have traditionally respected a division of labour, which distances Extension agents from the research process and researchers from Extension practice.

In order to engage in successful research collaborations with Extension agents, the privileged position of researchers within the research/Extension partnership must be challenged. Mutual respect, confidence and trust must be built between Extension and research colleagues in order for Extension/research collaborations to be effective. Extension colleagues are co-equals with different areas of expertise; their time and voice must be honoured. Researchers must avoid the temptation to "leave their Extension partners behind" once they have gleaned the necessary local insights and do not want to trouble with the challenges of continuing to incorporate a practical Extension perspective in subsequent writing and analysis.

Reality complicates theory; and theory, by definition, attempts to simplify reality. Too close an engagement with local practice removes the distance often thought necessary to assure research objectivity. Knowledge is socially constructed and the collaborative research process capitalises on the value of differing interpretations of the same events. Such a close articulation with local actors inevitably pulls researchers into the action phase. This "erosion of research distance" is both the challenge and promise of participatory action research.

On the Extension side, effort must be made to lower the costs of engaging in academic scrutiny of one's practice. Time is a limited commodity and the value of time spent on research must be recognised at an institutional level. The benefits of seeing one's experience through a researcher's lens can yield new insights for future practice. However, if the research collaboration focuses on the past and ignores insight for the future, then the benefits to extension practice will be greatly reduced. Thus such collaborations strengthen the connections between research and practices are clear. Both research and extension stand to gain from new insights. However hidden costs to extension effectiveness and research independence should be assessed.

'Education Facilitator' to 'Information Provider'

Extension is in the process of expanding its delivery methods through the use of the latest technology. Innovative delivery efforts include satellite/television link-up, computer-assisted learning, distance learning, and dial-access information service. These delivery methods pose two potential problems: first, they can reduce the direct interaction between information provider and information user; and second, they tend to maximise educational information and minimise the processing of learning experiences by users. So in such a situation extension needs to change its mission from an institution that facilitates learning by clients to one that only disseminates information? By looking at "innovative" Extension delivery methods, we can judge if, in fact, Extension is changing its mission.

This is not to say that these delivery methods cannot facilitate learning. However, they require conscious effort by the educator to ensure that we do not confuse facilitating education with providing information.

In facilitating education, we include the opportunity for learners to process the educational information and make sense of the learning experience. In providing information, on the other hand, we only deliver information to the learner. We need to be careful that Extension professionals do not become caught up in the development and use of new delivery methods to provide information and forget their role in helping adults in their learning efforts.

How can delivery methods be used to facilitate learning in addition to providing information? The following three points may serve to answer this question.

Provide Continuity in Learning

Since contacts with adults may be brief, we need to use this limited opportunity to place the educational message into the learner's overall learning process by providing ideas on how inquiry can continue through readings, discussion, and classes. In this way, information is provided, but learning is also encouraged to continue. Any programme, regardless of its format, should encourage the learner to continue learning after the programme is finished.

Give Opportunity for Processing Material

In any type of delivery method, opportunity for reflection or integration of the material should be included. For example, one advertises a "workshop" on a particular topic, but it actually becomes a two-hour lecture with several Extension speakers and no opportunity for learner involvement.

Study Personal Styles of Learning

Extension people are continually engaged in learning. What things facilitate your learning? What hinders your learning? By studying our own learning, we might get a better picture of what works and does not work so we can adjust the learning experiences and the delivery methods we use with clientele.

In conclusion, it is much easier and faster to simply provide information than to take the time and effort needed to facilitate learning. Yet, to truly be seen as adult educators and serve the mission of "helping people to help themselves," we must take the time and effort.

Work Force in the Changed Scenario: Is Extension Agriculture Ready?

The effectiveness of Extension has always depended on its human resource base. Most of its budget—80 per cent to 90 per cent—goes for salary and benefits for its extension professionals nation-wide. Does Extension have the human resource base to maintain its competitive position in the delivery of agricultural and natural resources technology? Particularly in a situation where there is:

- Competition from private sector for workers.
- Cluster of development programmes with no single focus.
- A reward and recognition system that encourages individual competition rather than teamwork.
- A need for workers with more sophisticated skills in high technology.
- Reduction in staff due to budget cuts.
- Lack of training.

What can be done to identify and recruit some of the best and brightest young minds to agricultural, natural resources, and extension careers? Some suggestions:

- Improve the image of agriculture and natural resource careers.
- Pay competitive salaries.

- Target youth early to remove biases against agriculture.
- Analyse how young people make career decisions and determine what factors motivate them.
- Tap their interest in being part of the solution to environmental problems.
- Let people know this sector works on contemporary concerns and people issues.
- Reclassify agent positions to suit future needs.

Human Resource Development Policies must be set to restructure staffing patterns, enhance salaries and benefits, increase training and technical support, and implement a more family-oriented workplace. Without such efforts, Extension will be unable to maintain its competitive position in delivering research-based agricultural and natural resources knowledge and practices.

Public and Private Sector Linkages

By now it is clear that the private players entry in the field of agricultural extension is inevitable, is also important that the situation be utilised to its optimum benefits. Therefore, the advantage of the situation can be taken by developing partnership between the two, *i.e.*, public and private extension system. The phenomenon of transformation of traditional agriculture into commercial enterprise is mainly continued to big farmers. The resource poor small farmers development is an extremely complex process. According to an estimate during this new millennium out of 100 million operational holdings 90 million are small holdings. Further 70 per cent of the sown area in the country come under rainfed agriculture. These resource poor farmers in villages cannot afford to pay up for the services they receive. The time has not yet come to completely withdraw the public extension services. The Public extension system needs to focus its attention on the small and resource poor farmers and the overall rural development. Hence, a dynamic strategy is urgently needed to integrate public and private extension system for rural development in India. This is a challenge and needs to be tackled immediately.

Are We Up to It?

We would better be. If we are not, in the very near future we would not have a professional foundation from which to work. Anyone assuming they will be working in Extension in the same way they do today beyond the next 5 to 10 years is just not paying attention.

REFERENCES

Deshler David and Merril Ewert (1995). "Participatory Action Research: Tradition and Major Assumptions," *The PAR Tool Box,* Part #001, http://www.parnet.org/tools.

Eldon, M. and Levin, M. (1991). "Co-generative Learning: Bringing Participation into Action Research," in W.F. Whyte (ed.), *Participatory Action Research*, Newbury Park, CA: Sage, pp. 51-84.

PART V

SOCIO-ECONOMIC AND ENVIRONMENTAL IMPACT OF AGRIBUSINESS

44

FAMILY FARMING OF PIG AND POULTRY IN SOUTH BRAZIL
Impact on Socio-Economic and Environmental Sustainability

— *Airton Spies, Mal Wegener, S. Chamala* and *Bob Beeton*

There are 2,03,000 family in Santa Catarina but 90 per cent are smaller than 50 ha in area (IBGE, 1997). Intensive pig and poultry industries contribute 40 per cent of the gross value of agricultural production in this State. The vertically integrated production system, in which producers contract the production to an agribusiness firm, began in 1972 and was responsible for 80 per cent of the 7.83 million pigs and 95 per cent of the 600 million chickens produced in Santa Catarina in 2000. For poultry, the meat processing firms supply all of the ration, chicks, technical and veterinarian advice and determine the production programme. Farmers are suppliers of labour and capital. Inputs for pig production can be either fully or partially supplied by the firms, depending on the farmer's level of maize production and supply of breeding animals. These industries are undergoing major structural changes, with a trend towards concentration of production on large specialised farms, rather than on a large number of smaller operations. Accordingly, in 1986, 60,000 pig farms produced 2,50,000 tonnes of carcass weight while, 10 years later, 5,00,000 tonnes were produced from only 25,000 farms.

The barriers to long-term sustainability of family farms in these industries arise because the current management approach is centred on technical and economic efficiency, but ignore the social and environmental dimensions. To continue to grow, the industry needs continuing access to export markets, where strong pressure is being exerted against environmentally unsound production practices. Market prices do not yet reflect the true cost of production, nor do they account for environmental and social costs. High cost, and lack of capital to access sustainable technologies for items such as waste treatment systems, are preventing producers using sustainable practices. Some operators are still focused more on short-term profits rather than long-term sustainability, ignoring the limits of the natural environment, and their social responsibility. This chapter therefore, explores the following themes:

- the process of establishing and developing the vertically integrated pig and poultry industries in SC;
- the contribution of the pig and poultry industries to socio-economic development in SC;
- the environmental impacts of intensive production systems in which volume of waste exceed the assimilation capacity of the local environment; and
- the barriers to adoption of sustainable practices for waste management.

Social and Economic Impact

Despite having approximately 1.1 per cent of Brazil's total area (or 95,000 km^2), the State of Santa Catarina is the largest pig and poultry producer, accounting for 28 per cent of total pig and 21 per cent of total poultry production in the year 2000 (instituto Cepa/SC 2000a). These activities are extremely important in the State's socio-economic context due to the large number of jobs and the flow-on effects they have in the economy. Activities directly and indirectly related to production and processing of pigs and chicken contribute to the survival of rural communities. Pig and chicken production are closely linked to cropping activities, such as maize and soybean, as these are the most important components of the animal rations (Talamini, 1977).

Thousands of families make their living from these enterprises and from growing crops that are inputs to the pig and poultry production chains. This has a positive impact on the survival of schools, health services, and cultural centres. Brazil's five largest pig and poultry meat processors are based in Santa Catarina.

Pigs

In the case of pig production, it is a typical activity for farms with total land area of 15 to 35 ha. Pig production contributed 19.3 per cent of the gross value of agricultural production in Santa Catarina in 2000, with a turnover of US $1.05 billion. Pig production employs 55,000 people directly and creates 1,20,000 indirect jobs (Kafruni, 2001a).

Poultry

Brazil is the second largest poultry exporter in the world after USA and Santa Catarina currently contributes 80 per cent of the exported production. The poultry production chain has a turnover of US $1.2 billion per year in Santa Catarina and employs directly 32,000 people (Kafruni, 2001a). Moreover, an estimated 65,000 people are involved in other activities related to the poultry production chain (Altman, 1997).

Characteristics of Contracts in the Vertically Production System

The development of agribusiness essentially aims to maximise profits by organising the production chain in an efficient way. To fulfil this objective, the pig and poultry industries in South Brazil, have entered into contracts that bind the key links of the production chain together in a vertically integrated structure. The meat processing and marketing firms became suppliers of materials, genetics, and technical advice to pig and poultry producers who, in turn, supply capital and labour. The system operates to a tight programme, determined by the firms, in order to ensure a constant supply of animals according to market demands to the abattoirs. This has made a major contribution to technical and economic efficiency, which assured constant supply, reduced costs, and increased product quality. Farmers are constantly pushed by the meat processing firms' extension advisers to adopt new technology and only those who are able to operate within the technological standards set by the firms are kept in the system. Fig. 44.1 shows the agents, processes and products involved in the vertically integrated pig and poultry production chain in Santa Catarina. The dotted square indicates the stages and agents that have the highest local impact.

As the pig and poultry industries became export oriented, the producers faced the need to match world standards in terms of quality and price. This is the major driver for the dynamic state of the industries and its continuous improvement. Currently, 80 per cent of the pig production in Santa Catarina is produced by vertically integrated contracts (ACCS, 2001).

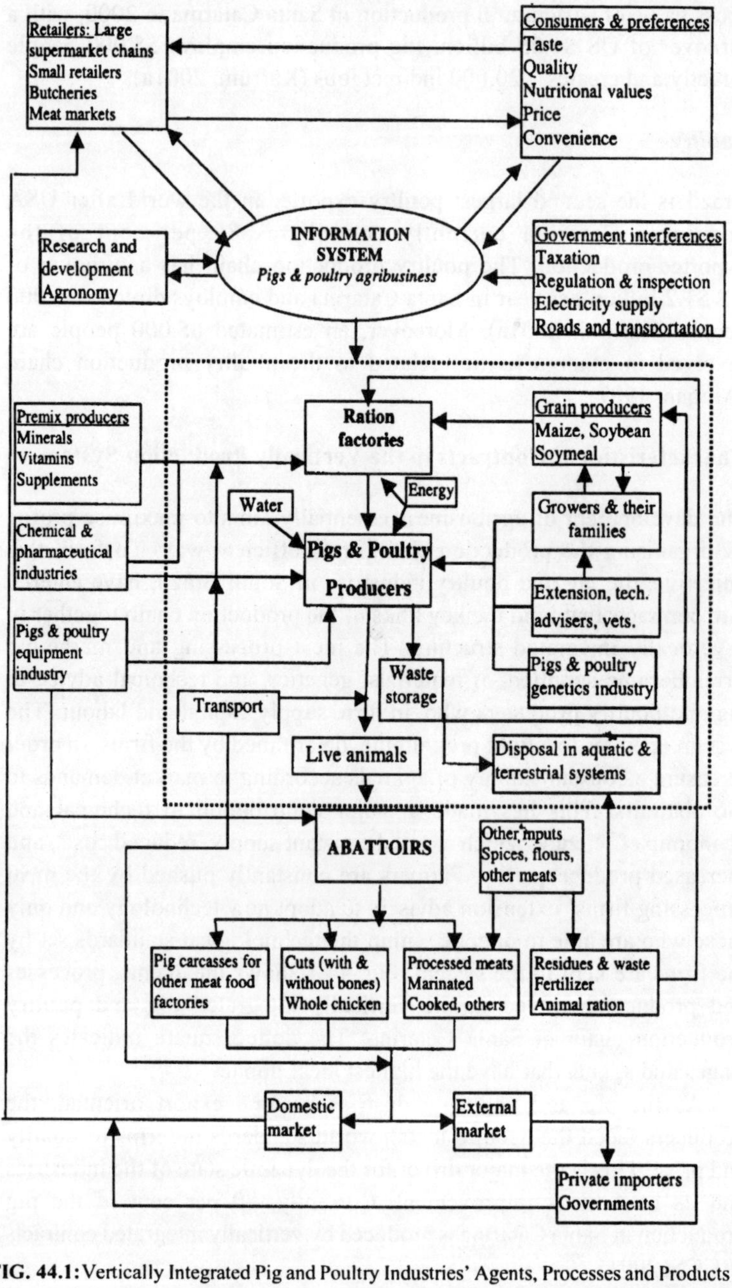

FIG. 44.1: Vertically Integrated Pig and Poultry Industries' Agents, Processes and Products

Vertically Integrated Pigs Production

For pig production in Santa Catarina, three different production system can be identified. They vary according to the level of specialisation and the level of inputs that the producer and the firm supply. Ration for pigs on average is composed by 77 per cent maize, 20 per cent soymeal, and 3 per cent of a nucleus of minerals and vitamins (Barbosa *et al.*, 1988) and therefore the producers may contribute with the supply of maize produced at their farm. The three production systems are:

(a) *Farrowing-to-finish*—In this system, the producer has breeding stock (sows, boars and replacement animals), produces the piglets, and finishes them for the market. These finished animals are sold to the meat processing firms at 90 to 130 kg live weight. In 1996, 9,766 producers, or 63 per cent of the total, adopted this system. The trend is for the number of producers in the farrowing-to-finish system to diminish, as the firms prefer the specialised system.

(b) *Breeders*—These producers have breeding stock and are dedicated to produce piglets, which are later transferred to finishers, when they reach 22 kg live weight. In 1996, there were 2,248 producers (14.5% of the total) operating as breeders only. These producers need more capital but normally have plenty of labour at the farm level and are willing to dedicate unlimited time to their operation, including working at nights and weekends.

(c) *Finishers*—These producers do not have breeding animals, as they receive the piglets at 22 kg live weight from the breeders. Their infrastructure is simple, as they basically supply ration, water and do the cleaning of the barns. In 1996, 3,440 or 22.2 per cent of the pig producers were specialised finishers.

Vertically Integrated Poultry Production

For poultry, the vertically integrated system is the most common choice for commercial producers in Santa Catarina. The market is organised in a tight production programme in such way that the meat firms do not buy live animals from independent producers. Over the years, almost all but the contracted commercial producers disappeared. There are however a few small abattoirs that have their own production and therefore do not operate a vertically integrated system. Their production is estimated to represent 5 per cent of the total in Santa Catarina (Canever *et al.*, 1997). In the poultry industry, the supply of ration and all other cosumbale inputs, except water, gas, electricity and bedding is provided by the meat firms.

Poultry producers are paid according to performance of each batch by a formula that takes into consideration the feed conversion rate, mortality, average live weight and age at killing. The formula can be summarised as:

$$\text{BEI} = \frac{[(100 - \text{MOR}) \times \text{ALW} \times 100]}{\text{FCR} \times \text{AGE}}$$

where BEI = batch efficiency index (in points)
ALW = average live weight
FCR = feed conversion rate, which is total kg of ration/total kg of live weight
MOR = mortality rate (in %)
AGE = age at killing (in days)

The Brazilian poultry industry has access to figures for the national average of BEI, which in 2001 was approximately 250 points. Table 44.1 shows the evolution of some of the indicators that are part of the formula in Santa Catarina, while Table 44.2 shows figures from the national level. If the producer achieves average performance, he/she may be paid at an average remuneration per bird delivered. If the BEI is above or below average, the producers may receive a premium or a discount, respectively. From the producers' point of view, they have the opportunity to increase their profits by achieving good technical performance and delivering a higher number of birds. This is one of the major reasons for intensification in the production of meat chickens. The "goal posts" are constantly moving and new benchmarks are continuously being set for producers.

The age of killing varies according to the market. On average, chickens are killed at 46 days, with feed conversion of 1.88 to 1.0. Daily ration consumption varies according to the chicken's live weight. Water consumption, according to Avila (2001) is 2 to 3 litres per kg of ration, depending on the seasons and temperature. A weighted average for the three stages of production shows the following ingredients in the poultry ration: Maize (55.5%); soymeal (35.0%); nucleus (4.0%); vegetable oil (5.3%) and DL-methionine (0.2%).

Intensive poultry production was always based on contracts with the meat industries and the scale was based on the standard 1,200 m^2 shed when the industry began in its current form in Santa Catarina

30 years ago. As revealed in Table 44.1, the number of producers increased in similar proportion to total production. However, the most recent trend is that meat companies now prefer to allocate a second production unit to those farms that have already shown to be good performers and which are located close to the abattoirs.

TABLE 44.1: Key Technical Coefficients for Vertically Integrated Production Systems in Santa Catarina

Coefficient	*Unit*	*1985*	*1998*
Feed conversion	kg ration/kg LW	2.21	1.88
Mortality	(%)	3.00	3.40
Live-weight	kg	2.10	1.98
Carcass weight	kg	1.68	1.58
Age to killing	days	48	46

Source: Instituto CEPA/Santa Catarina (2000b). Primary data from the meat industries.

History of the Development Process of the Vertically Integrated Pig and Poultry Industries in Santa Catarina

The development of the vertically integrated production system in the pig and poultry industries began in 1972 in Santa Catarina. However, the first step towards the contract based supply system was taken in 1964, when the company Sadia implemented a rural extension and farm advisory service and began to supply breeding pigs to small farmers. In 1969, they already had 1,200 producers involved in their technical advisory network (Altman, 1997). The meat processing firms where demanding pigs with less fat, as the market for animal fat dropped dramatically after the introduction of vegetable cooking oil. This brought two immediate benefits for the mean processing firms: better quality animals, and a more consistent supply.

The poultry industry began its development in the 1950s, mainly in Sao Paulo State. In the 1970s, the Brazilian chicken meat industry passed through a profound restructuring process and, since then, poultry industry development occurred in the southern region, particularly in Santa Catarina. This was driven by the meat firms, which already had competitive pig operations in this region, the availability of grain such as maize, and supply of soybeans. The quality of labour and the climatic conditions also made a contribution to expansion. Finally, the increase in poultry meat consumption and the availability of rural credit for long-term investments in rural areas had a major positive impact. This allowed the introduction of improved genetic material as well as new technologies.

Table 44.2 shows some of the technical indicators for poultry and how they improved over time in Brazil. The access to improved genetics and technology was responsible for significant gains in efficiency in feed conversion, weight gain, and age of killing.

TABLE 44.2: Evolution of Key Technical Indicators for the Poultry Industry in Brazil

Year	*Average Weight (kg)*	*Feed Conversion (kg feed/kg LW)*	*Age to Killing (days)*
1930	1.50	3.50	105
1940	1.55	3.00	98
1950	1.60	2.50	70
1960	1.60	2.25	56
1970	1.60	2.00	49
1980	1.70	2.00	49
1984	1.86	1.98	45
1989	1.94	1.98	45
2001*	2.24	1.78	41

Source: Adapted from Canever *et al.* (1997) after Andrade (1990).
* Estimates.

Size of the Pig and Poultry Industries in Santa Catarina

Santa Catarina produced 7.83 million pigs in 2000, from a stock of 4.53 million animals. There were 9,000 poultry farmers and production reached 1.1 million tonnes of chicken carcasses in 2000 (Instituto Cepa/SC 2000b). Poultry production in Brazil in 2000 reached 5.976 million tonnes and exports 9,06,000 tonnes, making Brazil the second largest producer and exporter of poultry meat in the world. Tables 44.3 and 44.4 show recent figures of the size and key technical performance indicators of the pig and poultry industries respectively. The exports of chicken to Europe have increased substantially since the mad-cow disease and foot-and-mouth disease crisis. While pork exports benefited from a new market in Russia as well as the Mercosur after Santa Catarina achieved the status of foot-and-mouth disease free zone. In addition to the increase in exports, the domestic consumption has risen substantially since the Brazilian economy was reformed in 1994 and average purchase power of the population improved.

Recent Developments and Trends

The pig and poultry industries in Santa Catarina have undergone major structural changes over recent years. In the case of pig production, the

TABLE 44.3: Evolution of the Numbers of Vertically Integrated Pig and Poultry Producers in Santa Catarina—1969 to 2001

Year	*No. of Pig Producers*	*No. of Poultry Producers*
1969*	1,200	—
1978	2,500	—
1986	23,936	4,450
1988	29,100	6,500
1996	15,800	9,128
2001**	12,500	6,500

* Only receiving technical advice.
** Estimates by ACCS (2001) personal communication.

TABLE 44.4: Production of Pigs in Brazil and Santa Catarina—1996 to 2000

Year	*Brazil*		*Santa Catarina*	
	Commercial Herds No. of Animals (millions)#	*Production ('000 tonnes carcass)*	*Pigs Slaughtered (millions)*	*Production ('000 tonnes)*
1996	26.6	1,900	6.74	571
1997	26.2	1,800	6.41	513
1998	27.3	2,000	7.07	562
1999*	27.8	2,100	7.36	586
2000*	28.9	2,200	7.83	624

* Estimates.
\# The total number of pigs in Brazil was estimated at 35 million according to Santos Filho *et al.* (1999). This includes the production for family subsistence, and represents 3.9 per cent of the total number of pigs in the world.
Source: Adapted from Instituto CEPA/SC (2000a).

trend is towards increased size, as well as the specialisation in either breeding or finishing operations. New improved genetic material is also widely used, with new pig breeds such as the MS58 and MS60 being examples of success. These animals have very low fat content, producing a carcass with 60 per cent of meat muscle, against 50 per cent for conventional breeds. The use of artificial insemination is facilitating farmer's access to improved genetic material as well as reducing costs (Kafruni, 2001b). Embrapa, the Brazilian National Agricultural Research Company, operates the National Centre for Pig and Poultry Research located in Santa Catarina and is contributing to the development of these industries, in cooperation with the private meat companies, particularly the cooperatives.

It is clear that the pig and poultry industries are adjusting to present circumstances by having fewer producers, but with larger scale operations, to benefit from lower costs. In 1986, over 60,000 pig farms produced 2,50,000 tonnes of carcass weight while 10 years later, 5,00,000 tonnes were produced by only 25,000 farms in Santa Catarina (Instituto CEPA/SC 2000b; ACCS, 1998). The remaining producers are predominantly those located close to the ration factories and abattoirs. The poultry industry is also changing, towards larger units, some with fully controlled environment (air conditioned) while the trend is for the meat firms to allocate new contracts preferably to the existing good performers rather than to new producers.

The total socio-economic and environmental consequences of this process are still unknown. The industries have improved their technological efficiency, and are highly competitive in world markets, but their long-term sustainability is not assured, due to negative externalities. Industry stockholders frequently raise problems such as poor waste management, water contamination, offensive odours, and increased populations of black flies (*Simulium* spp.), as critical issues. volumes of waste produced in some areas are beyond the thresholds for assimilation capacity of the local environment (Spies *et al.* 2001). It is clear that production in a vertically integrated system from large-scale producers is displacing thousands of small farmers, and concentrating production geographically, but it also contributed significantly to its economic efficiency. Figs. 44.2 and 44.3 reproduced from Icepa (2000a) respectively show the municipalities that have the highest concentration of pigs and poultry in Santa Catarina.

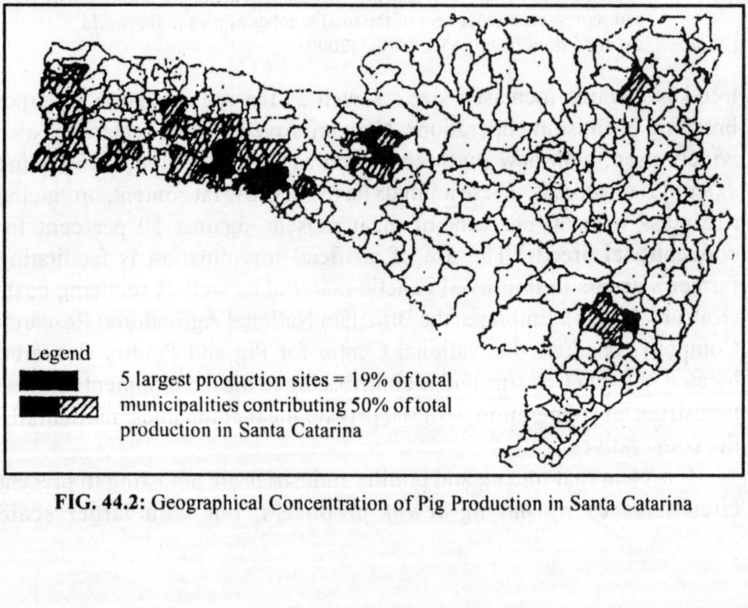

FIG. 44.2: Geographical Concentration of Pig Production in Santa Catarina

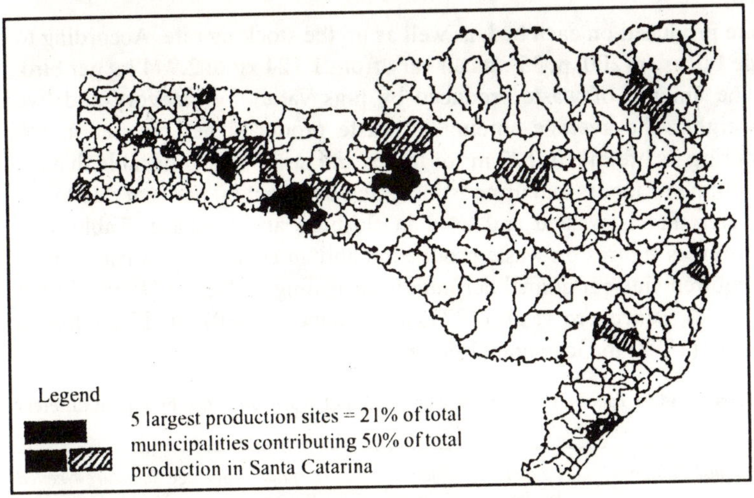

FIG. 44.3: Geographical Concentration of Poultry Production in Santa Catarina

Environmental Impact of Current Production Systems

Environmental degradation has become a major concern in Santa Catarina State because of excessive amounts of animal waste and the lack of suitable land for crops, which has forced farmers to dispose of large amounts of manure on small areas, causing soil erosion and pollution. The region has a rich river system, given its sloping landscape and the rainfall pattern of approximately 1,700 mm per year. Water contamination with nitrates above 10 mg/*l* was found in 37 per cent of the samples tested in Concórdia region between 1985 and 1998 and 82 per cent of water samples tested have shown *faecal coliform* organisms (SDA, 1998). The relatively high number of pigs and chickens in the small geographic area of 25,000 km^2 of the western region of Santa Catarina, implies production of a large quantity of manure to be disposed onto each hectare of cropping and pasture land. In addition to the high level of inorganic fertiliser used in cropping, the large amount of manure leads to excess nitrogen and phosphate flows into the soil and water, causing eutrophication. Therefore, it is believed that the amount of waste disposed in the region is beyond the assimilation capacity of the environment.

Waste Production and Management

The production of chicken litter depends on the number of batches that are produced on each bed, as well as on the stocking rate. According to de Lucas Jr (200, p. 35), it can vary from 1.124 kg to 2,944 kg per bird. The amount of waste produced by pigs varies by category and live weight. The average amount of waste produced per pig per month during its lifetime is 0.27 m^3 of faeces and urine. It is noticeable that the volume of faeces and urine is increased by about 60 per cent due to the water that is wasted and used in flushing and washing. Table 44.5 provides figures on waste production and an estimate of waste storage requirements per animal per month, according to Tietjen (1966), Loher (1974), Sancevero *et al.* (1979) and Konzen (1980) cited by Oliveira *et al.* (1993) in Gosmann (1997, p. 8).

TABLE 44.5: Faeces, Urine and combined Liquid Waste produced by Pigs at different stages

Animal Category	*Faeces kg/day*	*Faeces + Urine kg/day*	*Liquid Waste (Incl. water) l/day*	*Storage Structure Needed (m^3/animal/month)*	
				Faeces + Urine	*Liquid Waste*
Pigs 25-100 kg	2.30	4.90	7.00	0.16	0.25
Sows, pregnant	3.60	11.00	16.00	0.34	0.48
Lactating sows	6.40	18.00	27.00	0.52	0.81
Boars	3.00	6.00	9.00	0.18	0.28
Piglets	0.35	0.95	1.40	0.04	0.05
Average	2.35	5.80	8.60	0.17	0.27

Source: Adapted from Gosmann (1997, p. 8).

Pig Waste Management Systems used in Santa Catarina

Gosmann (1997) found that 88 per cent of the pig manure produced in the Western region of Santa Catarina is used as fertiliser on the pig producers' own farmland. He also found that 44 per cent of a sample of 163 producers in 1997 had bio-ponds, 47 per cent had conventional ponds, 7 per cent had other combined systems or biodigestors, and only one property (less than 1%) had no treatment. Most of the liquid manure is transferred from the ponds to cropping land or pastures by tanks pulled by tractors or other methods. Some producers have pumps installed so they can pump the manure directly to the cropping fields. The consequences of this were shown in a study by Pagani da Silva (2000), which detected an amount of animal waste in the Fragosos Creek catchment in Santa Catarina three times the 45 m^3/ha/year assimilation capacity of the local agricultural land available.

Poultry Waste Management Systemsused Used in Santa Catarina

Poultry waste is considered easier to manage because it is dry. In Santa Catarina, there is a market for poultry litter, so it can be sold to be used on cropping land or on other farms. It is also easier to store and there is less risk of it leaking to water streams, when compared to the liquid manure from pigs. Poultry producers are keeping the same bedding for up to five batches of chickens. For a 1,200 m^2 chicken shed, there is a need for 70 m^3 of woodchips, saw dust or straw initially, and then, for each of the next four batches, 18 m^3 of new material is added. After the fifth batch, the manure is completely removed and used as organic fertiliser.

Discussion and Conclusion

There is a need for changes to production methods to achieve sustainable pig and poultry production and sustainable family farming systems in Santa Catarina. As these industries are gradually becoming more exposed to world markets, new standards of environmental performance need to be achieved. Fig. 44.4 shows how the perception of successful agribusiness has evolved over time. The current system is still mainly focusing on client (or consumer) satisfaction but, to be sustainable, the industries need to take a broader view, which includes stakeholder[1] satisfaction. As environmentally sensitive industries, the pig and poultry operations need to be aware of pressures that come from the market, as well as their reputation within the communities. The failure to manage

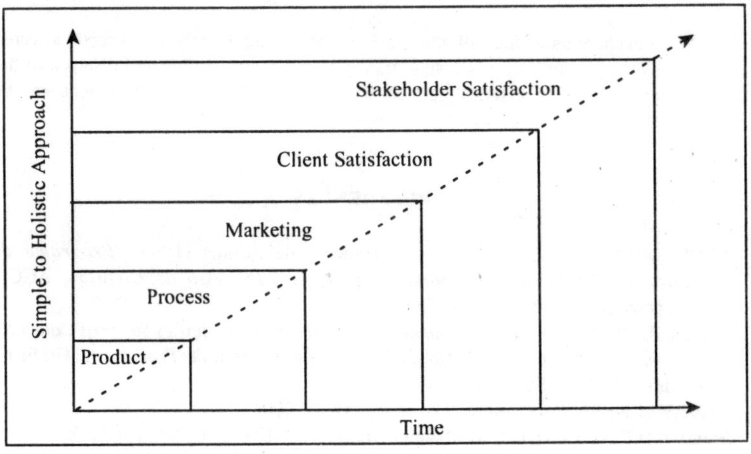

FIG. 44.4: Factors of Success in Agribusiness Performance

and minimise community reactions and outrage can expose the industries to reputation risks. A company's "licence to operate" and compliance with regulations will not be sufficient in the years to come. Operators need increasingly to demonstrate progress in keeping up with less tangible community expectations (Pigott, 2000).

The vertically integrated pig and poultry industries in Santa Catarina are an example of successful agribusiness development, but they are now faced by the challenges of building a sustainable future in a competitive market in the environmental and social issues are increasingly part of the equation. These industries play a key role in building sustainable family farming systems in Santa Catarina, because they are able to intensify the economic performance of small farms. However, the recent restructuring process, in which production is being concentrated among fewer producers and in a smaller geographic area, is not contributing to its long-term sustainability, nor does it help to build on the sustainability of the traditional family farming systems of Santa Catarina. The role of public research, development and extension (R, D&E) and policy-makers is to foster the process of integrating the interests of family farming systems with the market driven initiatives of the vertically integrated pig and poultry industries. This cooperation is likely to be successful in the pursuit of sustainable industries and family farming systems.

NOTE

1. Stakeholders include all interested parties that are directly or indirectly affected by the production, processing, transportation, consumption and disposal of the products. This includes the local communities where the environment may be damaged by impacts such as odours, pollution, flies, etc.

REFERENCES

ACCS—Associação Catarinense de Criadores de Suínos (1998). *Programa de Sustentabilidade da Suinocultura Familiar em Processo de Exclusão*, ACCS, Concórdia, pp. 1-11, Discussion paper.

Altman, R. (1997). *A Agricultura Familiar e os Contratos—Reflexões Sobre os Contratos de Integraçã, a Concentração da Produção e a Seleçã de Produtores*, Gráfica e Editora Pallotti, Florianópolis.

Avila, V., CNPSA. Pers. comm., Concórdia, 16.09, 2001.

Barbosa, H.P., de Lima, G.J. M.M. and Ferreira, A.S. (1988). "Estimativa da Quantidade de Ração Necessária para a Produção de um Suníno com 100 Kg de Peso Vivó", *Comunicado Técnico*, Vol. 133, No. March, pp. 1-3.

Canever, M.D., Talamini, D.J.D. Campos, A.C. and Santos Filho, I.J.d. (1997). *A Cadeia Produtiva do Frango de Corte no Brasil a na Argentina*, Embrapa, Concórdia.

de Lucas Jr, J. and Santos, T.M.B. (2000). "Aproveitamento de Resíduos da Indústria avícola par a Produção de Biogás", in *Simpósio Sobre Resíduos da Produção Avícola*, CNPSA, EMBRAPA, Concórdia, pp. 27-43.

Gosmann, H.A. (1997). Estudos Comparativos com Bioesterqueira e Esterqueira Para Armazenamento e Valorizção dos dejetos de Suínos. Unpublished Masters thesis, Departamento de Engenharia Sanitária e Ambiental, Universidade Federal de Santa Catarina, Florianópolis, p. 126.

IBGE—Instituto Brasileiro de Geografia e Estatística (1997). *Censo Agropecuário de 1995/1996, Santa Catarina*, IBGE, Rio de Janeiro, pp. 286.

Instituto Cepa/SC (2000a). *SC-AGRO 2000*, Instituto Cepa/SC. Available at: CD Rom.

_______ (2000b). *Síntese Anual da Agricultura de Santa Catarina—1998-1999*, Instituto Cepa, Florianópolis.

Kafruni, S. (2001a). "Modernizção muda Cara do Oeste", in *Diário Catarinense, 9-10-01*, Chapecó, pp. 1-4.

_______ (2001b). "Parceria Revoluciona a Suinocultura", in *Diário Catarinense*, Concórdia, pp. 1-3.

Pagani da Silva, A. (2000). *Diagnóstico Socio-econômico e Ambiental Aspectos Sobre a Sustentabilidade da Bacia Hidrográfica dos Fragosos*, Concórdia/SC. Unpublished Masters thesis, Engenharia Ambiental, UFSC, Florianópolis.

Pigott, C. (2000). "Managing Community Reactions to Project Developments", in *3rd Queensland Environmental Conference: Sustainable Environmental Solutions for Business and Government*, Environmental Engineering Society-Qld Chapter, Brisbane.

Santos Filho, I.J.d., Santos, N.A.d., Canaver, M.D., Sousa, I.S.F.d. and Vieira, L.F. (1999), "O Cluster Suinícola do oeste de Santa Catarina", in Haddad, P.R. (Ed.), *A Competitividade do agronegócio e o Desenvolvimento Regional no Brasil: Estudo de Clusters*, EMBRAPA/CNPq, Brasília, pp. 125-80.

SDA—Secretaria de Estado do Desenvolvimento rural e da Agricultura de SC (1998). *Projeto Microbacias-2 (documento básico)*, SDA, Florianópolis.

Spies, A., Spies, C.I.S., Chamala, S., Wegener, M. and Beeton, R. (2001). "The Need for Environmental Extension Education in Facilitating Sustainability of Pig and Poultry Industries in Santa Catarina State, Brazil", in *APEN 2001 International Conference*, APEN, Toowoomba, Australia, pp. 175-82.

Talamini, D.J.D. (1977). *Análise Econômica dos Sistemas de Produção de Suínos—Santa Catarina*, Unpublished Masters thesis, Faculdade de Ciências Econômica, IEPE, Universidade Federal do Rio Grande do Sul, Porto Alegre, p. 85.

45

WATERSHED MANAGEMENT
Impact on Lives of Farmers and Landless Labourers

— *R.S. Palande* and *G.K. Sawant*

Allotment of house sites, providing required clothings and food, minimum household material by creating employment was the points of the 20-point programme of the Government of India. In general, the marginal, small and big farmers and the landless labourers leads a sub-standard life as far as their consumption of food, household material possession. Similarly, they are poorly clothed and poorly housed in scarcity areas in particularly. To overcome these serious problems the effect of watershed management as a fundamental avenue of agribusiness in scarcity area would be one of the best solution. Keeping in view the study entitled "Socio-Economic impact of various recommended aspects of watershed management on the lives of beneficiary farmers and the landless labourers" were conducted with the following specific objectives are:

> To study the changes occurred in living conditions amongst the beneficiaries in respect to changes occurred in type of house, clothing pattern, material possessions and type of food consumed due to National Watershed Development Programme.

Methodology

A research project sanctioned and funded by Government of Maharashtra State under their National Watershed Programme was studied in Khospuri Watershed in Ahmednagar district on the lives of beneficiary farmers and the landless labourers with special reference to the changes in living conditions occurred due to Watershed Development Programme. A list

TABLE 45.1: Distribution of Beneficiaries according to Impact on Watershed Management on their lives

Sl. No.	Categories	Marginal Farmers (N = 229)				% Increase or Decrease	Small Farmers (N = 127)				% Increase or Decrease
		Before		After			Before		After		
		No.	%	No.	%		No.	%	No.	%	
(1)	(2)	(3)	(4)	(5)	(6)	(7)	(8)	(9)	(10)	(11)	(12)
1.	*Type of House*										
	(i) Hut	139	61	61	27	–34	55	43	37	29	–14
	(ii) Kaccha	53	23	98	43	+20	43	34	75	59	+25
	(iii) Pucca	37	16	70	31	+14	29	23	38	30	+7
2.	*Clothing Pattern*										
	(i) Half of the requirement	183	80	103	45	–35	56	44	23	18	–26
	(ii) Full requirement	46	20	89	39	+19	71	56	104	82	+26
3.	*Type of Clothing*										
	(i) Cotton	187	82	103	45	–37	102	80	81	64	–16
	(ii) Synthetic (Terrycot + Nylon)	42	18	26	55	+37	25	20	46	36	+16
4.	*Changes in Material Possession*										
	(i) Radio	53	23	103	45	+22	34	27	67	53	+26
	(ii) Chair	41	18	132	58	+40	22	17	29	23	+6
	(iii) Table	38	17	67	29	+12	11	9	38	30	+21
	(iv) Fan	—	—	—	—	—	3	2	17	13	–11
	(v) Cycle	56	24	118	52	+28	28	22	39	31	+9
	(vi) Motor cycle	—	—	11	5	+5	1	1	7	6	+5
	(vii) T.V.	—	—	19	8	+8	—	—	3	2	+2

(*Contd.*)

TABLE 45.1 (*Contd.*)

(1)	(2)	(3)	(4)	(5)	(6)	(7)	(8)	(9)	(10)	(11)	(12)
5.	*Type of Food Consumed*										
(i)	Cereals	229	100	229	100	—	127	100	127	100	—
(ii)	Pulses	229	100	229	100	—	229	100	229	100	—
(iii)	Vegetable	128	56	197	86	+30	65	67	110	87	+20
(iv)	Eggs, fish and meat	145	63	189	83	+20	89	70	110	87	+17
(v)	Oil and fats	103	45	158	69	+24	57	45	88	69	+24
(vi)	Fruits	40	17	68	30	+13	34	27	58	46	+19
(vii)	Milk and milk products	41	18	53	23	+5	38	30	63	50	+20
(viii)	Sugar and Jaggery	229	100	229	100	—	127	100	127	100	—

(*Contd.*)

TABLE 45.1 (*Contd.*)

Sl. No.	*Categories*	*Big Farmers (N = 103)*				*% Increase or Decrease*	*Landless Labours (N = 22)*				*% Increase or Decrease*
		Before		*After*			*Before*		*After*		
		No.	*%*	*No.*	*%*		*No.*	*%*	*No.*	*%*	
(1)	(2)	(13)	(14)	(15)	(16)	(17)	(18)	(19)	(20)	(21)	(22)
1.	*Type of House*										
	(i) Hut	63	61	39	38	–23	20	91	15	68	–23
	(ii) Kaccha	30	29	49	48	+19	2	9	7	32	+23
	(iii) Pucca	17	17	39	38	+21	—	—	—	—	—
2.	*Clothing Pattern*										
	(i) Half of the requirement	89	86	53	51	–35	22	100	14	64	–36
	(ii) Full requirement	30	29	50	49	+20	—	—	8	36	+36
3.	*Type of Clothing*										
	(i) Cotton	94	91	67	65	–26	22	100	19	86	–14
	(ii) Synthetic (Terrycot + Nylon)	9	9	36	35	+26	—	—	3	14	+14
4.	*Changes in Material Possession*										
	(i) Radio	23	22	58	56	+34	1	5	5	23	+18
	(ii) Chair	18	17	65	63	+46	1	5	3	14	+9
	(iii) Table	14	14	31	30	+16	—	—	1	5	+5
	(iv) Fan	23	22	42	41	+19	—	—	—	—	—
	(v) Cycle	36	35	72	70	+35	3	14	13	59	+45
	(vi) Motor cycle	2	2	9	9	+7	—	—	—	—	—
	(vii) T.V.	1	1	15	15	+14	—	—	—	—	—

(*Contd.*)

TABLE 45.1 (*Contd.*)

(1)	(2)	(13)	(14)	(15)	(16)	(17)	(18)	(19)	(20)	(21)	(22)
5.	*Type of Food Consumed*										
(i)	Cereals	103	100	103	100	—	22	100	22	100	—
(ii)	Pulses	229	100	229	100	—	229	100	229	100	—
(iii)	Vegetable	68	66	87	84	+18	5	23	9	41	+18
(iv)	Eggs, fish and meat	38	37	59	57	+20	22	100	22	100	—
(v)	Oil and fats	38	37	54	52	+15	7	32	13	59	+27
(vi)	Fruits	18	17	32	31	+14	3	14	3	14	—
(vii)	Milk and milk products	31	30	48	47	+17	3	14	3	14	—
(viii)	Sugar and Jaggery	103	100	103	100	—	22	100	22	100	—

of total beneficiary farmers was prepared from the records of *Talathi* of Khospuri village. Out of 900 beneficiaries, fifty per cent (459) were selected by using random sampling method. Thus the sample consisted marginal farmers (229), small farmers (127) and big farmers (103). However, in all 22 landless labourers of the village who benefited indirectly by NWDP are also studied. Thus the data were collected from 481 farm families and landless labourers by conducting personal interview method with the help of interview schedule prepared in view of the objectives framed for the study. The evaluatory research design was used.

Findings

Table 45.1 clearly indicates the considerable changes which occurred in the lives of beneficiaries of the watershed management programme are given below:

1. Due to impact of different watershed management aspects the appreciable changes in dwelling was observed, *i.e.*, from hut to Kaccha and from Kaccha to Pucca house was noticed amongst the marginal, small and big farmers and the landless labourers too.
2. The efforts made in Watershed Management changed the food habits of beneficiary farmers and the landless labourers by making their food more richer by including protective foods.
3. The study revealed that the changes in clothing items pattern of the beneficiaries were noticed in all the categories of beneficiaries due to increased income and employment opportunities.
4. The study also pointed out that the marginal, small and big farmers and the landless labourers had improved their material possessions.

Conclusion

The management of different recommended aspects of watershed brought considerable changes in the lives of all categories of beneficiaries. The marginal and small farmers comparatively made more changes in their food items and habits than the big farmers and the landless labourers. The study also indicates to consider watershed as a best source of agribusiness to bring all-round development of the farmers as well as landless labourers in particular.

46

INTEGRATED WATERSHED DEVELOPMENT MANAGEMENT

— *R.S. Palande* and *G.K. Sawant*

As per the scope of agribusiness activity the resource supply sector found to be more important in accelerating the growth of agro-business based industry. The management process of watershed includes diagnostic analysis resource base. The watershed management activity help to increase the demand for various resources required by production activity. In view of the efforts were made to study the extent of changes that are taken place in diagnostic characters and management objective. The study in regard to "Socio-economic impact on the lives of beneficiary farmers", was taken into the district of Ahmednagar (M.S.) by selecting 459 beneficiaries in Khospuri watershed, with the following specific objective:

> To study the physical changes occurred amongst the beneficiaries in respect of farm power, means of irrigation, use of manures and fertilisers, annual income, cropping pattern, animal possession, social participation and socio-economic status due to National Watershed Development Programme.

Methodology

A list of total beneficiary farmers was prepared from the records of *Talathi* of Khospuri village. Out of 900 beneficiaries, fifty per cent (459) were selected by using random sampling method. Thus the sample consisted marginal farmers (229), small farmers (127) and big farmers (103) were studied. Thus the data were collected from 459 farm families by conducting personal interview method with the help of interview schedule prepared in view of the objectives framed for the study. The evaluatory research design was used.

Findings

Data in Table 46.1 indicate the impact of Watershed Management on the lives of beneficiary farmers are given below:

1. Marginal, small and big farmers used improved farm power *viz.*, bullock pairs/electrical pump etc. which helped them in increased irrigation potential of their areas.
2. The definite change in the use of improved seeds/seedrill, recommended spacing and seed treatment amongst the marginal, small and big farmers was noticed. However, the significant change was noticed amongst the small farmers than the marginal farmers.
3. Watershed management has shown great impact on the lives of marginal, small and big farmers in increasing their irrigation potential and this brought more area under irrigation.
4. Marginal, small and big farmers reported an increased application of Farm Yard Manure alongwith nitrogenous and complex fertilisers on their farm.
5. Before watershed management programme 39 per cent marginal farmers, 38 per cent small farmers and 34 per cent big farmers had the annual income between Rs. 10,001 to Rs. 25,000. After Watershed Management Programme the farmer beneficiaries raised their income from 59, 61 and 49 per cent by registering 20, 23 and 15 per cent increased, respectively.
6. After Watershed Management Programme by and large helped all the farmer beneficiaries in growing more number of crops per year, *i.e.*, Kharif, Rabi and Summer crops.
7. It was also noticed that the beneficiaries of all the categories by and large benefited in respect of animal possession.
8. Study revealed that all the beneficiaries showed increased participation in the formal and in the informal organisations.
9. Study definitely helped the beneficiary farmers in improving their socio-economic status.

TABLE 46.1: Distribution of Beneficiaries according to Watershed Management Impact on their lives

Sl. No.	*Area and Means of Irrigation*	*Categories: Marginal Farmers (N = 229)*				*% Increase or Decrease*
		Before		*After*		
		No.	*%*	*No.*	*%*	
(1)	(2)	(3)	(4)	(5)	(6)	(7)
1.	*Farm Power*					
(i)	Bullock	115	50	180	79	+29
(ii)	Pump sets (Oil Engine/Elec. Motor	110	48	153	67	+19
2.	*Adoption*					
(i)	Improved seedrill	39	17	89	39	+22
(ii)	Recommended spacing	—	—	56	24	+11
(iii)	Seed treatment	—	—	56	24	+24
3.	*Means of Irrigation*					
(i)	Land under well irrigation	36	16	53	23	+7
(ii)	Well with pump set (Oil Engine/Elec. Motor)	37	16	61	27	+11
4.	*Use of Manures and Fertiliser*					
(i)	FYM	49	21	92	40	+22
(ii)	Nitrogenous	36	16	88	38	+22
5.	*Increased Annual Income*					
(i)	Rs. 10,001 to Rs. 25,000	90	39	136	59	+20
6.	*Cropping Pattern Changed*					
(i)	*Kharif*					
	(a) Bajra	109	48	173	76	+28
	(b) Sunflower	48	19	57	25	+6
(ii)	*Rabi*					
	(a) Jowar	89	39	143	62	+13
	(b) Wheat	48	21	106	46	+25
(iii)	*Summer*					
	(a) Groundnut	38	17	66	28	+12
	(b) Mung	69	30	117	51	+21
7.	*Animal Possession*					
(i)	Cow	49	21	111	48	+37
(ii)	Bullock	68	30	95	41	+11
(iii)	Goat	65	28	129	56	+28
8.	*Social Participation*					
(i)	No participation	63	28	19	8	–20
(ii)	Better participation	47	21	88	38	+17
9.	*Socio-economic Status*					
(i)	Lower (Score below 75)	117	51	75	33	–18
(ii)	Middle (Score between 76 to 150)	109	48	149	65	+17
(iii)	Higher (Score above 150)	3	1	5	2	+1

(Contd.)

TABLE 46.1 (*Contd.*)

Sl. No.	*Area and Means of Irrigation*	*Categories: Small Farmers (N = 127)*				*% Increase or Decrease*
		Before		*After*		
		No.	*%*	*No.*	*%*	
(1)	(2)	(8)	(9)	(10)	(11)	(12)
1.	*Farm Power*					
	(i) Bullock	49	39	69	54	+15
	(ii) Pump sets (Oil Engine/Elec. Motor	34	37	88	69	+42
2.	*Adoption*					
	(i) Improved seedrill	19	15	49	39	+24
	(ii) Recommended spacing	—	—	17	10	17
	(iii) Seed treatment	—	—	17	10	+10
3.	*Means of Irrigation*					
	(i) Land under well irrigation	36	28	35	43	+15
	(ii) Well with pump set (Oil Engine/Elec. Motor)	17	10	38	30	+20
4.	*Use of Manures and Fertiliser*					
	(i) FYM	36	28	61	48	+20
	(ii) Nitrogenous	14	11	50	39	+28
5.	*Increased Annual Income*					
	(i) Rs. 10,001 to Rs. 25,000	48	38	77	61	+23
6.	*Cropping Pattern Changed*					
	(i) *Kharif*					
	(a) Bajra	65	51	87	69	+18
	(b) Sunflower	18	14	33	26	+12
	(ii) *Rabi*					
	(a) Jowar	65	51	95	75	+24
	(b) Wheat	33	26	57	45	+19
	(iii) *Summer*					
	(a) Groundnut	19	15	40	31	+16
	(b) Mung	38	30	54	43	+13
7.	*Animal Possession*					
	(i) Cow	29	23	48	38	+15
	(ii) Bullock	41	32	63	50	+18
	(iii) Goat	39	31	68	54	+23
8.	*Social Participation*					
	(i) No participation	46	36	31	24	–12
	(ii) Better participation	2	2	25	20	+18
9.	*Socio-economic Status*					
	(i) Lower (Score below 75)	78	61	53	42	–19
	(ii) Middle (Score between 76 to 150)	48	38	73	57	+19
	(iii) Higher (Score above 150)	1	1	1	1	—

(*Contd.*)

TABLE 46.1 (*Contd.*)

Sl. No.	*Area and Means of Irrigation*	*Category: Big Farmers (N = 103)*				*% Increase or Decrease*
		Before		*After*		
		No.	*%*	*No.*	*%*	
(1)	(2)	(13)	(14)	(15)	(16)	(17)
1.	*Farm Power*					
	(i) Bullock	34	33	40	48	+15
	(ii) Pump sets (Oil Engine/Elec. Motor	28	27	60	58	+31
2.	*Adoption*					
	(i) Improved seedrill	7	7	13	13	+6
	(ii) Recommended spacing	8	8	36	35	+26
	(iii) Seed treatment	8	8	36	35	+27
3.	*Means of Irrigation*					
	(i) Land under well irrigation	39	38	49	48	+10
	(ii) Well with pump set (Oil Engine/Elec. Motor)	30	29	51	50	+21
4.	*Use of Manures and Fertiliser*					
	(i) FYM	23	22	48	47	+25
	(ii) Nitrogenous	21	20	48	47	+27
5.	*Increased Annual Income*					
	(i) Rs. 10,001 to Rs. 25,000	35	34	50	49	+15
6.	*Cropping Pattern Changed*					
	(i) *Kharif*					
	(a) Bajra	58	56	70	68	+12
	(b) Sunflower	12	12	23	22	+10
	(ii) *Rabi*					
	(a) Jowar	38	37	53	51	+14
	(b) Wheat	11	11	24	23	+12
	(iii) *Summer*					
	(a) Groundnut	7	7	22	21	+14
	(b) Mung	43	42	68	67	+25
7.	*Animal Possession*					
	(i) Cow	15	15	32	31	+16
	(ii) Bullock	29	28	38	37	+9
	(iii) Goat	26	25	53	51	+26
8.	*Social Participation*					
	(i) No participation	20	15	11	11	–8
	(ii) Better participation	25	24	25	24	Nil
9.	*Socio-economic Status*					
	(i) Lower (Score below 75)	59	57	28	27	–30
	(ii) Middle (Score between 76 to 150)	34	33	58	56	+23
	(iii) Higher (Score above 150)	10	10	17	17	+7

Conclusion

On the basis of available data on different aspects of watershed management it can be safely inferred that the national watershed development programme has definitely helped the beneficiary farmers in improving there socio-economic status. It is, therefore, necessary on the part of planners and programme executors to provide the better infrastructural facilities and continuous monitoring which would further help to generate more income and employment opportunities.

INDEX